D1796596

TABLE OF CONTENTS

Abstract of Dissertation Presented to the Graduate School of the University of Florida in Partial Fulfillment of the Requirements for the Degree of Doctor of Philosophy

1,3-DIPOLAR CYCLOADDITIONS OF FLUORINATED ALLENES AND STUDIES OF FLUORINATED TRIMETHYLENEMETHANES

By

CONRAD BURKHOLDER

April 1984

Chairman: William R. Dolbier, Jr.

Major Department: Chemistry

The 1,3-dipolar cycloadditions of 1,1-difluoroallene 1, fluoroallene 2, and 1,1-difluoro-3-methyl-1,2-butadiene 3 are reported. Dipoles studied are diazomethane 4, 2-diazopropane 5, diazofluorene 6, diphenyldiazomethane 7, diphenylnitrone 8, C-phenyl-N-methylnitrone 9, triphenylnitrone 10, mesitylnitrile oxide 11, and tetracyanocarbonyl ylide 12.

Reaction of 1 with 4 is regiospecific giving 4-(difluoromethylene)-1-pyrazoline. Similar regioselectivity is observed for dipoles 5 through 12 except for 6 and 7, which give mainly 5-(difluoromethylene)-1-pyrazolines. Regiochemistry is controlled by both electronic and steric interactions.

Reaction of 2 with 4 gives only 4-(fluoromethylene)-1-pyrazoline. Reaction of 2 with 8 gives an 86 to 14 ratio of (E)-and (Z)-4-(fluoromethylene)isoxazolidines. Reactions of 3 with 4, 8, and 9 are regiospecific for the 4-(difluoromethylene)-cycloadducts.

The deazetations of the fluorinated 4-methylene-1-pyrazolines by photolysis and thermolysis are studied in the gas phase and in solution. The kinetic product ratios of fluorinated methylenecyclopropanes are consistent with trimethylenemethane biradical intermediates. Thermodynamic parameters are determined for the gas-phase equilibria of the products.

The (difluoromethylene)cyclopropane from 6 gives cycloadducts with oxygen at room temperature and acrylonitrile at 70°. The distribution of products is consistent with formation of a highly stabilized trimethylenemethane biradical which is trapped.

Regiochemistry is also studied for reactions of 1 with acrylonitrile 13, methacrylonitrile 14, 1,1-dichloro-2,2-difluoroethylene 15, 1,2-bis-(methylene)cyclobutane 16, 2,3-dicyanobutadiene 17, and diphenylisobenzofuran. Olefins 13, 14, and 15 give 1,2-cycloadducts with fluorines mainly on the ring, a result consistent with a biradical mechanism. Dienes 16 and 17 give competing biradical and concerted reactions. Fluoroallene reacts non-regiospecifically with 15. Allene 3 with cyclopentadiene gives the Diels-Alder cycloadduct with an exocyclic difluoromethylene group.

SECTION I
INTRODUCTION

Allene Cycloadditions

Though most unconjugated olefins are unreactive in cycloadditions, it has been found that allene 1 dimerizes at elevated temperatures.[1-5] At less than 140° the only product is the head to head dimer 2. However at higher temperatures (200°) the head to tail dimer 3 is also observed, in addition to large amounts of oligomers.[6]

2 $CH_2=C=CH_2$ (1) ⟶ 2 + 3

CH_2 CH_2 CH_2 H_2C

1 2 3

Allene also forms methylenecyclobutanes with a variety of activated olefins including acrylonitrile, dichlorodifluoroethylene, chlorotrifluoroethylene and tetrafluoroethylene.[7-10] Through the stereochemical studies of Taylor, Warburton, and Wright,[9,10] the nonstereoselectivity of these [2+2] cycloadditions has been demonstrated. Consequently, a stepwise biradical mechanism has become generally accepted for the formation of cyclobutanes.

$CH_2=CH-CN$ (4) + $CH_2=C=CH_2$ ⟶ 3-methylenecyclobutanecarbonitrile (5) 60%

Under forcing conditions, allene will give Diels-Alder reactions with cyclopentadiene,[11] hexachlorocyclopentadiene,[12] and isoprene.[13]

cyclopentadiene (6) + $CH_2=C=CH_2$ —200-230°⟶ 5-methylenebicyclo[2.2.1]hept-2-ene (7) 49%

The measurement of secondary deuterium isotope effects for 1,1-dideuterioallene has confirmed that the methylenecyclobutane formation is biradical, while the Diels-Alder reaction is concerted.[12]

Tetrafluoroallene Cycloadditions

The first fluorine-substituted allene to be made was tetrafluoroallene 8.[14] It was noted by Jacobs and

Bauer[15] that tetrafluoroallene polymerizes at room temperature to give an unsaturated polymer containing difluoromethylene groups.[16] However, when an inhibitor was present and the neat tetrafluoroallene was heated at 40° for 20 hours, a good yield of dimer 9 was obtained.

$$2\ CF_2{=}C{=}CF_2\ (\underline{8}) \xrightarrow[20h]{40^\circ} \underline{9}\quad 83\%$$

In a more recent publication[17] Banks, Haszeldine and Taylor studied the behavior of tetrafluoroallene in ionic reactions and confirmed that the only dimer formed was the head to head isomer. In the dimerization of substituted allenes, only the head to head isomer is ever observed.

Until recently, only one cycloaddition of tetrafluoroallene had been reported.[18] This was the reaction of hexafluoro-2-butyne 10 to give cycloadduct 11.

$$CF_2{=}C{=}CF_2 + CF_3{-}C{\equiv}C{-}CF_3\ (\underline{10}) \xrightarrow{80^\circ} \underline{11}$$

After abandoning the study of tetrafluoroallene for

many years, Blackwell, Haszeldine, and Taylor[19] are currently working on 1,3-dipolar cycloadditions of tetrafluoroallene.

Trifluoroallene and 1,3-difluoroallene, though both known compounds,[20,21] have not been studied except for their spectra.

Difluoroallene Cycloadditions

The most well-studied fluorinated allene is 1,1-difluoroallene 12. Its cycloaddition reactions were first noted by Knoth and Coffman[22] and later studied by Piedrahita.[23]

Knoth and Coffman found that at elevated temperature, difluoroallene forms a head to head dimer 13 having one vinyl group and one difluorovinyl group. No other isomers were reported.

$$2\ CF_2{=}C{=}CH_2\ (\mathbf{12}) \xrightarrow{260\text{-}298^\circ} \mathbf{13}\ \ 30\%$$

(13: F_2-substituted cyclobutane bearing $=CH_2$ and $=CF_2$ groups)

They also found that difluoroallene reacts with acrylonitrile, maleic anhydride, and ethyl azidodiformate to give methylenecyclobutanes 14, 16, and 18.

$$CH_2{=}CHCN + CF_2{=}C{=}CH_2 \xrightarrow[8h]{150^\circ} \mathbf{14}\ (43\%)$$

$$\mathbf{15} + CF_2{=}C{=}CH_2 \xrightarrow[2)\ H_2O\ \text{heat}]{1)\ 150^\circ\ 8h} \mathbf{16}\ (16\%)$$

$$\mathbf{17} + CF_2{=}C{=}CH_2 \xrightarrow{120\text{-}192^\circ} \mathbf{18}\ (6.8\%)$$

When allowed to react with tetrafluoroethylene at 20° a diadduct of difluoroallene 19 was formed, probably arising from reaction of tetrafluoroethylene with the difluoroallene dimer.

$$2\ CF_2{=}C{=}CH_2 + CF_2{=}CF_2 \xrightarrow[8h]{20^\circ} \mathbf{19}\ (9.2\%)$$

However, the most interesting reaction studied by Knoth and Coffman was the cycloaddition of difluoroallene with cyclopentadiene to give 20 as the only product in nearly quantitative yield.

$$\text{cyclopentadiene} + CF_2{=}C{=}CH_2 \xrightarrow{20^\circ} \underline{20}\ (=CF_2)\quad 90\%$$

A careful study by Dolbier and Piedrahita[24] has confirmed the results of Knoth and Coffman regarding the cyclopentadiene reaction. In addition, a reexamination of the acrylonitrile reaction with difluoroallene revealed two isomeric products 14 and 21 in the ratio of approximately four to one.

$$CH_2{=}CH{-}CN + CF_2{=}C{=}CH_2 \xrightarrow[9h]{125^\circ} \underline{14} + \underline{21}\quad 4:1$$

The reactions of difluoroallene with a variety of 1,3-dienes all give complete regiospecificity for the [2+4] cycloaddition. That is to say, all Diels-Alder products have an exocyclic difluoromethylene group.

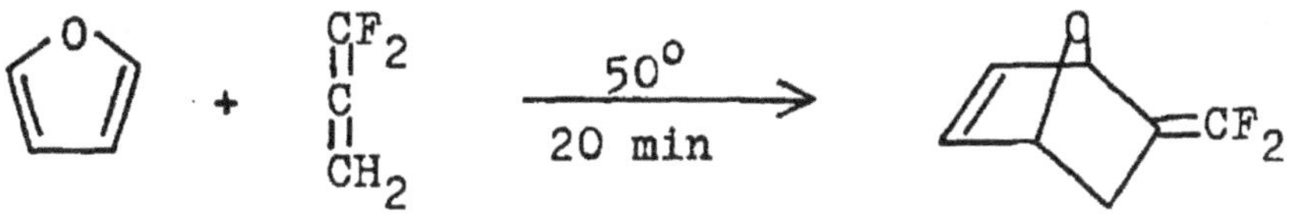

50°, 20 min

22 → 23

100°, 2.5h

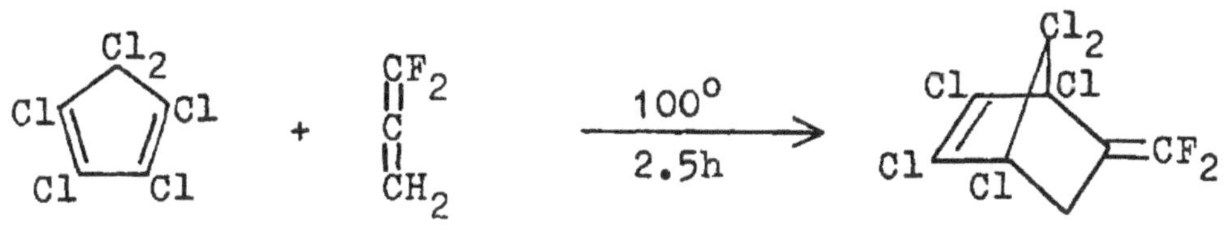

24 → 25

For the reaction of difluoroallene with butadiene, two isomers were obtained, with cyclohexene 27 formation favored over cyclobutane 28 formation by 63 to 37.

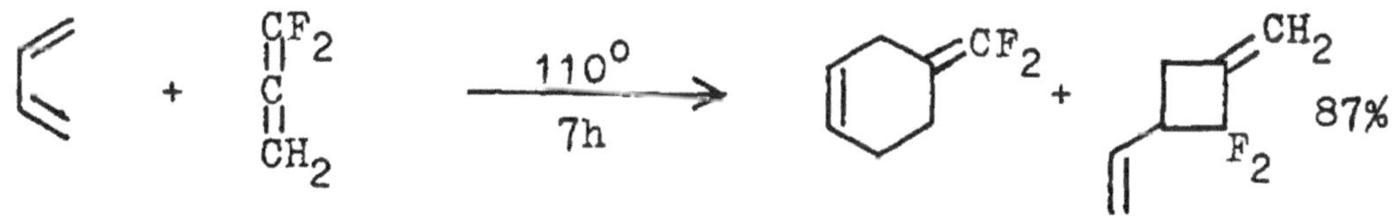

110°, 7h

26 → 27 63:37 28 87%

The formation of cyclohexene 27 as the major product is best explained by assuming there is competition between a concerted [2+4] cycloaddition and a biradical [2+2] cycloaddition. The fact that the major reaction product is the [2+4] cycloadduct 27 argues strongly for a concerted [2+4] cycloaddition mechanism for the following reason.

It has long been known that butadiene 26 exists in two equilibrating conformations designated as cisoid 26c and transoid 26t. Addition of difluoroallene to the cisoid butadiene leads to a cisoid biradical, while addition to the transoid butadiene gives a transoid biradical. Due to the fact that rotation of an allylic radical out of conjugation requires around 14.3 kcal/mole,[25] the cisoid and transoid biradicals are not interconvertible. According to O'Neal and Benson,[26] the energy barrier to closure of a biradical to methylenecyclobutane is around 6.6 kcal/mole. Therefore it may be concluded that closure is totally predominant over rotation of the allyl radical.

While the cisoid biradical can close to give any of the four possible products 27, 28, 29, or 30, the

H_2C CF_2 H_2C H_2 CF_2 + CH_2 F_2

26c cisoid biradical 27 29

H_2C H_2C CF_2 H H_2 CH_2 + CF_2 F_2

26t transoid biradical 28 30

transoid biradical can close to give only methylenecyclobutanes 28 and 30. One would expect attack on the transoid butadiene to be at least as easy as on the cisoid. If this is the case, then following the reasoning of Bartlett and coworkers[27] by the biradical mechanism, the ratio of [2+4] to [2+2] cycloadducts can never exceed the ratio of cisoid to transoid diene in the conformational equilibrium. Accordingly, one way in which competition of concerted and stepwise cycloaddition might show itself would be by cyclohexene to cyclobutane ratios in excess of the known cisoid to transoid ratio for the diene. This is only true if the initial addition is irreversible. Though the addition of various ethylenes is reversible, it has been determined that addition of allenes to 1,3-dienes is irreversible.[24]

In the equilibrium mixture, transoid butadiene is favored over cisoid by about 2.3 kcal/mole.[28] The fraction of cisoid butadiene in the mixture is estimated to be around 9% using Bartlett and coworkers' empirical equation.[29] However, the fraction of cyclohexene 27 in the product mixture is 63%. Thus the competing concerted cycloaddition is detected by the product ratio.

In addition, the totally biradical mechanism fails to explain why the fluorine-substituted double bond of difluoroallene reacts preferentially in [2+2] cycloadditions and not at all in [2+4] cycloadditions.

This regiospecificity of the difluoroallene [2+4] cycloadditions can be successfully rationalized by application of the Frontier Molecular Orbital Theory of cycloadditions. The Highest Occupied Molecular Orbital(HOMO) of difluoroallene is located at the fluorine-substituted double bond, while the Lowest Unoccupied Molecular Orbital(LUMO) is at the other double bond.[30] Since it is the LUMO of difluoroallene and the HOMO of the diene which is the major interaction, reaction occurs exclusively at the unsubstituted double bond of difluoroallene, where the LUMO is located.

In contrast, the [2+2] cycloadditions of difluoroallene are undoubtedly stepwise biradical reactions. In all cases, the major [2+2] cycloadduct has the fluorines on the ring rather than on the double bond. This is consistent with a biradical intermediate which closes preferentially to give the more stable isomer, that is with fluorines on sp^3 carbon rather than sp^2 carbon.[31]

Fluoroallene Cycloadditions

Fluoroallene[32] reacts very similarly to difluoroallene to give regiospecific [2+4] cycloaddition at the unsubstituted double bond. Thus reaction of cyclopentadiene gave 32 and 33 in a one to one ratio. Reaction of fluoroallene with furan at 50° produced two products 34 and 35 in a ratio of 2.3 to one. Though

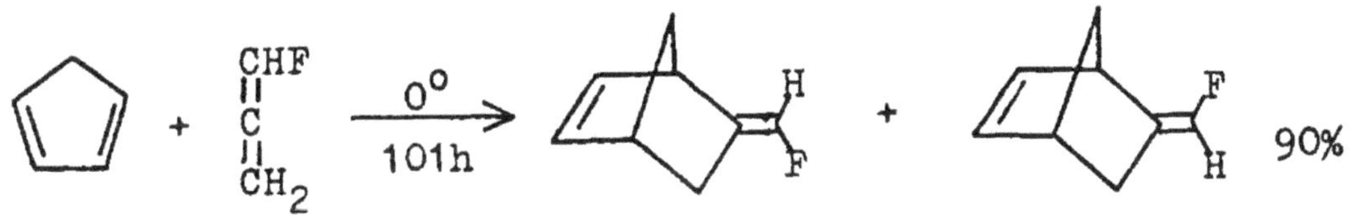

31 32 1:1 33

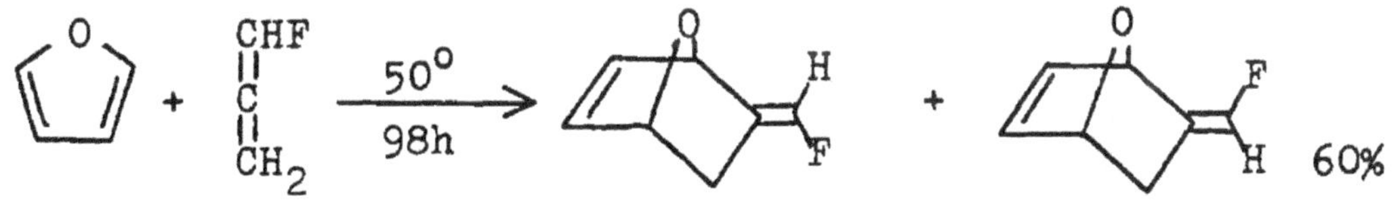

34 1:2.3 35

fluoroallene is less reactive than difluoroallene, it appears a single fluorine is still sufficient to completely control the regiospecificity of the [2+4] cycloadditions.[33]

In the reaction of fluoroallene with butadiene, there is again competition between concerted [2+4] and biradical [2+2] cycloaddition. The formation of

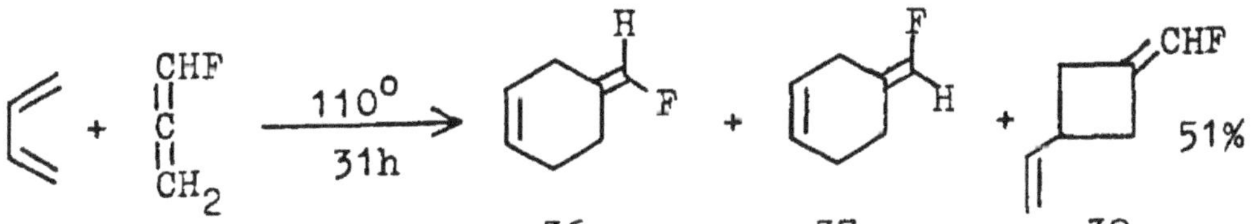

36 37 38

2:2:1

cyclohexenes 36 and 37 is apparently regiospecific. It is interesting to observe that the cyclohexene to cyclobutane ratio is 80 to 20. Apparently both the concerted and biradical reactions are decreased relative to the more reactive difluoroallene.

As expected, reaction of fluoroallene with acrylonitrile gave a biradical [2+2] cycloaddition producing a mixture of 39, 40, and 41. The ratio of

CN + CHF=C=CH_2 —135°, 21h→ 39 + 40 + 41 45%

1:4.4:1

these products is one to 4.4 to one. Once again, the more thermodynamically stable isomer[31] is the major product, only this time it is 40 with a single fluorine on the double bond.

With this background, the stage is set for exploring reaction mechanisms which may be concerted or biradical in nature. The basic idea is to use difluoroallene as a mechanistic probe for concertedness. The assumption is that a concerted reaction will give one product with an exocyclic difluoromethylene group, while a biradical reaction will give two products, the major product with a cyclic difluoromethylene. So from a simple product analysis, the reaction mechanism can be determined.

SECTION II
[2+4] CYCLOADDITIONS

The validity of using difluoroallene as a mechanistic probe is only as good as the reliability of the underlying assumptions:

1. Concerted reactions give products with the fluorines only on sp^2 carbon.

2. Biradical reactions give products with the fluorines on both sp^2 and sp^3 carbons.

The first assumption has considerable experimental evidence to back it up. The second assumption will be considered in the next section. The reactions of difluoroallene with six different dienes have been reported.[24] In all cases, [2+4] cycloaddition occurs exclusively at the unsubstituted allene double bond. Let us consider just one of these reactions, the reaction of difluoroallene with butadiene. The products isolated from the reaction mixture are reported to be 27 and 28 in the indicated ratio. This product ratio

CF_2=C=CH_2 ; 110°, 7h ; =CF_2 ; =CH_2, F_2 ; 87%

26   27   63:37   28

has been rationalized as coming from two distinct reaction pathways, the concerted pathway giving rise to 27 and a biradical pathway forming 28. Noticeably absent from the product mixture are 29 and 30, both of which could, in principle, be formed from the biradical pathway.

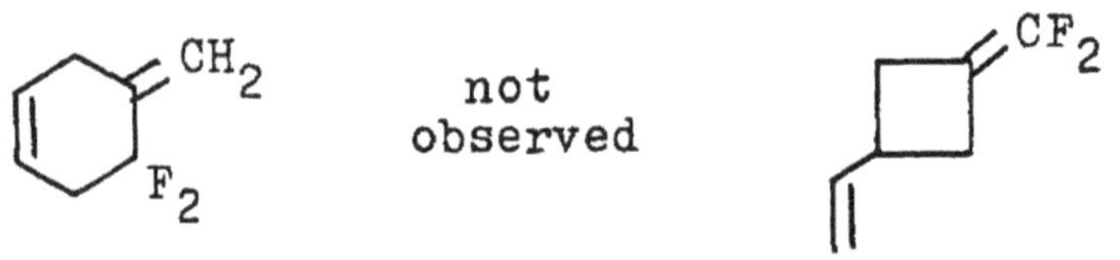

CH_2 F_2 29 not observed CF_2 30

In order to observe products of the type 29 and 30, it is necessary to slow down the concerted pathway which produces most of the product. One way to do this would be to modify the diene appropriately.

Bis-methylenecyclobutane 2 is a diene held in a rigid cisoid conformation. It has been observed that the reaction of 2 with 1,1-dichloro-2,2-difluoroethylene[34] gives only the biradical product 43. In contrast,

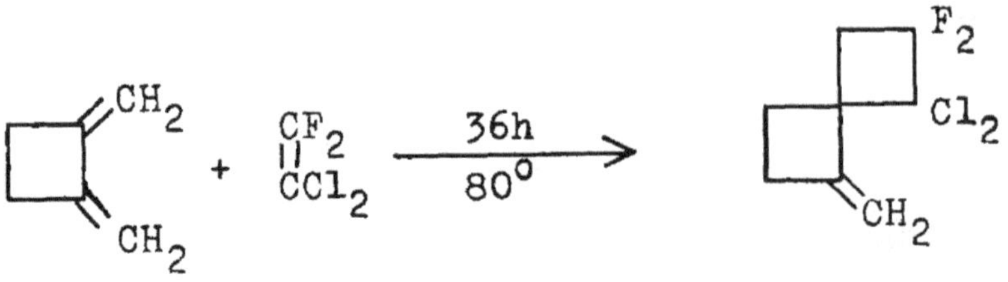

CH_2 CH_2 2 + CF_2 CCl_2 42 —36h, 80°→ F_2 Cl_2 CH_2 43

dichlorodifluoroethylene reacts with cyclopentadiene to give a mixture of biradical and concerted products.[35]

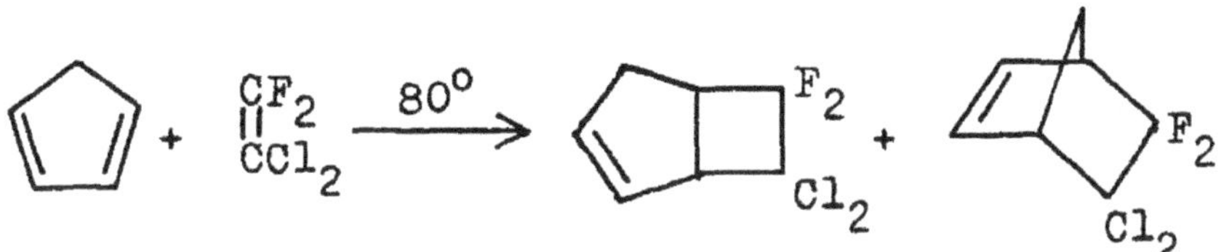

$$\text{cyclopentadiene} + CF_2{=}CCl_2 \xrightarrow{80^\circ} \underset{\underline{44}}{\text{bicyclo (}F_2, Cl_2\text{)}} + \underset{\underline{45}}{\text{norbornene (}F_2, Cl_2\text{)}}$$

44 1:5 45

This difference in relative reactivity for the concerted reaction is not due to a difference in the electron-donating ability of these two dienes. Their ionization potentials are nearly equal being 8.66 eV for 2 and 8.77 eV for cyclopentadiene.[36,37] The decreased reactivity of bis-methylenecyclobutane 2 in concerted reactions has been rationalized by pointing out that the distance between the terminal methylenes of 2 is much larger than for cyclopentadiene or cisoid butadiene. This increased distance leads to an increase

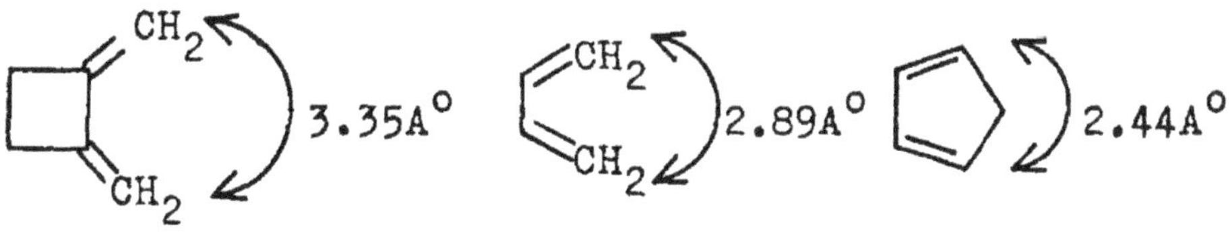

CH_2 ... CH_2 3.35A° — 2; CH_2 ... CH_2 2.89A° — 26; 2.44A° — 6

in activation energy for the concerted reaction, where the olefin must span the terminal methylenes of the diene. The rate of biradical reaction remains the same to a first approximation. On this basis, it was thought that reaction of difluoroallene with 2 might give only biradical reaction products.

Bis-methylenecyclobutane 2 has been prepared by several methods.[38-41] The most direct route to 2 is by gas phase dimerization of allene.[42] This simple approach suffers from two serious drawbacks. First of all the yield of allene dimers is only 20% at best, even after recycling the allene 10 times. Secondly, the dimer mixture contains 20% of 1,3-bis-methylenecyclobutane 3 so that pure 2 can only be obtained after preparative GLPC. This makes the allene dimerization unsuitable for larger preparations.

A convenient method for preparing bis-methylenecyclobutane 2 is by an easy five-step synthesis.[43]

CO_2H, CO_2H — 80° 1h, $SOCl_2$, 84% → COCl, COCl — -60°, $HNMe_2$, 95% → $C(=O)NMe_2$, $C(=O)NMe_2$

46 47 48

$LiAlH_4$, Et_2O, 81% → CH_2NMe_2, CH_2NMe_2 — H_2O_2, 97% → $CH_2N(O)Me_2$, $CH_2N(O)Me_2$

49 50

180-200°, 1mm, 49% → CH_2, CH_2 2

Starting with commercially available 1,2-cyclobutane-dicarboxylic acid 46, the dichloride 47 is prepared by refluxing in thionyl chloride to give an 84% isolated yield of 47 after distillation. The dichloride 47 was treated with dimethylamine at -60° to give a 95% yield of solid crystalline dicarboxamide 48. Reduction with $LiAlH_4$ in ether gave an 81% isolated yield of diamine 49, which was converted by oxidation with 30% H_2O_2 to the diamine N,N'-dioxide in 97% yield. The diamine N,N'-dioxide 50 is an insoluble, hygroscopic solid. It may be stored indefinitely under dry nitrogen and used as a convenient source of 1,2-bis-methylenecyclobutane 2. The bis-methylenecyclobutane 2 is obtained by pyrolysis of diamine N,N'-dioxide under full vacuum. Highly pure product is obtained by a simple distillation. Analysis by GLPC indicated a purity of 99.8%.

Although bis-methylenecyclobutane 2 does not react with difluoroallene at room temperature, after heating at 80° for 4 hours, two products 51 and 52 were isolated

CH_2 CH_2 + CF_2 C CH_2 $\xrightarrow[4h]{80^{o}}$ CF_2 + CH_2 F_2 CH_2 79%

2 51 76:24 52

by preparative GLPC in 79% yield. The product ratio obtained directly from the reaction mixture by integration of the GLPC peakes was 76 to 24. The analysis of the reaction mixture indicated no other products were present.

The mass spectra of 51 and 52 indicate they are isomeric difluoroallene adducts. The strong absorption in the IR for 51 at 1760 cm^{-1} is diagnostic of the difluoromethylene group as is the AB pattern in the ^{19}F NMR with a 60 Hz F-F coupling constant. The ^{1}H NMR of 51 indicates no olefinic protons are present. The ^{13}C NMR of 51 is consistent with the structure and all carbons are assigned including the difluoromethylene carbon which is a triplet at 152 ppm with a C-F coupling constant of 282 Hz.

The minor isomer 52 has four olefinic protons. These are observed in the ^{1}H NMR of 52 as two AB patterns with an integration of 4 to 6 for the olefinic to aliphatic protons. The ^{19}F NMR of 52 is a complex multiplet. All carbon resonances are assigned for the ^{13}C NMR and the CF_2 is observed as a triplet at 118 ppm with a C-F coupling of 284 Hz.

Considering the product ratio for the cycloaddition, it would appear that the relative rate of concerted reaction is even greater here than for butadiene.

However, it should be remembered that only 9% of butadiene is in the cisoid conformation. Since only cisoid butadiene can give the concerted [2+4] cycloadducts, a direct comparison between the two systems is not valid.

Because the relative amount of cisoid butadiene at 110° is only 9%, it must be concluded that the concerted reaction of difluoroallene has a faster relative rate than the product ratio of 63 to 37 would seem to indicate. If a comparison of product ratios is to be made, the product ratio from cisoid butadiene must be used. This ratio can be approximated by assuming the biradical reaction rates for cisoid and transoid butadiene are equivalent and then correcting the product ratio for the mole fraction of cisoid butadiene at 110°. Very simply, the observed product ratio of 63/37 is divided by the mole fraction of cisoid butadiene(0.0964 at 110°) to give a corrected ratio of 95/5. Consequently, when we compare the reactions of difluoroallene with cyclopentadiene, <u>cisoid</u> butadiene, and bis-methylenecyclobutane <u>2</u>, we see that the relative yield of concerted [2+4] cycloadducts decreases from 100% to 95% to 76% respectively.

In a further attempt to observe biradical products analogous to <u>29</u> and <u>30</u>, it was decided to slow down the concerted reaction pathway by using a diene with electron-withdrawing substituents. Dicyanobutadiene <u>60</u> appeared to be an excellent choice. It reacts with a variety of unactivated olefins to give good yields of cyclohexene adducts.[44,45] For example, dicyanobutadiene 60 can be generated *in situ* from the 1,2-dicyanocyclobutene 54. The reaction works well for cyclopentene and even for ethylene.

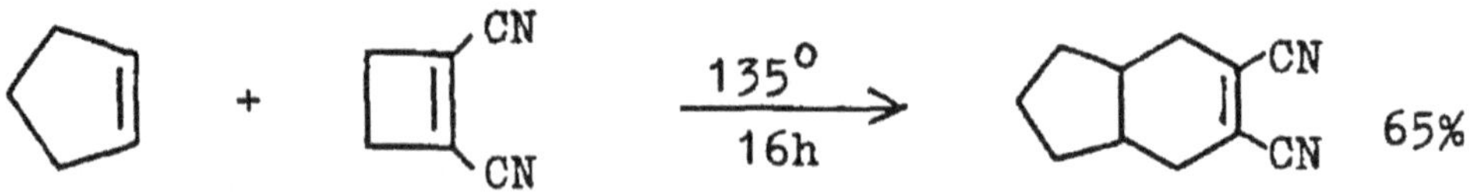

CN, CN, 135°, 16h, CN, CN, 65%

53 54 55

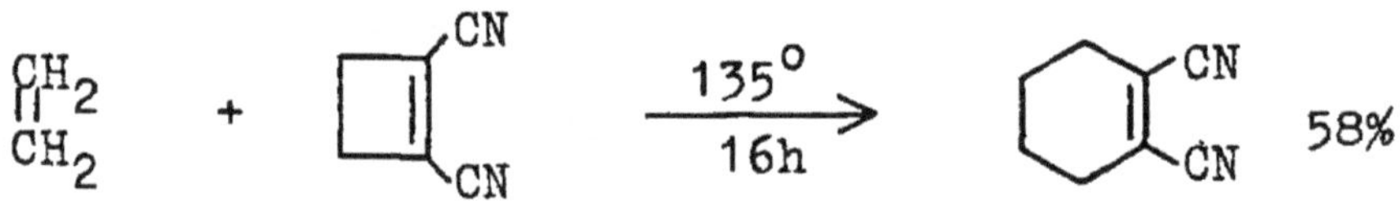

CH_2 CH_2 + CN CN 135° 16h CN CN 58%

56 54 57

Dicyanocyclobutene was prepared according to the literature[44] from commercial 1,2-dicyanocyclobutane in two easy steps. Following the procedure for *in situ*

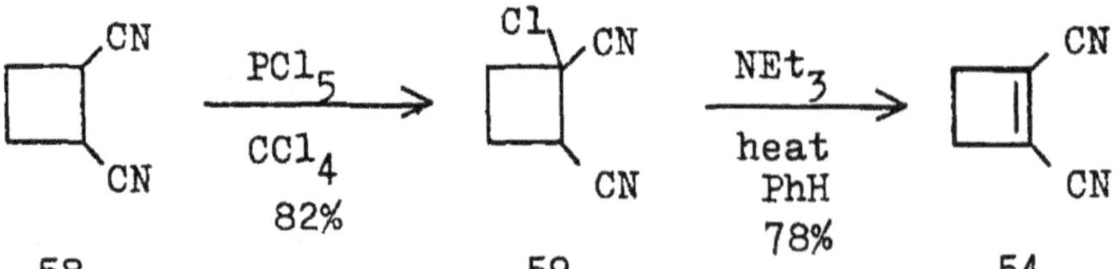

CN CN PCl_5 CCl_4 82% Cl CN CN NEt_3 heat PhH 78% CN CN

58 59 54

generation, a benzene solution of 54 was heated in a sealed tube with difluoroallene for 5 hours at 145°. The reaction mixture was found to contain difluoroallene oligomers and dicyanobutadiene dimer. Obviously, the method of in situ generation is unsuitable for reactions with olefins which are thermally unstable. Although the in situ method was reported to give the best results, it was decided to allow difluoroallene to react directly with 2,3-dicyanobutadiene 60.

According to the literature procedure,[45] 60 was prepared by thermolysis of dicyanocyclobutene 54 in an isolated yield of 83%. The crystalline product 60

CN CN $\xrightarrow[0.1\text{mm}]{395^\circ}$ CN CN 83%

54 60

is reported to polymerize slowly at room temperature, so it was stored at dry ice temperature until used.

Dicyanobutadiene 60 reacted slowly with difluoroallene at room temperature in solution. After 4 days, the reaction was complete and the products were isolated by preparative GLPC. Three primary products 61, 62, and 63 were isolated in 60% yield. In addition, three

NC NC + CF_2=C=CH_2 $\xrightarrow[4\text{days}]{28^\circ}$ NC CH_2 CN F_2 + NC CF_2 CN + NC CF_2 NC 60%

60 61 62 63

84:4:12

secondary products 64, 65, and 66 were obtained in isolated yields of 1%. These secondary products

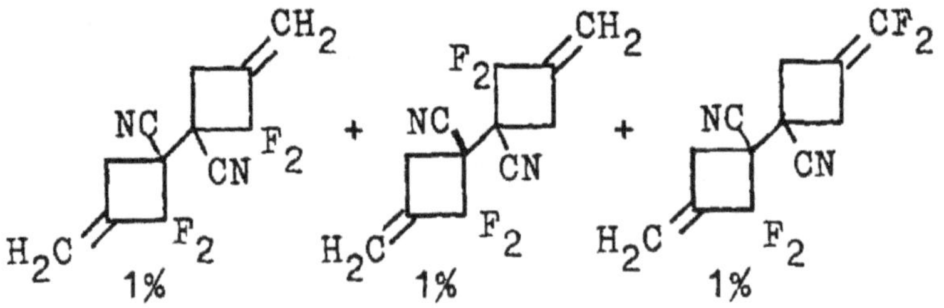

64 (1%), 65 (1%), 66 (1%)

were demonstrated to be diadducts of difluoroallene by their mass spectra and ^{13}C NMR spectra. The primary products were obtained in relative yields of 84%, 4%, and 12% as determined by integration of the ^{19}F NMR spectrum of the reaction mixture.

In view of the 12% relative yield of [2+4] cycloadduct 63, it is concluded that the reaction of dicyanobutadiene with difluoroallene was successful in decreasing the rate of concerted reaction. In fact, considering the electron-withdrawing cyano groups on the diene, it is rather surprising that any concerted product is observed at all. Even though the major reaction pathway is by a biradical, there was no trace of cyclohexene product with fluorines on the ring in the reaction mixture. Since the cisoid to transoid ratio is unknown for dicyanobutadiene, the relative rates of [2+2] and [2+4] cycloaddition for the *cisoid* diene cannot be estimated.

In order to further broaden the scope of the [2+4] reactions of difluoroallene, the reaction of diphenylisobenzofuran 67 was studied. The usual concerted product 68 was observed by ^{1}H NMR(see experimental section) however this initial product rearranged upon work-up to give 69 and 70 in a ratio

$$\mathbf{67} + CF_2{=}C{=}CH_2 \xrightarrow[4h]{28^\circ} \mathbf{68} \longrightarrow \mathbf{69} + \mathbf{70} \quad 32\%$$

of one to one. This rearrangement probably occurs by acid-catalyzed ring opening followed by dehydration. Product 69 comes from partial hydrolysis of the difluoromethylene group and 70 comes from attack of fluoride ion on the difluoromethylene group.

Compound 70 was also prepared by an alternative two-step synthesis. Cycloaddition of trifluoropropene with diphenylisobenzofuran gave cycloadducts 72 and 73 in the ratio of 90.4 to 9.6. Pure 72 was obtained by recrystallization. Dehydration of 72 with polyphosphoric acid gave compoud 70 which was the same as the product obtained form the difluoroallene reaction.

67 + 71 → (83°, 63h) → 72 + 73, 89%, 90.4:9.6

A sample of difluorodimethylallene 80 was available by a six-step synthesis starting with ethyl trifluoroacetate.[32]

CF_3CO_2Et + i-PrMgBr (74) → (Et_2O) → 75 → (AcCl, heat 4h, 93%) → 76

→ (500°, flow, 75%) → 77 → (Br_2, 79%) → 78 → (KOH, heat, 74%)

79 → (-85° BuLi, heptane, 28%) → $(H_3C)_2C{=}C{=}CF_2$ 80

It was found that difluorodimethylallene is less reactive than difluoroallene. The steric effect of the two methyl groups slows down the reaction. Also, the electronic effect of the methyl groups is to make difluorodimethylallene a less electron-deficient olefin than difluoroallene. Even so, the reaction

$$\text{cyclopentadiene} + \underset{\underline{80}}{(H_3C)_2C{=}C{=}CF_2} \xrightarrow[\text{4days}]{28^{\circ}} \underline{81}\ (38\%)$$

with cyclopentadiene is completely regiospecific giving only product 81.

These studies seem to indicate that the first assumption is valid. That is, difluoroallene does, in fact, give only products with fluorines on sp^2 carbon in concerted reactions. The inverse demand Diels-Alder reaction is not observed and neither are [2+4] cycloadducts from biradical pathways.

SECTION III
[2+2] CYCLOADDITIONS

The assumption that biradical cycloadditions of difluoroallene give two products with fluorines on sp^2 and sp^3 carbons is really based only on one reaction. Some additional [2+2] cycloadditions of difluoroallene needed to be studied. A good place to start seemed to be with the dimerization of difluoroallene, which has been reported[22] to give only one product 13.

Difluoroallene was kept in a sealed tube at room temperature for four days. After this time roughly half of the difluoroallene had reacted. In addition to polymeric material, three products were obtained.

$CF_2{=}C{=}CH_2$ —(28°, 4days)→ 13 + 82 + 83 (18.4%)

They proved to be 13, 82, and 83, which were present in relative yields of 33%, 17%, and 49%. The neat dimer 13 was very unstable to polymerization at room temperature and had to be kept at dry-ice temperature.

Not a trace of dimer 84 was seen in the reaction mixture. Presumably it is converted rapidly to trimer 83 under the reaction conditions by reaction with difluoroallene.

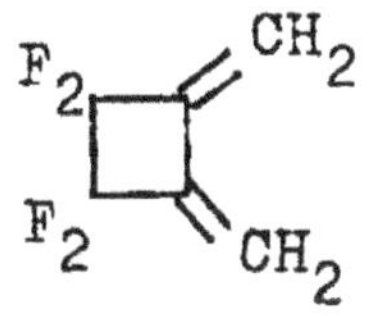

F2 CH2 F2 CH2 not observed

84

Due to the polymerization and the presence of minor products which were not isolated, the relative amounts of initial [2+2] cycloadducts cannot be determined accurately. However, it can be concluded that both 13 and 84 are formed, probably in comparable amounts.

The reaction of difluoroallene with acrylonitrile gave two products 14 and 21 in the ratio 77 to 23, as determined by GLPC integration of the reaction mixture. In addition, 2% of difluoroallene trimer 83 was isolated.

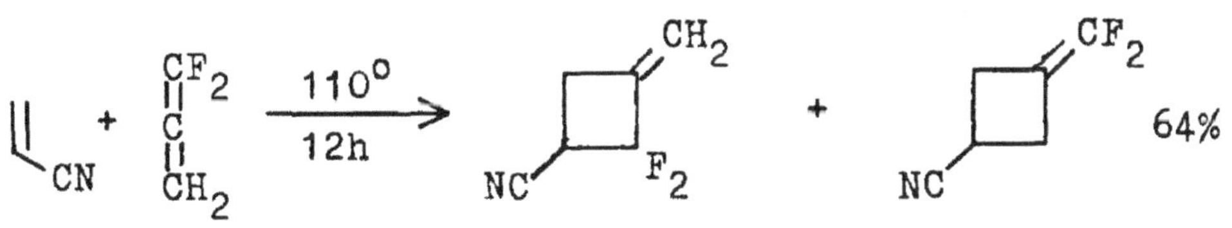

CN + CF_2=C=CH_2 $\xrightarrow[12h]{110^{\circ}}$ (NC, F2, CH2) 14 + (NC, CF2) 21 64%

14 77:23 21

For this reaction, a 10-fold excess of acrylonitrile was used. This large excess of acrylonitrile is needed because difluoroallene dimerizes more rapidly than it reacts with other olefins. For this reason difluoroallene does not react with olefins which are not activated towards biradical cycloadditions.

The reaction of difluoroallene with methacrylonitrile gave a 39% isolated yield of products 86 and 87 along with 9% of isolated trimer 83. This lower yield of methylenecyclobutane products may be due to the fact that only a 7-fold excess of methacrylonitrile was used. The higher yield of difluoroallene trimer 83 supports this view.

85 + $CF_2{=}C{=}CH_2$ → (110°, 24h) 86 + 87 (84:16) 39%

It was also found that dichlorodifluoroethylene reacted with difluoroallene at 135° to give cycloadducts 88 and 89 in a ratio of 27 to 73. This ratio, unlike the others, appeared to change with reaction time. When

$CF_2{=}CCl_2$ + $CF_2{=}C{=}CH_2$ → (134°, 27h) 88 + 89 (51:49) 20%

the reaction was carried out at the same temperature for only one hour, analysis of the reaction mixture by GLPC integration indicated a product ratio for 88 and 89 of 51 to 49. This ratio at lower conversion is taken to be the more reliable one.

Dichlorodifluoroethylene also reacts with fluoroallene under similar conditions. The ratio of cycloadducts 90, 91, and 92 was 25.3 to 49.3 to 25.4 for a reaction time of 8 hours at 134°. At 80° for 30 hours, the product ratio was very similar, being 23.4 to 51.1 to 25.5; however, the conversion was very low under these conditions.

$CF_2{=}CCl_2$ + $CHF{=}C{=}CH_2$ $\xrightarrow[8h]{134^\circ}$ 90 + 91 + 92 (45%)

90: F_2, Cl_2 cyclobutane with =CHF (H, F)
91: F_2, Cl_2, F cyclobutane with $=CH_2$
92: F_2, Cl_2 cyclobutane with =CFH (F, H)

25.3:49.3:25.4

A summary of the [2+2] cycloadditions of olefins with difluoroallene is presented in Table I. The methylenecyclobutane products 93 and 94 are obtained in the indicated ratios. Though a number of 1,3-diene cycloadditions have been studied previously, only the present reaction with dicyanobutadiene was done carefully enough to isolate a product with structure 94.

Table I. Summary of Product Ratios from Difluoroallene [2+2] Cycloadditions.

$X_2C{=}CYZ$			Temp.	X_2, YZ ring with =CH_2, F_2	X_2, YZ ring with =CF_2
X	Y	Z	(°C)	93	94
H	CN	$CH_2{=}CCN$	28°	95	5
H	CN	CH_3	110°	84	16
H	CN	H	110°	77	23
F	Cl	Cl	135°	51	49

There seems to be a rough correlation between the stability of the biradical intermediate and the product ratio. This correlation may be explained by applying the Hammond Postulate to closure of the biradical intermediate. Early transition states for biradical closure are less selective and product ratios of roughly 50:50 are observed. For longer-lived, more stable biradicals, the transition state is later and resembles the products. Therefore the thermodynamic preference of geminal fluorines for an sp^3 carbon leads to predominant closure of CF_2 to give fluorines on the cyclobutane ring. This theory is presented pictorially in Figure 1 where the total energy is presented as a function of the reaction coordinate.

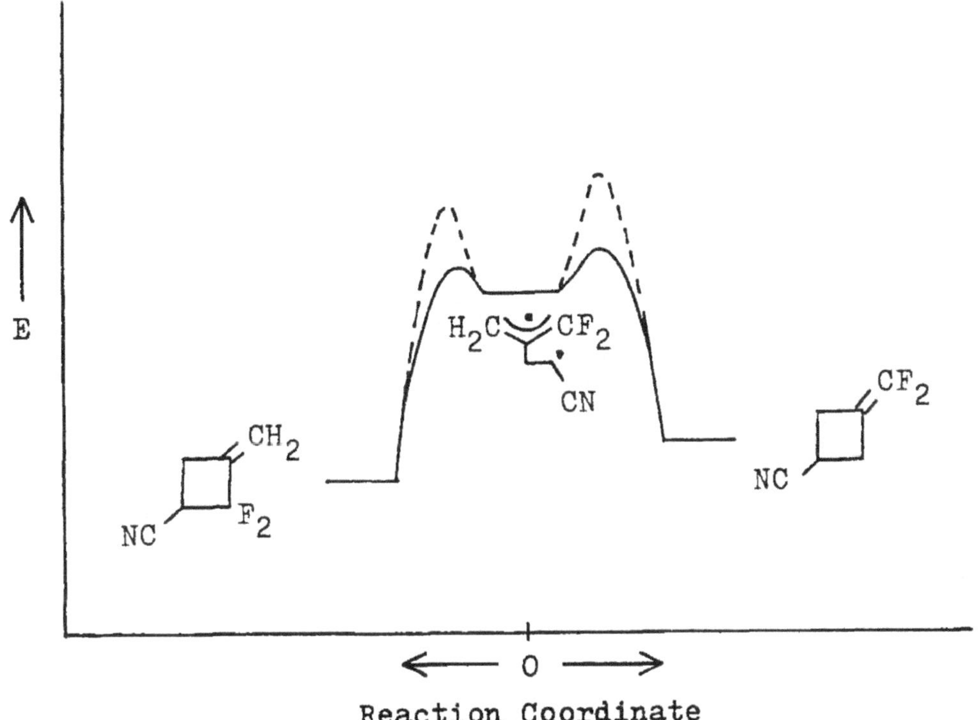

Figure 1. Potential Energy as a Function of the Reaction Coordinate for Closure of the Difluoroallyl Biradical. Earlier Transition State ——— Later Transition State - - -

This hypothesis explains the data rather nicely, but more experiments need to be done in order to confirm or refuteeit. It should be noted that the few results for cycloadditions of fluoroallene are also completely consistent with this hypothesis. Remembering that the thermodynamic preference for a single fluorine substituent is for sp^2 carbon, we predict an opposite trend for the product ratios from the closure of the fluoroallyl biradical. This seems to be the case as can be seen by looking at the fluoroallene product ratios in Table II.

Table II. Summary of Product Ratios from Fluoroallene [2+2] Cycloadditions.

X_2C=CYZ			Temp.	X_2/YZ cyclobutane, =CH_2, F	X_2/YZ cyclobutane, =CHF
X	Y	Z	(^{o}C)	95	96
H	CN	H	135°	31	69
F	Cl	Cl	134°	49	51

All the data indicate that biradical cycloadditions of difluoroallene give two products with fluorines on sp^2 and sp^3 carbons, as previously assumed. Though in certain cases, more highly stabilized biradicals may give almost entirely one product, the regioselectivity is exactly the opposite of that observed for concerted cycloadditions. Thus the two basic assumptions concerning cycloaddition regiochemistry for difluoroallene reactions are confirmed.

SECTION IV
1,3-DIPOLAR CYCLOADDITIONS

Diazoalkanes

Let us consider a typical 1,3-dipole, for example diazomethane. The two most commonly considered resonance forms are presented below, along with the less commonly considered singlet biradical resonance form.

$$H_2\overset{-}{C}-\overset{+}{N}\equiv N \longleftrightarrow H_2C=\overset{+}{N}=\overset{-}{N} \longleftrightarrow H_2\dot{C}-\ddot{N}=\dot{N}$$

Recent theoretical studies of 1,3-dipoles indicate that the biradical resonance form may be as good as, or even better than the charge-separated resonance forms.[46] According to a simplified theoretical model for diazomethane, it has a π system composed of three atomic p-orbitals, containing four electrons. This makes it isoelectronic with the allyl anion. There is also an orthogonal π bond not participating in this system as indicated in structure 97. The 4 π system in diazomethane is similar to the π system in cyclopentadiene, except that the four electrons are spread out over three atoms instead of four atoms as in dienes. For this reason,

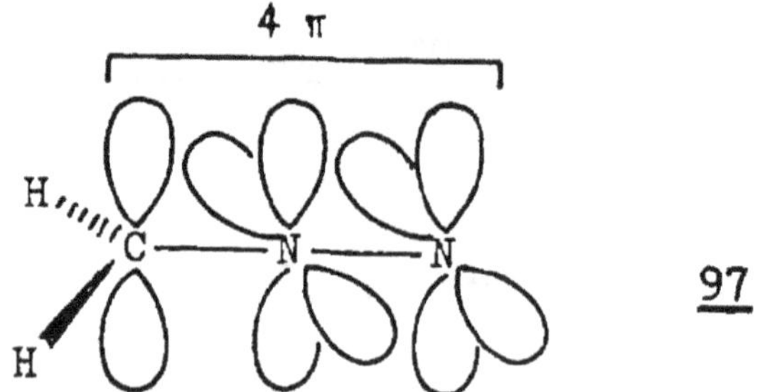

1,3-dipoles will give concerted cycloaddition reactions with olefins. The leading worker in this field is Huisgen who has written two reviews.[47,48] Reactions of substituted allenes have been studied by Battioni and coworkers[49,50] and some 1,3-dipolar cycloadditions of tetrafluoroallene have been reported by Blackwell, Haszeldine, and Taylor.[19]

Though all the evidence points to a concerted mechanism for 1,3-dipolar cycloadditions, it has been suggested that these reactions could be stepwise in nature,[51,52] proceeding by diazenyl biradical intermediate 98. The proposed mechanisms are depicted below.

Huisgen Firestone

CH_2 CH_2 Z → 99 ← 98

Since the reaction goes faster with electron-withdrawing groups on the olefin, it was suspected that difluoroallene, an electron-deficient olefin and good dieneophile, would react easily with 1,3-dipoles. Furthermore, difluoroallene could act as a mechanistic probe, because the Firestone intermediate in this case has a difluoroallyl system 100 which must close to give two products 101 and 102 as shown below.

100 ⟶ 101 + 102

From analogy with the [2+2] cycloadditions of difluoroallene, which are authentic biradical cycloadditions, the Firestone intermediate is expected to give 102 as the major product with perhaps a minor amount of 101. On the other hand, FMO Theory applied to 1,3-dipolar cycloadditions[53-55] leads us to expect that no 102 at all will be formed.

With due precautions, diazomethane was prepared as a yellow ether solution. Reaction with difluoroallene occurred rapidly as indicated by the immediate disappearance of the yellow diazo color. Only one product was formed as indicated by ^{19}F NMR analysis of the ether solution. This product was 4-(difluoromethylene)-1-pyrazoline 101. Careful distillation at reduced pressure gave pure pyrazoline 101 in 52% isolated yield. It was characterized spectroscopically. The NMR yield was determined to be 95% using

$$H_2C{=}N_2 \quad + \quad CF_2{=}C{=}CH_2 \xrightarrow[Et_2O]{0^\circ\ 5min} \text{101} \quad 52\%\ (95\%)$$

101

a slight excess of diazomethane and adding benzene as an internal standard. Pyrazoline 101 is unstable to air, light, and heat. It may be stored indefinitely at -78° in the absence of oxygen, but decomposes if kept at room temperature.

Since 101 is produced without even a trace of 102, it must be concluded that the Firestone intermediate is not involved in the reaction, even to a small extent. Thus the reaction appears to be completely concerted.

It is tempting to make generalizations about diazoalkane cycloadditions from this one regiospecific reaction with difluoroallene. In fact, it turns out that diazomethane is the only diazoalkane which reacts regiospecifically. Thus the reaction of 2-diazopropane with difluoroallene gave two regioisomers 104 and 105 along with some acetone azine. The formation of these

$$N_2{=}C(CH_3)_2 + CF_2{=}C{=}CH_2 \xrightarrow[Et_2O]{0^\circ\ 5min} \textbf{104} + \textbf{105}\ (99\%)$$

103   104   61:39   105

two isomers is once again consistent with the concerted mechanism and not with Firestone's biradical intermediate.

Though the NMR yield is 99% for the reaction, pure samples of regioisomers 104 and 105 could not be obtained due to instability to GLPC conditions and to silica gel chromatography. Although the ^{1}H NMR, ^{19}F NMR, and IR spectra of the mixture were all consistent with the structural assignments, further confirmation was desired. It was decided to synthesize pyrazoline 104 by a different route.

Analysis of all possible data available at the time indicated that if difluorodimethylallene could be synthesized, its reaction with diazomethane would give only regioisomer 104. In a series of six reactions, difluorodimethylallene was obtained from ethyl trifluoroacetate.[32]

It was found that difluorodimethylallene reacted regiospecifically with diazomethane as predicted to give a 74% isolated yield of pyrazoline 104. The pure material was obtained by distillation at reduced pressure and it was characterized spectroscopically.

$$H_2C{=}N_2 + (H_3C)_2C{=}C{=}CF_2 \xrightarrow[2h]{28^\circ} \textbf{104} \quad 74\% \ (90\%)$$

104

Thus the structure of 104 was confirmed; however, assignment of structure to compound 105 still relies on spectra of the reaction mixture.

Difluoroallene reacted more slowly with diphenyldiazomethane to give, once again, two regioisomers 107 and 108. The major isomer 108 was unstable in chloroform solution, but stable in acid-free solvents such as ether. Pyrazoline 108 was also unstable to silica gel chromatography but was able to be isolated

$$Ph_2C{=}N_2 + CF_2{=}C{=}CH_2 \xrightarrow[5h]{28^\circ} \textbf{107} + \textbf{108} \quad 95\%$$

106   107   14:86   108

and purified by rapid recrystallization from hexane. The minor isomer 107 was isolated in pure form by silica gel chromatography of the mother liquor from the recrystallization. The product ratio was determined by ^{1}H NMR and ^{19}F NMR analysis of the reaction mixture before work-up. The relative ratio of 107 to 108 was 14 to 86.

The reaction of diazofluorene 109 with difluoroallene gave a similar ratio of pyrazoline products 110 and 111. The relative yield was 28 to 72 by ^{1}H NMR integration of the reaction mixture. The isolated yield for the

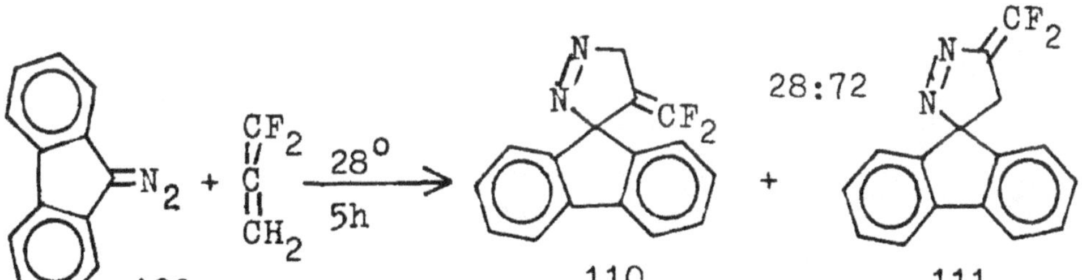

109 + $CF_2{=}C{=}CH_2$ —(28°, 5h)→ 110 + 111 (28:72, 99.8%)

crude product mixture was essentially quantitative. The minor isomer 110 was thermally very unstable, extruding nitrogen at room temperature during the course of the reaction to give 112 which was the

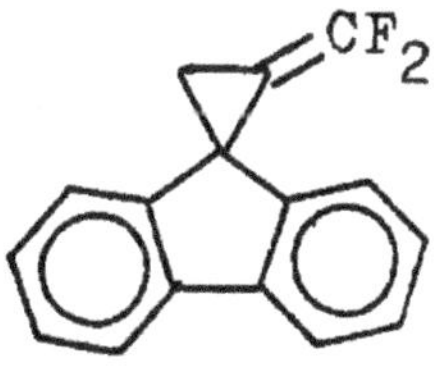

112

product isolated. Pyrazoline 110 could be observed in the ^{1}H NMR spectrum during the reaction, but by the time reaction was complete, it had all been transformed to 112. The ^{1}H NMR chemical shift and coupling constant to fluorine in 110 for the methylene protons are completely consistent with the proposed structure as can be seen by looking at Table III.

Table III. ^{1}H NMR Chemical Shifts in ppm and Coupling Constants to Fluorine in Hz for the Methylene Protons of 4-(Difluoromethylene)- and 4-(Fluoromethylene)-1-pyrazolines.

Pyrazoline	δ CH_2 (ppm)	J_{HF} (Hz)
101	5.1	3.7
104	5.19	3.8
107	5.34	3.5
110*	5.65	3.7
113	5.16, 5.06	2.8

*observed in reaction mixture but not isolated.

Pyrazoline 111 could not be purified by recrystallization as had been 108 and it was just as unstable to silica gel chromatography. However, spectra of the mixture containing 72% pyrazoline 111 were consistent with the proposed structure as is seen by looking at Table IV.

Conjugation of the difluoromethylene group with the azo group appears to lower the $=CF_2$ stretching

Table IV. ^{1}H NMR Chemical Shifts in ppm and Coupling Constants to Fluorine in Hz for the Methylene Protons of 5-(Difluoromethylene)-1-pyrazolines.

Pyrazoline	δ CH_2 (ppm)	J_{HF} (Hz)
105*	2.1	3.7
108	3.02	3.6
111*	2.97	3.8

*compound not isolated in pure form.

frequency in the IR spectrum of 105, 108, and 111 to 1756, 1757, and 1756 cm^{-1} respectively. More typical $=CF_2$ stretching frequencies are observed for 102, 104, and 107, being 1797, 1780, and 1780 cm^{-1} respectively.

Since a sample of 4-(fluoromethylene)-1-pyrazoline 113 was desired for some further studies, fluoroallene was allowed to react with diazomethane. The single product 113 was isolated by distillation at reduced pressure in 64% yield.

$$CH_2N_2 + CHF{=}C{=}CH_2 \xrightarrow[10min]{28^o} \text{113} \quad 64\%$$

113

Why are some of these cycloadditions regiospecific, while others are regioselective for the opposite isomer?

The FMO Theory fails to explain this dramatic reversal of regioselectivity. Calculations indicate that diazomethane, diazopropane, and diphenyldiazomethane all have virtually identical orbital coefficients for their HOMO.[56] This leads to the incorrect prediction that their regiochemistry in cycloadditions should be the same.

One possible explanation could be polar effects in the cycloaddition transition state. When two polar compounds such as diazomethane and difluoroallene react they may line up so that their dipoles minimize the polarity of the transition state. This is an attractive hypothesis at first glance. It is, in fact, easily tested. If the transition states leading to the two different products have different polarities, the product ratio should be changed by changing the solvent polarity.

The effect of solvent polarity was determined for reaction of diphenyldiazomethane with difluoroallene. Product ratios were determined by ^{1}H NMR integration and by ^{19}F NMR integration. The results are presented in Table V. Going from DMSO to pentane, there is no significant difference in the product ratio of pyrazolines <u>107</u> and <u>108</u>. Hydrogen bonding solvents were not used in this study. In view of this result, the dipole argument deserves to be discarded.

Table V. Solvent Study of Product Ratio for Reaction of Difluoroallene with Diphenyldiazomethane.

Solvent	Dielectric Constant	Product Ratio	
		107	108
DMSO-d_6	46.7	14.5	85.5
Acetone-d_6	20.7	12.9	87.1
Ether	4.3	12.0	88.0
CCl_4	2.2	16.5	83.5
Hexane	1.9	14.3	85.7
Pentane	1.8	15.6	84.4

The observed regiochemistry for the diazoalkane cycloadditions, which is summarized in Table VI, is easily rationalized in terms of two opposed effects. They are an electronic effect and a steric effect.

It is proposed that the electronic preference is for the carbon end of the 1,3-dipole to interact with the central carbon of difluoroallene. This electronic preference favors formation of pyrazolines 114. When the two ends of the diazoalkane have similar sizes, as with diazomethane, the electronic preference is dominant. However, substituted diazoalkanes have a much larger carbon end and the sterically more favored product 115 starts to show up in the reaction mixture.

Table VI. Summary of Product Ratios for Reaction of Difluoroallene with Diazoalkanes.

Dipole R_2CN_2 R	Temp. (°C)	114	115
H	0°	100	0
CH_3	0°	61	39
Fl=	28°	28	72
Ph	28°	14	86

With large aromatic substituents on the 1,3-dipole, the sterically favored pyrazoline 115 becomes the major product. This reversal of regiochemistry has been observed before for cycloadditions of diazoalkanes with thiete 1,1-dioxide.[56]

A survey of the literature reveals that allene itself reacts with diazomethane[57] and diazopropane,[57,58] though at much slower rates than difluoroallene. The product ratios are similar to the difluoroallene cycloadditions. Thus the same steric reversal of

$$H_2CN_2 + CH_2{=}C{=}CH_2 \xrightarrow[24h]{\text{ambient}} \text{4-methylene-1-pyrazoline}$$

$$(H_3C)_2C{=}N_2 + CH_2{=}C{=}CH_2 \xrightarrow[3h]{0^{\circ}} \text{4-methylene-3,3-dimethyl-1-pyrazoline} + \text{4-methylene-5,5-dimethyl-1-pyrazoline}$$

80:20

regiochemistry appears to be present for allene as for difluoroallene. This indicates that the orbital coefficients of the respective LUMO's for allene and for difluoroallene are probably similar in magnitude.

What is the origin of the electronic preference in these cycloadditions? To answer this question we may use the FMO Theory. To do this, it is necessary to know the orbital coefficients and energies for the 1,3-dipoles and for difluoroallene. Calculations for 1,3-dipoles have been reported in the literature.[53,55] Calculations for difluoroallene have also been reported[30] but the orbital coefficients necessary for predicting regiochemistry of cycloadditions have not been presented. Consequently, the only way to explain the observed electronic preference without making any unwarranted assumptions, was to calculate the shapes of the difluoroallene molecular orbitals.

Extended Huckel calculations were performed in collaboration with Randy Winchester. Using the experimental geometry for difluoroallene derived from its microwave spectrum[59] gave the results presented in Figure 2. The energies are the Extended Huckel energies and are, of course, not reliable. However, the shapes of the molecular orbitals are accurately represented by the Extended Huckel Method.

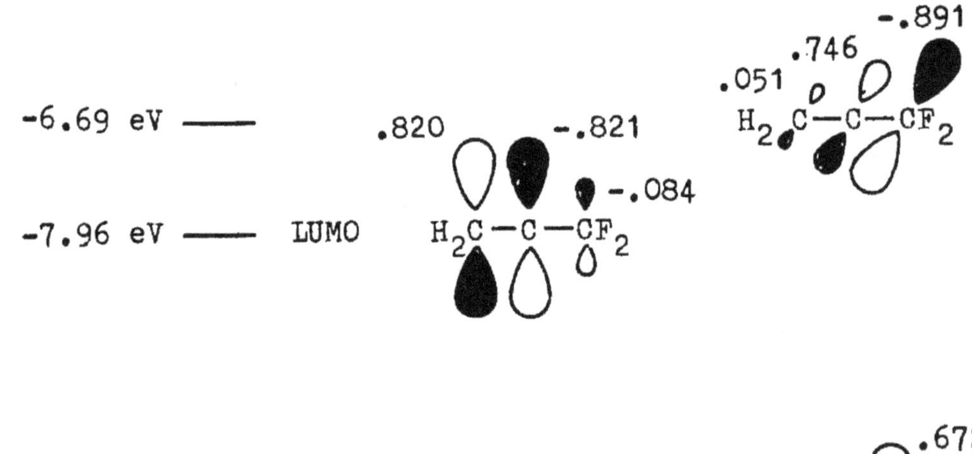

Figure 2. Extended Huckel MO Energies and AO Coefficients for Difluoroallene.

A similar calculation for diazomethane gave the results presented in Figure 3. Again, the energies are the Extended Huckel energies.

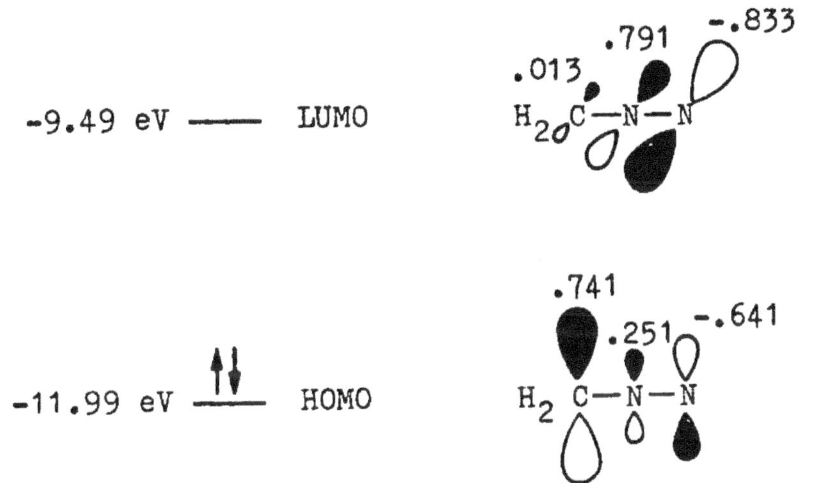

Figure 3. Extended Huckel MO Energies and AO Coefficients for Diazomethane.

An accurate estimation of the Frontier Molecular Orbital energies can be obtained from experimental observations. The energy of the HOMO is obtained from the Ionization Potential and the energy of the LUMO is available from the Electron Affinity. The estimated FMO energies for difluoroallene[30] and diazomethane[53] are presented in Figure 4 and the controlling Frontier Orbital interaction is identified with a line. The energy separation between the LUMO of difluoroallene and the HOMO of diazomethane is 8.19 eV, while the energy difference for the HOMO of difluoroallene and the LUMO of diazomethane is 10.3 eV. Thus the controlling interaction is the LUMO of difluoroallene and the HOMO of diazomethane.

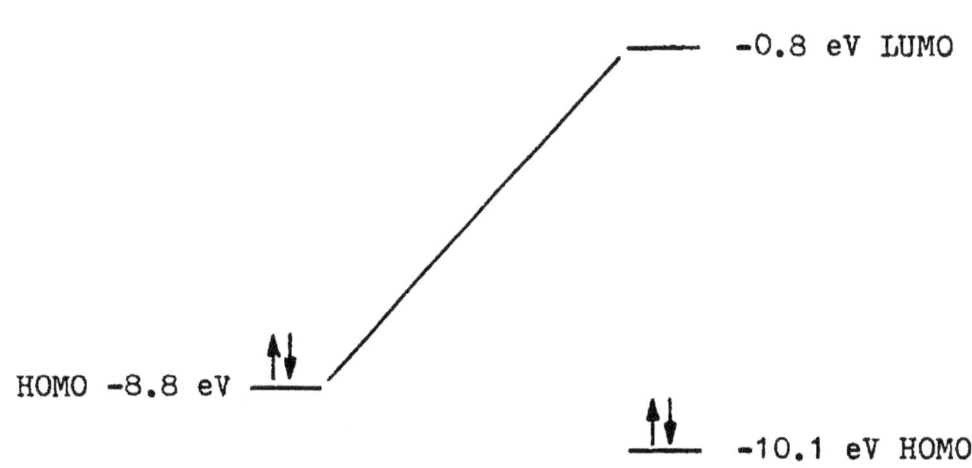

Figure 4. Experimental HOMO and LUMO Energies for Difluoroallene and for Diazomethane in eV.

For the sake of simplicity, the Atomic Orbital coefficients of fluorine and hydrogen are not presented in Figures 2 and 3. Let us consider the actual shapes of these Molecular Orbitals. The HOMO of difluoroallene is localized on C_1 and C_2, while the LUMO is orthogonal and localized on C_2 and C_3. One can see, however, that the classical view of an allene as having two orthogonal, non-interacting π bonds is an oversimplification. Let us look more closely at the

LUMO of difluoroallene, which is the Frontier Orbital of interest in explaining the regiochemistry of concerted cycloadditions.

It is important here to include all the Atomic Orbital contributions if an accurate picture of the LUMO is desired. Figure 5 is a computer-simulated representation of the LUMO for difluoroallene, which includes the contributions from Atomic Orbitals of all the atoms. It should be obvious that the

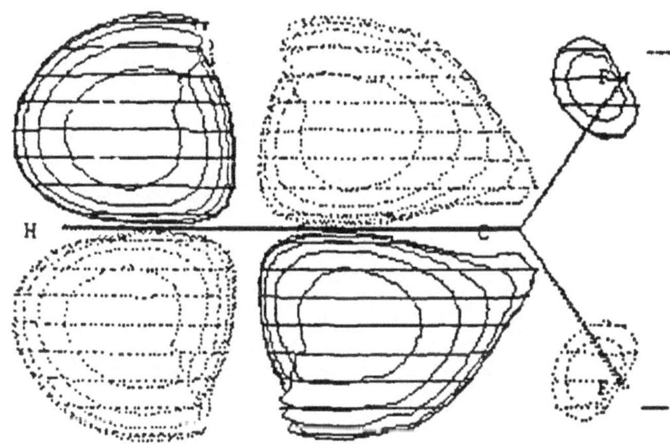

Figure 5. Contour Surface for the Wavefunction of the Difluoroallene LUMO.

contributions to the wavefunction from carbon and fluorine Atomic Orbitals of the CF_2 determine the shape of the LUMO. Specifically, the bonding interaction of the C_1 p-orbital with the C_2 p-orbital results in reinforcement of the wave function at C_2. What this means is that even though the AO coefficients for C_3 and C_2 are virtually identical(0.820 and 0.821), the value of the wave function is larger at C_2 for the LUMO.

The kind of orbital contribution we see here is generally called a "secondary interaction" and is usually ignored when predicting regiochemistry of cycloadditions. In the case of the 1,3-dipolar cycloadditions with difluoroallene, ignoring the "secondary interaction" of the CF_2 leads to the incorrect prediction that no regioselectivity would be observed. In fact, as we shall see, it is the "secondary interaction" of the CF_2 which is responsible for determining regioselectivity in the 1,3-dipolar cycloadditions.

In order to demonstrate this conclusion, a perturbation calculation was performed for the two modes of cycloaddition of diazomethane with difluoroallene. What this method involves is a calculation for the change in energy which accompanies the interaction of two molecules involved in a cycloaddition reaction. This interaction energy ΔE is given by the second-order perturbation expression presented in Equation 1.[60,61] This equation deals only with the stabilization due to orbital overlap and not at all with coulombic attractions or closed shell repulsions. The lack of solvent effect for the cycloaddition indicates that coulombic attractions are not important in determining regiochemistry. We assume that the closed shell repulsions(steric effect) for diazomethane can be neglected since the experimentally observed product is the more sterically hindered one. In Equation 1, γ_{rs} is the interorbital interaction

$$\Delta E = 2\left(\sum_{R}^{occ}\sum_{S}^{unocc} - \sum_{R}^{unocc}\sum_{S}^{occ}\right)\frac{(\sum_{rs} c_r\, c_s\, \gamma_{rs})^2}{E_R - E_S} \quad \text{Eq(1)}$$

integral for Atomic Orbitals r and s in Molecular Orbitals R and S, C_r is the Atomic Orbital coefficient at atom r in Molecular Orbital R, C_s is the Atomic Orbital coefficient at atom s in Molecular Orbital S, E_R is the energy of Molecular Orbital R, and E_S is the energy of Molecular Orbital S.

For our model, we will consider only the Frontier Orbital interaction of the HOMO of the 1,3-dipole and the LUMO of difluoroallene. In addition, γ_{rs} will be approximated in the usual manner by the overlap integral S_{rs} between r and s. Thus the expression for the interaction energy reduces to the following expression in Equation 2, where k is a proportionality constant. The validity of this

$$\Delta E_{HO-LU} = k \frac{(\sum_{rs} C_r C_s S_{rs})^2}{E_{HO} - E_{LU}} \quad Eq(2)$$

expression for diazomethane cycloadditions has been experimentally confirmed by Geittner, Huisgen, and Sustmann.[62]

Perturbation calculations using the two geometries presented in Figure 6 were performed by the Extended Huckel Method. For this model, diazomethane is still

Type I	Type II
N—N—CH_2	H_2C—N—N
H_2C—C—CF_2	H_2C—C—CF_2

Figure 6. Geometries for the Two Modes of Cycloaddition of Difluoroallene and Diazomethane.

linear and it is positioned exactly between C_2 and C_3 of difluoroallene at a distance of 2.7 A°. The values obtained for the numerator of Equation 2 are -0.0620 for Type I addition and -0.0456 for Type II addition.

These results predict Type I addition to be favored and are in agreement with the experimental fact that only Type I addition occurs. This is interpreted as evidence that "secondary interactions" of the difluoroallene CF_2 control the regiochemistry for diazomethane cycloaddition.

In conclusion, it is the reinforcement of the wavefunction at the central carbon of difluoroallene by CF_2 "secondary interactions" which leads to the electronic preference in diazoalkane cycloadditions. Normally one thinks of such "secondary interactions" as unimportant, but in this case they result in complete regiospecificity. As we shall see in the next section, these "secondary interactions" are even more important for nitrone cycloadditions.

Nitrones

Another class of 1,3-dipoles is the nitrones. The cycloaddition reactions of nitrones have been investigated extensively by Huisgen, Seidl, and Bruning.[63] Reaction of nitrones with substituted allenes[49] and with tetrafluoroallene[19] have been reported. The two most common and easily prepared nitrones are diphenylnitrone and C-phenyl-N-methylnitrone. They are both stable, crystalline solids.

It was found that difluoroallene reacts easily at room temperature with these two nitrones in high yield. The reactions are regiospecific and the stable products were purified by silica gel chromatography.

$$\mathrm{PhCH{=}\overset{+}{N}(O^-)Ph} + \mathrm{CF_2{=}C{=}CH_2} \xrightarrow[3h]{28^\circ} \underline{116} \quad 97\%$$

$$\mathrm{PhCH{=}\overset{+}{N}(O^-)CH_3} + \mathrm{CF_2{=}C{=}CH_2} \xrightarrow[3h]{28^\circ} \underline{117} \quad 95\%$$

Just as in the case of diazoalkanes, the regiospecificity of these reactions rules out the Firestone biradical intermediate.

In order to see if steric effects are important in nitrone cycloadditions, difluoroallene was allowed to react with triphenylnitrone 118. The reaction proceeded at a much reduced rate and mostly a single regioisomer 119 was produced. However, another primary reaction product, which was not isolated, was present.

From this result, it appears that nitrones have an even stronger electronic preference for the carbon

$$\underset{\underline{118}}{Ph_2C{=}\overset{+}{N}(O^-)Ph} + CF_2{=}C{=}CH_2 \xrightarrow[\text{5 days}]{28^o} \underset{\underline{119}}{\text{2-Ph-3,3-}Ph_2\text{-4-(}{=}CF_2\text{)-isoxazolidine}} \quad 27\%$$

of the dipole to react with the central allene carbon. Although the Atomic Orbital coefficients for the HOMO of nitrones are roughly equal[53,64] being 0.68 for C, 0.15 for N, and -0.71 for O, two factors lead to better overlap of the carbon in the transition state. The first factor is the reinforcement of the HOMO wave function on carbon by a bonding interaction with the central nitrogen. The second factor is the greater long-range overlap for carbon than for oxygen. Due to the stronger electronic preference of nitrones, steric effects seem to be unimportant in determining regiochemistry, at least in the reactions with difluoroallene.

As expected, difluorodimethylallene also reacted regiospecifically with diphenylnitrone and C-phenyl-N-methylnitrone. Reacton rates were slower than with difluoroallene. The stable products <u>120</u> and <u>121</u> were isolated by silica gel chromatography.

$$\text{PhCH}=\overset{+}{\text{N}}(\text{O}^-)\text{Ph} + \text{CF}_2=\text{C}=\text{C}(\text{CH}_3)_2 \xrightarrow[\text{40h}]{28^\circ} \underline{120}\ \ 97\%$$

$$\text{PhCH}=\overset{+}{\text{N}}(\text{O}^-)\text{CH}_3 + \text{CF}_2=\text{C}=\text{C}(\text{CH}_3)_2 \xrightarrow[\text{9 days}]{28^\circ} \underline{121}\ \ 40\%$$

Only one reaction of fluoroallene with a nitrone was studied, but it turned out to give a most interesting result. Fluoroallene reacted slowly with diphenylnitrone to give two isomers <u>122</u> and <u>123</u> in excellent yield. The reaction has the same regiospecificity

$$\text{PhCH}=\overset{+}{\text{N}}(\text{O}^-)\text{Ph} + \text{CHF}=\text{C}=\text{CH}_2 \xrightarrow[\text{36h}]{28^\circ} \underline{122} + \underline{123}\ \ 99.7\%$$

<u>122</u> 86:14 <u>123</u>

observed previously for difluoroallene, but in this case two products were formed in an 86 to 14 ratio. This ratio was determined by ^{1}H NMR and ^{19}F NMR integration of the reaction mixture. The two isomers were separated by silica gel chromatography and their structures determined spectroscopically. The distinction between 122 and 123 is made on the basis of homoallylic H-H coupling assigned by decoupling experiments. Coupling of the methine proton to the olefinic proton is 1.71 Hz in 122 and 1.58 Hz in 123. In view of literature reports[65] that the transoid homoallylic coupling is larger than the cisoid homoallylic coupling, structure 122 is assigned to the major isomer with the 1.71 Hz coupling. Coupling of the methylene protons to the olefinic proton is 1.77 Hz in 122 and 2.17 Hz in 123, again consistent with the structures.

Further evidence for structures 122 and 123 is provided by the ^{13}C NMR spectra of the two isomers.

Table VII. ^{1}H NMR Homoallylic H-H Coupling Constants and 13-C NMR Allylic C-F Coupling Constants in Hz for 4-(Fluoromethylene)isoxazolidines 122 and 123.

Compound	H-H Coupling (Hz)		C-F Coupling (Hz)	
	$J(CH_2)$	$J(CH)$	$J(CH_2)$	$J(CH)$
122	1.77	1.71	6.1	<1
123	2.17	1.58	<1	5.5

There is a large trans C-F coupling of 5.5 Hz to the methine carbon in 123 and a similarly large trans C-F coupling of 6.1 Hz to the methylene carbon in 122. When the fluorine is cis to these carbons, the coupling is negligible as indicted in Table VII.

The regioselectivity for fluoroallene cycloaddition is exactly opposite to what is expected for a steric effect. The 1,3-dipole approaches predominantly from the side of the allene which has fluorine oriented toward the 1,3-dipole to give 122 as the major product. In situations where the fluorine is replaced by a methyl or methoxy group the stereochemistry is just the reverse.[49,50] This unusual effect of fluorine substitution is still unexplained and deserves further study.both experimentally and theoretically.

The structural assignments of the nitrone cycloadducts are mostly straightforward. All of the adducts show CH and CH_2 proton chemical shifts consistent with heteroatom substitution as can be seen by looking at Table VIII. The IR stretches at 1787 to 1774 cm^{-1} are indicative of the $=CF_2$ group. In the ^{19}F NMR, the F-F coupling constants of 50 to 60 Hz confirm the presence of a $=CF_2$ group with non-equivalent fluorines.[66] For isomers 122 and 123, the fluoromethylene stretching frequencies in the IR spectra are 1713 and 1714 cm^{-1}. The ^{19}F NMR and ^1H NMR display H-F coupling constants of

82.0 and 81.3 Hz for 122 and 123 which unquestionably indicates monofluoromethylene groups.[66]

Table VIII. Summary of Spectral Data for Nitrone Cycloadducts. 1-H NMR Chemical Shifts in ppm and Coupling Constants to Fluorine in Hz for the Ring Protons. Also F-F Coupling Constants from the 19-F NMR and IR Stretching Frequencies of $=CF_2$ and =CHF.

Compound	δ CH (ppm)	δ CH_2 (ppm)	J_{HF} (Hz)	J_{FF} (Hz)	IR (cm^{-1})
115	5.31	4.61	3.3	54.8	1787
117	4.56	4.61	—	56.1	1785
119	none	4.91	3.6	51.5	—
120	5.26	none	3.0	58.6	1774
121	4.15	none	4.3	59.8	1777
122	5.52	4.58	82.0	none	1713
123	5.10	4.87 4.77	81.3	none	1714

Interestingly, the CH and CH_2 proton chemical shifts in 122 and 123 are consistent with a deshielding effect on these protons when the fluorine is cis. This amounts to 0.42 ppm deshielding of the CH in 122 and 0.24 ppm deshielding of the CH_2 in 123. There is, however, no apparent chemical shift difference in the ^{13}C NMR spectra.

Nitrile Oxides

Nitrile oxides are generally unstable 1,3-dipoles. Benzonitrile oxide must be prepared and used in situ. However, Grundmann and Dean[67] have found that more sterically hindered nitrile oxides can be easily prepared and isolated as stable solids. For example, mesityl nitrile oxide 124 is very stable and can be stored indefinitely at room temperature.

Mesityl nitrile oxide reacts rapidly with difluoroallene at room temperature to give mainly 125 and some diadduct 126. The failure of 125 to react with excess mesityl nitrile oxide proves that the diadduct 126 does not come from 125. The diadduct may be formed

H_3C–(2,6-dimethylphenyl)–$C\equiv\overset{+}{N}-\overset{-}{O}$ + $CF_2=C=CH_2$ $\xrightarrow[10\ min]{28^{\circ}}$ 125 + 126

98%

124   125   77:23   126

from rapid reaction of regioisomer 127 with more nitrile oxide. There is ample precedent for the diadduct formation in the work of Battioni and coworkers.[50] In addition, the ^1H NMR spectrum of diadduct 126 is nearly identical to the ^1H NMR spectrum which has been reported for the analogous compound without fluorines.[68] The two methylene protons of 126 have a chemical shift of 3.41 ppm, while the unfluorinated compound has four methylene protons at 3.51 ppm.

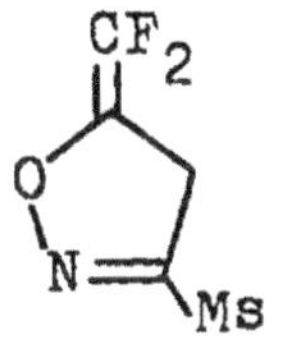

CF_2 O N Ms

127

The neat primary adduct 125 is unstable at room temperature and decomposes when kept overnight. The extensive work of Huisgen and coworkers[69] with nitrile oxide cycloadditions has indicated that steric effects are not very important here. However, very little can be concluded from the single reaction with difluoroallene.

Carbonyl Ylides

Tetracyanoethylene oxide 128 is a readily available source of carbonyl ylide.[70] When 128 is heated in solution, it is in equilibrium with a small amount of the ring-opened tetracyanocarbonyl ylide 129. If an olefin is present which can trap the carbonyl ylide, the reaction goes to completion.

It has been claimed that tetracyanocarbonyl ylide is an electron deficient 1,3-dipole. If this is true,

NC, CN, NC, CN (oxirane) ⇌ heat NC, NC C̄—Ȯ=C CN, CN

128 129

the carbonyl ylide might be expected to give a cycloaddition product with the fluorinated double bond of difluoroallene. Such a reaction would be an inverse demand 1,3-dipolar cycloaddition.

Tetracyanocarbonyl ylide when generated thermolytically from 128 reacts with difluoroallene to give one product 130. This solid compound is very

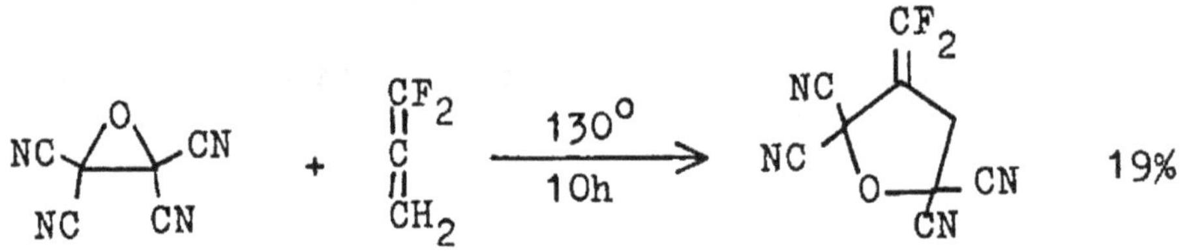

128 + CF_2=C=CH_2 → (130°, 10h) 130 19%

unstable to water and its spectra were determined in dry solvents.

From the regiospecificity of the reaction, it is apparent that the carbonyl ylide reacts in the normal manner with difluoroallene. The controlling

Molecular Orbital interaction is between the HOMO of the 1,3-dipole and the LUMO of difluoroallene.

Conclusions and Observations

The order of reactivity in concerted reactions such as the Diels-Alder reaction and 1,3-dipolar cycloadditions is as indicated. Difluoroallene is

$$CH_2{=}C{=}CH_2 < CF_2{=}C{=}CMe_2 < CHF{=}C{=}CH_2 < CF_2{=}C{=}CH_2$$

the most reactive and allene is the least reactive. Fluoroallene and difluorodimethylallene are nearly equal in reactivity and both are less reactive than difluoroallene. In biradical [2+2] cycloadditions, difluoroallene is also much more reactive than allene.

The regiochemistry of the 1,3-dipolar cycloadditions is determined mainly by "secondary orbital interactions" of the difluoroallene CF_2. In the case of diazoalkanes, the steric effect of substituents on the 1,3-dipole is very important in determining the regiochemistry of the products. However, reactions with other 1,3-dipoles are less sensitive to steric effects.

In conclusion, it should be pointed out that although the 1,3-dipolar cycloadditions studied have proven to be completely concerted, this does not disprove the biradical mechanism for other dipolarophiles. There is overwhelming evidence that most 1,3-dipolar cycloadditions are concerted, but the biradical pathway should be close in energy. The reason biradical products have not been observed is probably because these mechanistic studies with difluoroallene have been biased towards observation of the concerted pathway by the selection of 1,3-dipoles which react readily at room temperature. If the biradical pathway is to be observed, less reactive 1,3-dipoles would need to be investigated at higher reaction temperatures.

SECTION V
TRIMETHYLENEMETHANE STUDIES

Pyrazoline Decompositions

When a methylenecyclopropane is heated, typically around 200-250°, it is in equilibrium with a small concentration of trimethylenemethane biradical. Let us consider a substituted methylenecyclopropane, for example difluoromethylenecyclopropane 131. When the isomer with fluorines on the ring is heated, the fluorines appear to migrate to the double bond. This

CH_2 F_2 ⇌ CH_2 $H_2C\cdot$ $\cdot CF_2$ ⇌ CF_2

131 132 133

rearrangement occurs through the difluorotrimethylene-methane biradical 132. If the structure of the biradical is considered, there are three possible conformations. In the planar conformation 132, four p-orbitals interact in a π sense. Rotation of a methylene by 90° gives an orthogonal conformation 134, which in principle has an isolated methylene radical and a difluoroallyl radical. Rotation of the CF_2 by 90° gives another

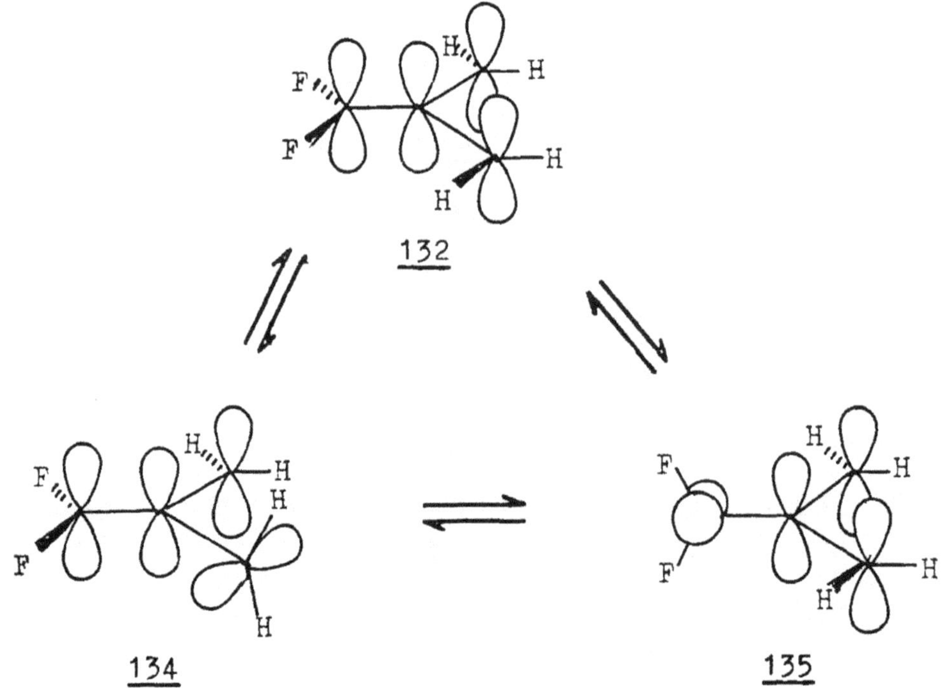

orthogonal conformation 135, which has an isolated difluoromethylene radical and an allyl radical. Any or all of these three conformations may exist in the triplet or singlet spin states.

There has been a remarkable amount of theoretical interest in the trimethylenemethane biradical. The calculations all agree that the planar triplet conformation is lowest in energy.[71] Experimental observations have confirmed these predictions.[71,72] It is, however, the singlet difluorotrimethylenemethane which closes to give the difluoromethylenecyclopropane products. It is probable that for the singlet spin state, all three conformations are in equilibrium or partially equilibrated and all are important in determining the kinetic product ratios for closure of the biradical.

The structure of the biradical intermediate, of course, has no effect on the thermodynamic ratio of difluoromethylenecyclopropanes 131 and 133 at equilibrium. The thermal equilibrium of 131 and 133 was studied in the gas phase by Dolbier and Fielder.[73] They obtained standard enthalpy ΔH^o and standard entropy ΔS^o values of -1.9 kcal/mole and -0.48 eu. The geminal fluorines are favored on the double bond in the equilibrium. The ratio of 131 to 133 at 236° is reported to be 16 to 84.

CH_2 F_2 131 $\underset{16:84}{\overset{236^o}{\rightleftharpoons}}$ CF_2 133

$$\Delta H^o = -1.9 \pm 0.07 \text{ kcal/mole}$$
$$\Delta S^o = -0.48 \pm 0.1 \text{ eu}$$

An alternative method of generating difluorotrimethylenemethane would be by photolysis or thermolysis of 4-(difluoromethylene)-1-pyrazoline 101. Loss of nitrogen from cyclic azoalkanes is generally considered to be a concerted process giving a free biradical.[74] Much of the work in the field of pyrazoline thermolysis has been reported by Crawford and coworkers.[75-77] Unfortunately, they have recently diverged from the accepted mechanism of pyrazoline thermolysis,[78,79] which has led to considerable confusion in the area. For this reason, the study of loss of nitrogen from 4-(difluoromethylene)-1-pyrazoline was conducted with extreme care when determining the product ratios.

Nitrogen extrusion from 4-(difluoromethylene)-1-pyrazoline 101 was studied both by thermolysis and by photolysis. The experiments were carried out in the gas phase, both at low pressure and at higher pressures. Argon was used to obtain the higher pressures since the vapor pressure of pyrazoline 101 was only 5 mm at room temperature. Experiments were also carried out in CCl_4 solution. The reported product ratios do not change with respect to time of photolysis or thermolysis and represent the true kinetic product ratios.

The product ratios for selected high pressures, low pressures and solution are presented in Table IX. These product ratios were found to be remarkably dependent on the conditions for photolysis and thermolysis.

In the photolysis and to a lesser extent in the thermolysis, loss of nitrogen produces a vibrationally excited difluorotrimethylenemethane biradical which

CF_2=C(ring, N=N) 101 —heat or light→ 131 (CH₂=cyclopropane-F₂) + 133 (CF_2=cyclopropane)

101 131 60:40 133

Table IX. Product Ratios for Photolysis and Thermolysis of 4-(Difluoromethyene)-1-pyrazoline 101.

Conditions	131	133	allene
162°, 5mm	54.2	45.8	0.0
162°, 700mm	54.4	45.6	0.0
150°, soln.	59.7	40.3	0.0
light, 2mm	30.1	31.6	38.3
light, 660mm	56.6	42.2	1.2
light, soln.	61.6	38.4	0.0

closes to give excited methylenecyclopropanes. There are two "hot molecule" reactions involved. The highest energy process is loss of difluorocarbene to produce allene. This high energy fragmentation pathway is observed only upon photolysis in the gas phase and not at all in solution or upon thermolysis. It is rapidly quenched in the gas phase by added argon.

The second "hot molecule" process is isomerization of the products. Isomer 133 is more thermodynamically stable so this process leads to an increase in the amount of 133 in the product mixture. This is a lower energy process and is less easily quenched by addition of inert gas.

For the case of 4-(difluoromethylene)-1-pyrazoline 101, the product ratios from photolysis and thermolysis converge to similar values in solution. The most reliable product ratio is taken to be the solution thermolysis result. Thus the product ratio for kinetic closure of difluorotrimethylenemethane 132 is 60 to 40. A completely statistical closure would give a ratio of 67 to 33. This statistical ratio is assuming a planar conformation 132. The presence of orthogonal conformations 134 and 135 in equilibrium would alter this ratio, as would preferential rotation and closure of CH_2 or CF_2. Surprisingly, there is virtually no effect of geminal fluorines on the kinetic product ratio for closure of difluorotrimethylenemethane.

An identical study was carried out for 4-(fluoromethylene)-1-pyrazoline 113 and the results are presented in Table X. In this case, no allene was produced upon gas phase photolysis, even at low pressures. This is due to the fact that fluorocarbene is much less stable than difluorocarbene. The "hot" fluoromethylenecyclopropane products 136 and 137 are apparently more difficult to quench than the difluoromethylenecyclopropanes. In this case the product ratios for photolysis

113 $\xrightarrow{\text{heat or light}}$ 136 + 137 (90:10)

Table X. Product Ratios for Photolysis and Thermolysis of 4-(Fluoromethylene)-1-pyrazoline 113.

Conditions	136	137
168°, 1mm	81.7	18.3
168°, 740mm	86.8	13.2
140°, soln.	90.5	9.5
light, 2mm	48.0	52.0
light, 770mm	49.8	50.2
light, soln.	64.6	35.4

and thermolysis approach each other, but do not fully converge even in solution. One can think of two possible explanations for why they do not converge to the same product ratio.

The first explanation would be a temperature effect. The solution photolysis is carried out at 38° while the thermolysis is at 140°. Such a large temperature difference could account for the difference in the product ratios. This explanation seems unlikely in view of the fact that difluorotrimethylenemethane shows no temperature dependence for the product ratios.

The second explanation is that photolysis produces a different distribution of partially-equilibrated orthogonal conformations for the fluorotrimethylenemethane biradical. Specifically, photolysis would produce less conformation with orthogonal CHF and thus less 136 in the product mixture. Although there is precedent for vibrationally-excited intermediates in solution,[80] it is not necessary for this argument to invoke vibrational excitation. That is to say, the biradical conformations are in their ground states, but not in complete equilibrium due to competing ring closure to products.

The most reliable kinetic ratio for 136 and 137 is taken to be 90.5 to 9.5 from the solution thermolysis. Comparing this ratio to the ideal statistical ratio of 67 to 33, it is apparent that a single fluorine has a large effect which is opposite to that of geminal fluorines.

The thermodynamic equilibrium for 136 and 137 was studied at five different temperatures and the standard enthalpy and entropy were determined.[81]

The equilibrium ratio for 136 and 137 at 243° was 14 to 88. The ΔH^{o} value of -2.6 kcal/mole is quite

$$\underline{136} \underset{}{\overset{243^{o}}{\rightleftharpoons}} \underline{137} \quad 14:88$$

$$\Delta H^{o} = -2.6 \pm 0.06 \text{ kcal/mole}$$

$$\Delta S^{o} = -1.0 \pm 0.02 \text{ eu}$$

close to the ΔH^{o} value for fluoropropene isomerization which is -2.7 kcal/mole.[31] This is the first indication that there is no destabilization of a cyclopropane by a single fluorine substituent.

Since 4-(difluoromethylene)-3,3-dimethyl-1-pyrazoline 104 was available, the kinetic product ratios for loss of nitrogen were studied for this more complex system. The results are presented in Table XI.

$$\underline{104} \xrightarrow{\text{heat or light}} \underline{138} : \underline{139} : \underline{140} \quad 54:15:31$$

Table XI. Product Ratios for Photolysis and Thermolysis of 4-(Difluoromethylene)-3,3-dimethyl-1-pyrazoline 104.

Conditions	138	139	140	$H_2C{=}C{=}CMe_2$ 141
162°, 3mm	41.7	16.7	41.6	0.0
162°, 280mm	42.3	16.3	41.4	0.0
155°, soln.	53.5	15.0	31.5	0.0
light, 1mm	43.4	11.8	37.6	7.1
light, 740mm	29.4	22.7	47.7	0.2
light, soln.	55.5	9.5	35.1	0.0

In this case difluorocarbene extrusion was once again observed upon gas phase photolysis. The solution product ratios tend to converge for thermolysis and photolysis. Taking the solution thermolysis to be the most reliable, the ratio of 138 to 139 to 140 is 53.5 to 15.0 to 31.5. For comparison, the thermodynamic equilibrium at 244° gave the ratio 72 to 2 to 26 starting with isomer 139, which was conveniently prepared from addition of difluorocarbene with commercially available dimethylallene 141.

Since isomer 139 is much less stable than 138 and 140, it was possible to determine an accurate kinetic ratio of 138 to 140 by starting with 139. At 201° the kinetic ratio of 138 to 140 was determined

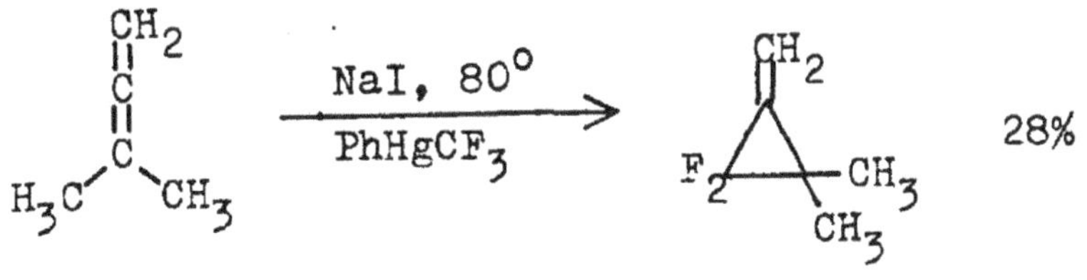

$$\underset{\textbf{141}}{(CH_3)_2C{=}C{=}CH_2} \xrightarrow[\text{PhHgCF}_3]{\text{NaI, 80°}} \underset{\textbf{139}}{\text{2,2-difluoro-3,3-dimethyl-1-methylenecyclopropane}} \quad 28\%$$

to be 52 to 48. This ratio is in agreement with the gas phase pyrazoline thermolysis ratio. The thermodynamic parameters for the system were determined.

$$\textbf{138} \rightleftharpoons \textbf{139} \qquad \Delta H^\circ = 3.25 \pm 0.072 \text{ kcal/mol} \qquad \Delta S^\circ = -1.2 \pm 0.03 \text{ eu}$$

$$\textbf{139} \rightleftharpoons \textbf{140} \qquad \Delta H^\circ = -2.68 \pm 0.07 \text{ kcal/mol} \qquad \Delta S^\circ = 0.32 \pm 0.03 \text{ eu}$$

$$\textbf{138} \rightleftharpoons \textbf{140} \qquad \Delta H^\circ = 0.58 \pm 0.01 \text{ kcal/mol} \qquad \Delta S^\circ = -0.89 \pm 0.006 \text{ eu}$$

A rate constant for disappearance of 139 at 200.8° was measured. It was $1.467 \pm 0.006 \times 10^{-3}\ s^{-1}$.

Comparing the thermodynamic parameters, the minor isomer 139 is less stable than expected considering the effect of geminal fluorines and of geminal methyls separately. In the first equilibrium, 138 and 139, the methyls are favored on the double bond by 3.25 kcal/mole. In the equilibrium without fluorines, the value is only 1.17 kcal/mole.[76] This is a net destabilization of 139 by 2.08 kcal/mole.

Carrying these kinds of studies further to the non-volatile pyrazolines looked promising at first. However, photolysis of 107 gave only one product 142. Isomer 143, if formed at all, is probably not stable at room temperature. The other possible isomer 144 would be stable, but is kinetically not accessible.[82]

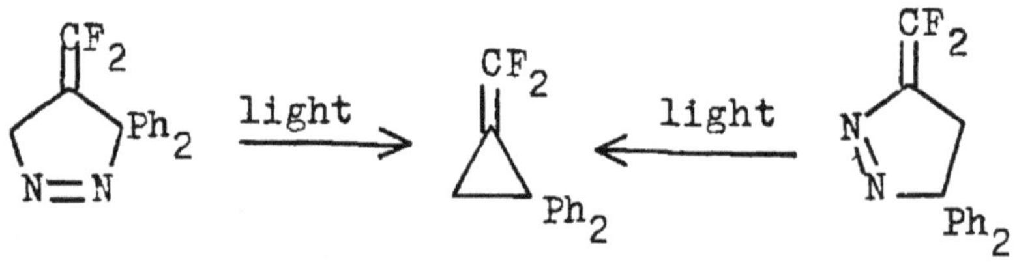

CF_2 / Ph_2 / N=N — light → CF_2 / Ph_2 ← light — CF_2 / N / N / Ph_2

107 142 108

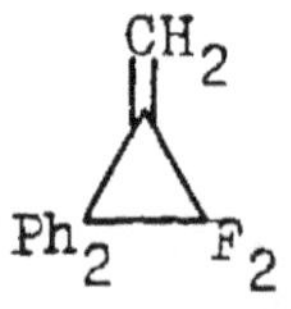

CH_2 / Ph_2 F_2 — not present — CPh_2 / F_2

143 144

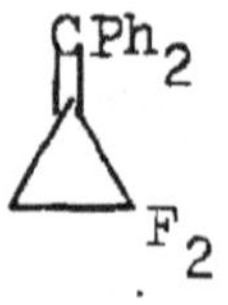

It was discovered that 2-(difluoromethylene)-1,1-diphenylcyclopropane 142 could also be obtained by photolysis of the major cycloaddition product, 5-(difluoromethylene)-3,3-diphenyl-1-pyrazoline 108. In this case, loss of nitrogen is probably stepwise and a trimethylenemethane biradical obviously is not involved. Photolysis of this isomer was slower, but eventually a good yield of 142 was obtained.

A better understanding of the chemistry involved was obtained by studying the pyrazolines from cycloaddition of diazofluorene and difluoroallene. As previously mentioned, pyrazoline 110 eliminates nitrogen at room temperature to give methylenecyclopropane 112. The best way to prepare 112 was to run the cycloaddition reaction and then irradiate the product mixture. The

CF_2 N N CF_2 28° light CF_2 N N

110 112 111

yield of 112 obtained in this way was 76% after chromatography. The methylenecyclopropane 112 turned out to be a very interesting compound, as we shall see in the next section.

Trapping Studies

During the purification of compound 112 by flash chromatography, there were trace impurities which were impossible to remove even though their R_f values were very different from 112. Small amounts of these compounds were isolated and analysis by mass spec. indicated an M^+ of 272. Comparing this value to the M^+ for 112 of 240, it is apparent that 112 has reacted with molecular oxygen and gained 32 mass units. To test this hypothesis, a solution of 112 was treated with a slow flow of oxygen at room temperature. The starting material disappeared slowly, to be replaced by the previously observed oxygen adducts. The three products were isolated by silica gel chromatography in 67% yield. The relative ratio of the products 145, 146, and 147 was determined by ^{19}F NMR integration to be 58.6 to 36.8 to 4.6 respectively.

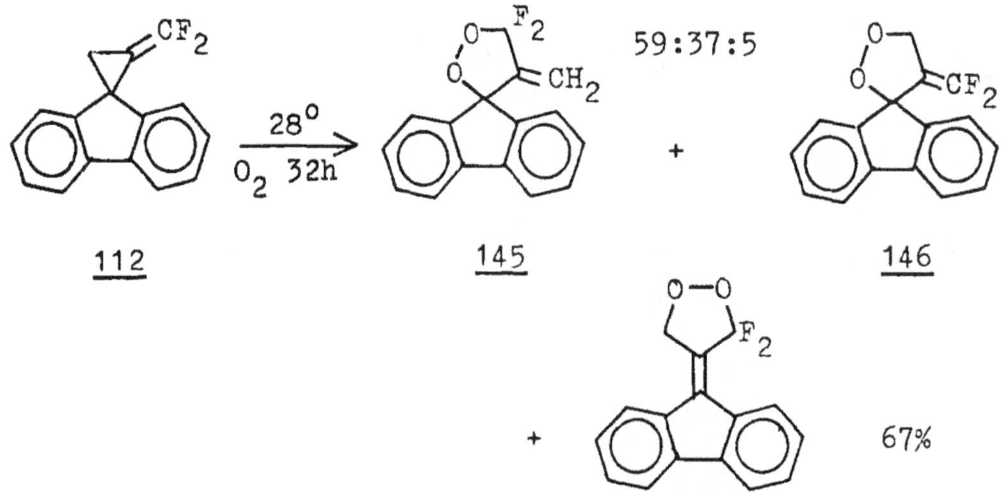

CF_2 112 $\xrightarrow[O_2\ 32h]{28^o}$ 145 (F_2, CH_2) + 146 (CF_2) + 147 (F_2)

59:37:5

67%

How is it possible for a normally stable olefin such as 112 to react with oxygen under such mild conditions? Usually, in order to form the trimethylenemethane biradical from a methylenecyclopropane, it requires temperatures in excess of 200°. In this case, the extensive delocalization of the biradical 148 and the added strain of the spiromethylenecyclopropane combine so that 112 is in equilibrium with a very small concentration of biradical 148 even at room temperature.

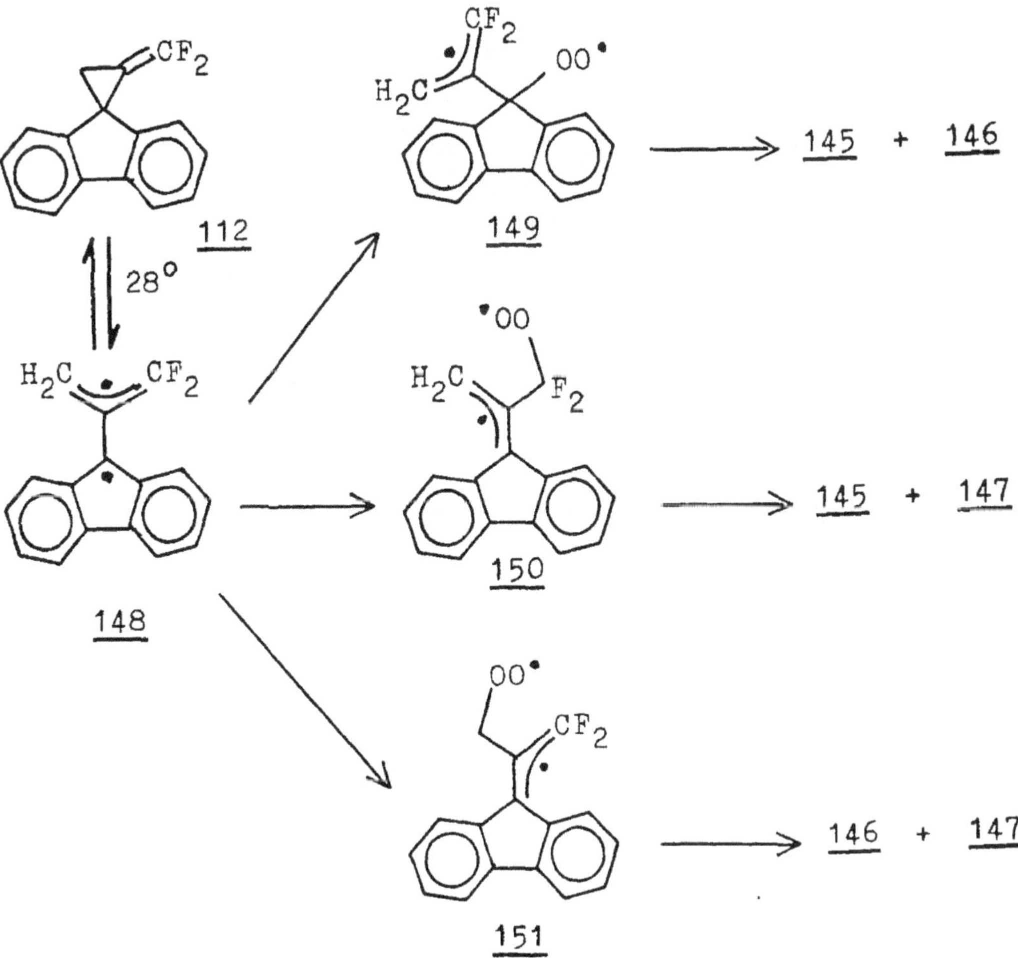

Oxygen is an incredibly efficient scavenger of monoradicals. It reacts with carbon-based radicals at diffusion-controlled rates where the rate of reaction is limited only by the viscosity of the solvent.[83-86] Even though the concentrations of oxygen and of biradical 148 are both small, there is a continuous source of the biradical and eventually the reaction goes to completion if enough oxygen is available. There has been one previous report of oxygen adducts from a biradical.[87]

The trapping of oxygen is most likely a stepwise process involving intermediates 149, 150, and 151. In the product mixture, 95% of the oxygen ends up bonded to the fluorenyl carbon as in 145 and 146 while only 5% gives product 147.

One possible explanation for this observed selectivity could in principle be that 149 is the predominant trapping intermediate. Use of an asymmetric olefin to trap intermediate 148 will tell which site of the biradical reacts first. For this experiment, acrylonitrile was chosen. Radicals react with acrylonitrile only at the terminal carbon due to the stabilization of the radical site by nitrile conjugation. Whichever carbon of biradical 148 reacts first in the trapping reaction must end up bonded to the terminal CH_2 of acrylonitrile in the products.

Methylenecyclopropane 112 reacted with excess acrylonitrile at 70°. The ^{19}F NMR of the reaction mixture indicated that only three of the possible six trapping products were formed. The products 152, 153, and 154 were isolated in 75% yield by flash chromatography. The product ratio was as indicated.

68:16:16

112 $\xrightarrow[CH_2=CHCN]{70^\circ 6h}$ 152 + 153 + 154 75%

The ratio of acrylonitrile trapping products is roughly similar to the ratio of the corresponding oxygen trapping products. Very significantly, there were no acrylonitrile trapping products produced with the terminal CH_2 of acrylonitrile bonded to the fluorenyl carbon. It must be concluded from this fact that the fluorenyl carbon does not react first in the trapping reactions. That is to say, intermediates 150 and 151 are formed exclusively with no intermediate 149 being formed.

It is interesting to consider the ratio of CF_2 *versus* CH_2 reaction for the initial trapping. For reaction with oxygen, $k_F/k_H = 1.72$, meaning that the

CF_2 reacts 1.72 times faster than the CH_2 with oxygen. For reaction with acrylonitrile $k_F/k_H = 5.40$, showing that the less reactive trapping reagent is more selective. The relative rate of reaction of the fluorenyl carbon is much too slow to be detected.

Rather curiously, the relative rates of biradical closure forming the products are different. Closure to the fluorenyl carbon is dominant in this case with the ratio $k_{Fl}/k_H = 20.7$ for biradicals 150 and 151. The corresponding biradicals from acrylonitrile, 155 and 156 give a ratio $k_{Fl}/k_H = 5.20$. Thus the more stable oxygen-based biradicals are more selective in the ring-closure, product-forming steps.

It is proposed that biradical 148 has only one lowest energy conformation. For steric reasons, the

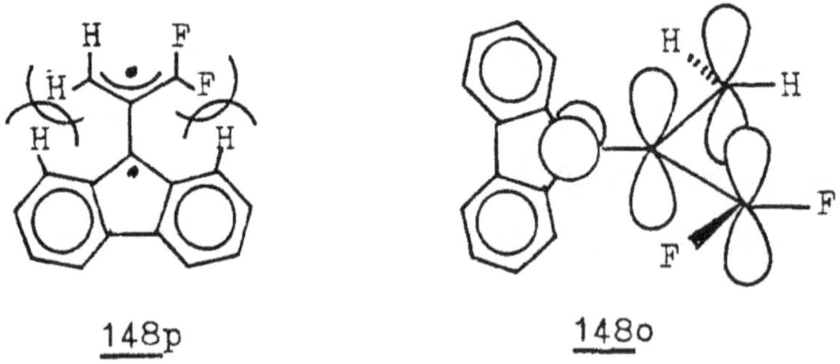

planar conformation 148p is simply not possible. If a model of 148 is made, the perihydrogens and the substituents of the allyl system must occupy the same space in order for 148 to be planar. Thus the only possible conformation of 148 is with the fluorenyl system orthogonal as in 148o.

The initially formed trapping products with acrylonitrile are 155 and 156. The conformations of these biradicals which are formed at first will contain a full double bond with an orthogonal fluorenyl radical.

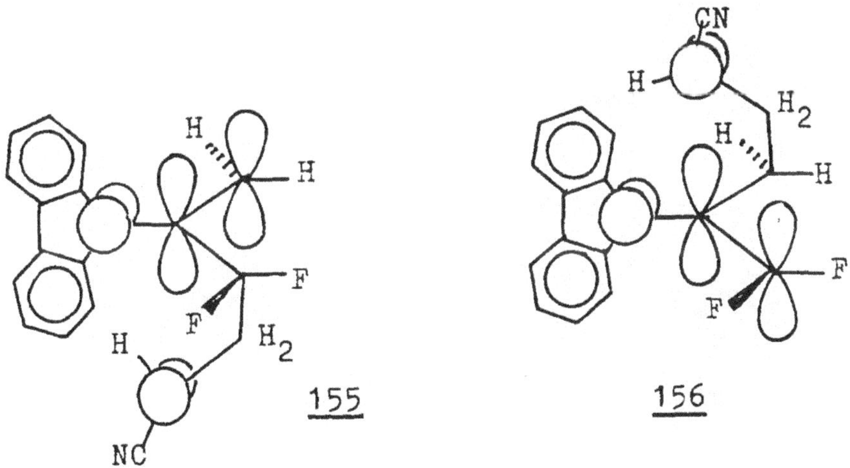

Thus direct combination of the radical centers competes with rotation of the fluorenyl radical into conjugation with the double bond. In this scheme, the fluorenyl carbon combines faster than it rotates leading to more 152 and 153 than 154. This theory nicely explains the product ratios, but more data are needed for confirmation. In particular, trapping with other olefins should be studied.

Preliminary results indicate that tetracyanoethylene reacts slowly at room temperature and that dimethyl fumarate also reacts at higher temperature. However, no attempt to isolate products has been made, for which reason the product ratios are not reported. The ratios do appear to be significantly different from the oxygen and acrylonitrile product ratios.

Is the ability to be trapped a special property of fluorinated biradicals, or is it a general reaction of highly stabilized biradicals? In order to answer this question, the analogous compound to 112 without fluorines was synthesized in embarrassingly low yield by photolysis of diazofluorene with excess allene. The numerous other products were not characterized, but after a couple chromatographies and a fractional crystallization, pure 157 was obtained and fully characterized.

Compound 157 was found to react with oxygen very slowly at room temperature in solution. Eventually the reaction went to completion and a 75% isolated yield of oxygen adducts 158 and 159 was obtained.

CH_2 $\xrightarrow[O_2,\ 7\ days]{28^\circ}$ CH_2 + 75%

157 158 92:8 159

Separation of pure 158 was obtained by recrystallization, however pure 159 could not be obtained by flash chromatography or fractional crystallization.

Thus the answer to our question is that the fluorine substitution has little if anything to do with the biradical trapping. Whether a spin correlation exists for the product distributions remains to be determined. Such a correlation has been postulated by Berson and coworkers[71,88,89] for trapping of 2-methylenecyclopenta-1,3-diyls.

SECTION VI
EXPERIMENTAL

Infrared spectra were determined on a Perkin-Elmer 283B spectrophotometer and absorption bands are reported in cm^{-1}. The spectra of neat liquids were determined as films between KBr plates. Solution spectra were determined with matched liquid cells (0.1 mm). Gas phase spectra were determined using a gas IR cell with KBr windows and a 5-cm path length using 10 to 20 mm pressure of sample.

The 60 MHz ^{1}H NMR spectra were determined on a Varian EM 360L spectrometer; 100 MHz spectra were taken with either a Varian FX-100 instrument or a Varian XL-100 instrument; 300 MHz spectra were determined using a Nicolet NT-300 spectrometer. Chemical shifts are reported in ppm downfield of internal TMS in $CDCl_3$ solution.

The ^{19}F NMR spectra were determined using the Varian XL-100 or the Nicolet NT-300 instruments. Chemical shifts are reported in ppm upfield of internal $CFCl_3$ in $CDCl_3$ solution.

The ^{13}C NMR spectra were determined on the Varian FX-100 or the Nicolet NT-300 instruments. Chemical shifts are reported in ppm downfield of internal TMS in $CDCl_3$ solution.

Coupling constants are reported in Hertz. Fluorine-hydrogen coupling constants are indicated by J_{HF}, fluorine-fluorine coupling constants by J_{FF}, carbon-fluorine coupling constants by J_{CF}, and proton coupling constants by J_{HH}. All assighnments of ^{13}C NMR resonances are made with the aid of off-resonance spectra or pulse-sequence spectra.

Mass spectra were determined on an AEI-MS 30 spectrometer at 70 eV. Exact masses were also determined on the AEI-MS 30.

The GLPC analyses and preparative separations were performed on a Varian Aerograph 90-P gas chromatograph with thermal conductivity detector. The gas phase samples were analysed using a Hewlett-Packard 5710A gas chromatograph with flame-ionization detector in conjunction with a Hewlett-Packard 3380S recording integrator.

Elemental analyses were performed by Atlantic Microlabs.

Melting points and micro-boiling points were determined with a Thomas-Hoover capillary melting point apparatus and are uncorrected.

The static gas-phase thermolysis apparatus consisted of a standard high-vacuum line connected to a thermolysis vessel submerged in a well-insulated, stirred, and heated, constant-temperature salt bath. A small-volume sampling area between the thermolysis vessel and the vacuum line allowed samples to be taken by expansion into an evacuated 25-mL gas sample bulb with teflon Rotoflo stopcock. After taking a sample, the bulb was pressurized with argon and then analyzed using a gas chromatograph equipped with gas injection valve. Three analyses by GLPC were made for each sample and the average value taken for the peak integrations.

Procedures

1,2-Cyclobutanedicarbonyl dichloride 47.

To a 250-mL, round-bottom, three-necked flask equipped with nitrogen inlet, reflux condenser and magnetic stirrer was added 200 g (1.68 moles) thionyl chloride and 24.6 g (0.171 mole) trans-1,2-cyclobutane dicarboxylic acid. The mixture was stirred and heated at reflux for 1 hour. The excess thionyl chloride was removed by simple distillation to give a brown liquid residue. The crude residue was purified by fractional distillation at reduced pressure using a 15-cm, vacuum-jacketed Vigreux column. A total of 29.5 g(84%) clear, colorless 47 was obtained: bp 43-55°/0.1 mm(lit.[90] bp 49.5-50.5^{o}/0.6-0.65 mm); IR(flim) 3010, 2970, 2882, 1795(s), 1036, 723 cm^{-1}; ^{1}H NMR(60 MHz) δ3.9(m,2H), 2.4(m,4H); mass spectrum gave 147(M^{+}-35,10), 145(30), 119(10), 117(30), 91(18), 89(52), 53(62), 36(100).

N,N,N',N'-Tetramethyl-1,2-cyclobutanedicarboxamide 48.

A 1000-mL, three-necked, round-bottom flask was equipped with a dry-ice condenser, low temperature thermometer, gas inlet addition tube, and magnetic stirrer. After flushing the system with dry nitrogen, 300 mL commercial anhydrous ether was added and the flask was cooled in a dry ice/isopropanol bath. Into the flask was condensed 100 g(2.22 moles) dimethylamine and the inlet tube was replaced with a pressure-equalizing addition funnel. Over a period of 50 minutes, 24.3 g(0.134 mole) dichloride 47 were added at an internal temperature of -60°. The mixture, containing a dense white precipitate, was allowed to come to room temperature.

To the mixture was added 200 mL water. The ether layer was separated and the aqueous layer was extracted with five 75-mL portions methylene chloride. The combined organic layers were dried with molecular sieves and concentrated by rotary evaporation at reduced pressure to give 25.4 g(95%) off-white solid which was recrystallized from 500 mL ether to give 21.6 g(81%) colorless rhombohedral crystals of diamide 48: mp 83.7-88.5^{o}; IR(CCl_4) 2947, 1645(s). 1400, 1143 cm^{-1}; ^{1}H NMR(60 MHz) δ3.7(m,2H), 3.0(s,6H), 2.9(s,6H), 2.1 (m,4H); mass spectrum gave 198(M^{+},15), 153(12), 126(42), 100(40), 72(100).

N,N,N',N'-Tetramethyl-1,2-cyclobutanedimethanamine 49.

A 2,000-mL, three-necked, round-bottom flask was equipped with a Friedrich's condenser, nitrogen inlet, pressure-equalizing addition funnel and efficient magnetic stirrer. To the flask was added 200 mL commercial anhydrous ether and 8.00 g(0.211 mole) lithium aluminum hydride. To the stirred mixture was added a solution of 21.6 g(0.109 mole) dicarboxamide 48 in 700 mL anhydrous ether at such a rate that a gentle reflux was maintained. Addition was completed in 70 minutes. After stirring overnight (16 h), the stirred reaction mixture was hydrolysed with extreme care using a solution of 200 g(0.709 mole) potassium sodium tartrate hydrate($KNaC_4H_4O_6 \cdot 4H_2O$) in 800 mL water. Evolution of hydrogen was vigorous at first and a white precipitate formed, which dissolved upon further addition of the tartrate solution. The ether layer was separated and the aqueous layer was extracted with four 100-mL portions of ether. The combined ether layers were dried over anhydrous sodium sulfate and concentrated by rotary evaporation at reduced pressure to roughly 125 mL. The remaining ether was removed by fractional distillation using a 15-cm, vacuum-jacketed Vigreux column. The remaining residue was purified by fractional distillation at reduced pressure to give 15.1 g(81%) clear, colorless distillate which was the diamine 49: bp 80-82°/13 mm; IR(film) 2980, 2950, 2870, 2825, 2775, 1460, 1267, 1045 846 cm^{-1}; ^{1}H NMR(60 MHz) δ 2.17(m); mass spectrum gave 125(M^{+}-45,15), 110(5), 84(56), 58(100).

N,N,N',N'-Tetramethyl-1,2-cyclobutanedimethanamine N,N'-dioxide 50.

To a 250-mL, three-necked, round-bottom flask equipped with pressure-equalizing addition funnel, thermometer, and magnetic stirrer, was added 60 g (0.529 mole) of 30% aqueous hydrogen peroxide. The flask was cooled in an ice-water bath and diamine 49 was added at such a rate that the temperature did not exceed 15°. Addition was completed in 15 minutes. The reaction mixture was stirred and allowed to come to room temperature overnight(17 h). The excess peroxide was decomposed by careful addition of 140 mg platinum black. After stirring for 1 hour, the dark grey mixture gave a negative test for peroxide with lead sulfide test paper. The platinum black was removed by suction filtration and the filtrate was concentrated by rotary evaporation at reduced pressure to give a gummy solid which was further dried under full vacuum at 90° to give 16.62 g(97%) off-white hygroscopic solid 50 which was stored under dry nitrogen: mass spectrum gave 168(M^+-34,0.4), 154(0.6), 139(0.5), 84 (17), 79(20), 58(100).

1,2-Bis-(methylene)cyclobutane 2.

To a 250-mL round-bottom flask equipped with a 20-cm glass tube connected to a vacuum trap was added 6.00 g(29.7 mmoles) finely ground diamine N,N'-dioxide 50, which was spread as thinly as possible on the bottom of the flask. After the system was evacuated, the trap was cooled with liquid nitrogen and the flask was heated at 170-200° for 5 hours. Vacuum was 0.1 mm.

The crude product in the vacuum trap was allowed to warm until it melted, then the cold liquid was washed with three 10-mL portions of 1 N aqueous HCl and then 10 mL water. A total of 1.18 g(49%) cloudy, pale brown liquid product was obtained.

To the crude product was added 50 mg hydroquinone and it was distilled using a short-path distillation apparatus to give 0.96 g(40%) clear, colorless liquid distillate which was collected in an ice-chilled receiving flask containing 50 mg hydroquinone. The spectra of 2 were identical to those previously reported:[41,43] bp 74-76^{o}(lit.[41] bp 73.8-74.2^{o}); IR(gas) 3100, 2995, 2960, 1770(w), 1660(w), 893, 880(s), 870 cm^{-1}; ^{1}H NMR(60 MHz) δ 5.07(s,2H), 4.63 (s,2H), 2.60(broad s,4H); mass spectrum gave M^{+} 80.0613 $\pm$ 0.00235(30 ppm), calcd for C_6H_8 80.06260 dev -0.0013(16 ppm).

3-(Difluoromethylene)bicyclo[4.2.0]oct-1(6)-ene 51 and 2,5-Bis(methylene)-1,1-difluorospiro[3.3]heptane 52.

Into a 5-mL glass tube containing 200 mg(2.50 mmoles) bis-(methylene)cyclobutane 2 was condensed 206 mg(2.71 mmoles) difluoroallene. The tube was sealed under vacuum and kept at room temperature for 2 days, then heated at 80° for 4 hours. The tube was cooled and opened. The products were isolated by prep GLPC(20 ft by 1/4 in 20% SE-30 at 120°, 80 mL/min) to give 223 mg(57%) of 51: bp 156.0-156.8^{o}; IR(film) 2923, 2845, 1760(s), 1690(w), 1438, 1270, 1236, 1210, 1080, 1007, 940 cm^{-1}; ^{1}H NMR(100 MHz) δ 2.59(m,2H), 2.46(m,4H), 2.24(m,2H), 2.03(m,2H);

^{19}F NMR(100 MHz) ϕ94.4 and 99.3(AB pattern J_{FF}=59.8 Hz); ^{13}C NMR(100 MHz) δ152.0(t, J_{CF}=281.6 Hz, CF_2), 141.9(C_6), 138.7(d, J_{CF}=2.4 Hz, C_1), 84.9(t, J_{CF}=18.9 Hz, C_3), 30.5 (CH_2), 30.2(CH_2), 25.1(t, J_{CF}=2.1 Hz, CH_2), 24.1(t, J_{CF}= 1.8 Hz), 21.2(dd, J_{CF}=3.1 and 1.2 Hz); mass spectrum gave M^+ 156.07406 $\pm$ 0.00137(9 ppm), calcd for $C_9H_{10}F_2$ 156.07506 dev -0.0010(6 ppm).

Analysis: Calcd, C 69.22, H 6.45

Found, C 69.05, H 6.48

There were also isolated 86 mg (22%) of 52: bp 142.5-143.0°; IR(film) 3090, 2997, 2960, 2935, 1680, 1428, 1278, 1090, 1017, 990, 918, 886 cm^{-1}; ^1H NMR(100 MHz) δ5.52(d of pentet, 1H, J_{gem}=0.98, J=2.9 Hz), 5.19 (d of pentet, 1H, J_{gem}=0.98, J=2.4 Hz), 4.99(m, 1H), 4.94(m, 1H), 2.8-2.4(complex m, 5H), 1.96(m, 1H); ^{19}F NMR(100 MHz) ϕ105.4(complex m); ^{13}C NMR(100 MHz) δ147.8 (t, J_{CF}=2.4 Hz, C_5), 144.0(t, J_{CF}=21.4 Hz, C_2), 118.4 (t, J_{CF}=283.8 Hz, CF_2), 112.1(=CH_2), 107.3(=CH_2), 56.8 (t, J_{CF}=22.6 Hz, C_4), 36.8(t, J_{CF}=8.5 Hz, C_3), 28.0 (C_6), 24.3(t, J_{CF}=4.0 Hz, C_7); mass spectrum gave M^+ 156.07383 $\pm$ 0.00169(11 ppm), calcd for $C_9H_{10}F_2$ 156.07506 dev -0.00123(8 ppm).

The combined yield of isolated products was 79%. Order of elution was 52 then 51. The GLPC relative yields were 76% for 51 and 24% for 52.

<u>1-(1-Cyanoethenyl)-2,2-difluoro-3-methylenecyclobutanecarbonitrile 61, 1-(1-Cyanoethenyl)-3-(difluoromethylene)cyclobutanecarbonitrile 62, 4-(Difluoromethylene)-1-cyclohexene-1,2-dicarbonitrile 63, 3,3'-Bis-(methylene)-2,2,2',2',-tetrafluoro-[1,1'-bicyclobutyl]-1,1'-dicarbonitrile 64, 65 and 2,2-Difluoro-3'-(difluoromethylene)-3-methylene[1,1'-bicyclobutyl]-1,1'-dicarbonitrile 66.</u>

Into a 30-mL glass tube containing a solution of 1.00 g (9.6 mmoles) 1,3-butadiene-2,3-dicarbonitrile[45] in 10 mL $CHCl_3$ were condensed 2.89 g (38.0 mmoles) of difluoroallene. The tube was sealed under vacuum and kept at room temperature for 4 days. The clear, colorless solution was concentrated by rotary evaporation at reduced pressure to approximately 2 mL and then filtered to remove a minute amount of insoluble material.

The products were isolated by prep GLPC (10 ft by 1/4 in 20% SE-30 at 160°, 60 mL/min); <u>61</u> and <u>62</u> eluted together, then <u>64</u> and <u>66</u> together, and finally <u>63</u> and <u>65</u> together. The yields were 1.22 g for <u>61</u> and <u>62</u>, 85 mg for <u>64</u> and <u>66</u>, and 168 mg for <u>63</u> and <u>65</u>. Further separation by prep GLPC (10 ft by 1/4 in 10% DEGS at 160°, 60 mL/min) gave 989 mg (57.2%) white solid <u>61</u>: mp 35.0-36.1° (sublimed); IR(CCl_4) 3120, 3000, 2250, 2230, 1930, 1870, 1790, 1690, 1616, 1283(s), 1164(s),

1125(s) cm^{-1}; ^{1}H NMR(100 MHz) δ 6.44 and 6.41(AB pattern, 2H, J=1.3 Hz), 5.85(hextet, 1H, J=2.4 Hz), 5.60(hextet, 1H, J=2.4 Hz), 3.27(non-first-order m, 2H); ^{19}F NMR (300 MHz) ϕ 91.9 and 98.4(AB pattern, J_{FF}=205 Hz); ^{13}C NMR(100 MHz) δ 138.6(t, J_{CF}=20.1 Hz, C_3), 136.9(=CH_2), 118.4(=CH_2), 115.7, 115.4(t, J_{CF}=293.6 Hz, CF_2), 114.5, 49.6(dd, J_{CF}=22.6 and 26.2 Hz, C_1), 35.4(t, J_{CF}=6.7 Hz, C_4); mass spectrum gave M^+ 180.04966 $\pm$ 0.0019(11 ppm), calcd for $C_9H_6N_2F_2$ 180.0499 dev -0.00024(1 ppm).

Analysis: Calcd, C 60.00, H 3.36, N 15.55

Found, C 59.70, H 3.43, N 15.46

There were also isolated 30 mg (1.7%) of white solid <u>62</u>: mp 32.3-34.0°; IR(CCl_4) 3120(w), 2943(w), 2245(w), 2230(w), 1794(s), 1617(w), 1277(s) cm^{-1}; ^{1}H NMR(60 MHz) δ 6.33(d, 1H, J=1 Hz), 6.25(d, 1H, J=1 Hz), 3.4(t, 4H, J_{HF}=4.0 Hz); ^{19}F NMR(300 MHz) ϕ 91.5(pentet, J_{HF}=3.9 Hz); ^{13}C NMR(100 MHz) δ 151.6(t, J_{CF}=286.3 Hz, =CF_2), 132.7 (=CH_2), 121.6, 119.0, 114.3, 78.7(t, J_{CF}=29.3 Hz, =CR_2), 35.9(CH_2); mass spectrum gave M^+ 180.0484 $\pm$ 0.0012(7 ppm), calcd for $C_9H_6N_2F_2$ 180.0499 dev -0.0015(8 ppm).

There were also isolated 25 mg (1.4%) of white solid <u>63</u>: ^{1}H NMR(100 MHz) δ 3.14(pentet, 2H, J=2.2 Hz), 2.5-2.4 (non-first-order m, 4H); ^{19}F NMR(100 MHz) ϕ 90.1 and 94.0 (AB pattern, pentet fine structure, J_{FF}=47.5 and J_{HF}=2.0 Hz).

There were also isolated 20 mg (0.81%) white solid <u>64</u>: mp 142-145.5°; IR($CHCl_3$) 2255(w), 1880(w), 1794(w), 1738(w),

1690(W), 1431, 1379(s), 1159(s), 1124(s), 1012, 936 cm^{-1}; ^1H NMR(100 MHz) δ5.9(m, 1H, J=2 Hz), 5.6(m, 1H, J=2 Hz), 3.2(non-first-order m, 2H); ^{19}F NMR(300 MHz) ϕ88.9 and 98.6(AB pattern, J_{FF}=211.1 Hz); ^{13}C NMR(300 MHz) δ138.4(t, J_{CF}=20.7 Hz, C_3), 119.4(=CH_2), 115.5(dd, J_{CF}=288.8 and J_{CF}=295.4 Hz, CF_2), 113.3(d, J_{CF}=3.2 Hz, CN), 48.4(t, J_{CF}=23 Hz, C_1), 34.2(dd, J_{CF}=6.0 and 8.7 Hz, C_4); mass spectrum gave M^+ 255.05466 ± 0.0020(8 ppm), calcd for $C_{12}H_7N_2F_4$ 255.05453 dev 0.00012(0.5 ppm).

There were also isolated 21 mg (0.85%) white solid <u>65</u>: mp 82-86.5°; IR(CCl_4) 1745(w), 1711(w), 1431(w), 1288(s), 1135, 907 cm^{-1}; ^1H NMR(60 MHz) δ 5.9(m, 1H, J=2.5 Hz), 5.6(m, 1H, J=2.5 Hz), 3.1(m, 2H); ^{19}F NMR(300 MHz) ϕ 90.7 and 98.8(AB pattern, J_{FF}=215.4 Hz); ^{13}C NMR (300 MHz) δ 138.4(t, J_{CF}=21.1 Hz, C_3), 119.2(=CH_2), 114.8 (t, J_{CF}=292 Hz, CF_2), 113.9(d, J_{CF}=2.7 Hz, CN), 47.9(dd, J_{CF}=22.2 and 27.3 Hz, C_1), 33.6(broad s, C_4); mass spectrum gave M^+ 255.05379 ± 0.0016(6 ppm), calcd for $C_{12}H_7N_2F_4$ 255.05453 dev -0.00074(3 ppm).

Finally there were isolated 28 mg (1.1%) white solid <u>66</u>: mp 78.6-80.0°; IR(CCl_4) 2940(w), 1794(s), 1430, 1278 (s), 1155, 1126, 1112 cm^{-1}; ^1H NMR(60 MHz) δ 5.9(pentet, 1H, J=3 Hz), 5.6(pentet, 1H, J=2.5 Hz), 3.3(m, 4H), 3.1(m, 2H); ^{19}F NMR(300 MHz) ϕ 89.9 and 101.2(AB pattern, 2F, J_{FF}=214.3 Hz), 90.8(pentet, 2F, J_{HF}=1.9 Hz); ^{13}C NMR(300 MHz)

δ 151.6(t, J_{CF}=287.0 Hz, =CF_2), 138.4(t, J_{CF}=20.5 Hz, C_3), 119.1(=CH_2), 118.8, 115.2(dd, J_{CF}=289.4 and J_{CF}=293.2 Hz, CF_2), 114.2(d, J_{CF}=3.6 Hz), 78.4(t, J_{CF}=32 Hz, $C_{3'}$), 51.2(t, J_{CF}=24.8 Hz, C_1), 33.7($C_{2'}$), 34.2(d, J_{CF}= 2.5 Hz, $C_{1'}$), 33.0(dd, J_{CF}=6.6 and J_{CF}=7.9 Hz, C_4); mass spectrum gave M^+ 255.05334 $\pm$ 0.0021(8 ppm), calcd for $C_{12}H_7N_2F_4$ 255.05453 dev -0.0012(5 ppm).

The total combined yield was 63%. The order of elution on DEGS was 62, 61, 66, 64, 65, then 63. The relative yields were 4.1%, 79.3%, 2.7%, 1.4%, 1.4% and 11.2% respectively as determined by GLPC integration, ^{19}F NMR integration and weighing. Compound 63 was somewhat unstable to GLPC conditions particularly on DEGS.

1,4-Diphenylnaphthalene-2-carbonyl fluoride 69 and 1,4-Diphenyl-2-trifluoromethylnaphthalene 70.

Into a 30-mL glass tube containing 2.00 g(7.40 mmoles) diphenylisobenzofuran[91] and 23 mL CCl_4 was condensed 1.7 g(22.4 mmoles) difluoroallene. The tube was sealed under vacuum. After 4 hours at room temperature the pale yellow solution was concentrated by rotary evaporation at reduced pressure to give 2.69 g of viscous oil. ^1H NMR analysis indicated only aromatic protons. Silica gel chromatography using a graded elution with hexane and CCl_4 gave 367 mg (15%) white solid 69: mp 135.0-135.4°; IR(CCl_4) 3070,

3038, 1824(s), 1804(s), 1595(w), 1383, 1238, 1202, 1076 (s), 977, 697 cm^{-1}; ^{1}H NMR(60 MHz) δ 8.0(s, 1H), 7.7-7.3 (m, 14H); ^{19}F NMR(100 MHz) ϕ -36.1(s); mass spectrum gave M^{+} 326.11014 $\pm$ 0.00345(11 ppm), calcd for $C_{23}H_{15}OF$ 326.11069 dev -0.00056(2 ppm).

Analysis: Calcd, C 84.64, H 4.63

Found, C 84.52, H 4.66

It also gave 445 mg (17%) white solid <u>70</u>: mp 128.8-130.0^{o}; IR(CCl_4) 3069, 3040, 2966, 2935, 2863, 1595(w), 1577(w), 1458, 1445, 1386, 1268(s), 1227, 1183, 1155(s), 1134(s), 1101, 896, 701 cm^{-1}; ^{1}H NMR(60 MHz) δ 7.8(s, 1H), 7.6-7.4(m, 14H); ^{19}F NMR(100 MHz) ϕ 56.7(s); mass spectrum gave M^{+} 348.11319 $\pm$ 0.00315(9 ppm), calcd for $C_{23}H_{15}F_3$ 348.11258 dev -0.00061(2 ppm).

Analysis: Calcd, C 79.30, H 4.34

Found, C 79.13, H 4.38

The combined yield of isolated products was 32%. The order of elution was <u>70</u> then <u>69</u>.

In a separate experiment, 31 mg (0.41 mmole) difluoroallene was condensed into an NMR tube containing 54 mg (0.20 mmole) diphenylisobenzofuran in 0.40 mL $CDCl_3$. The tube was sealed under vacuum and the reaction was monitored by ^{1}H NMR. The initial cycloadduct <u>68</u> displayed an AB pattern: δ 3.25(d of t, 1H, J_{HF}=3.4 and J_{HH}=13.6 Hz), 2.75(d of t, 1H, J_{HF}=2.4 and J_{HH}= 13.6 Hz).

<u>Endo-1,2,3,4-Tetrahydro-1,4-diphenyl-2-trifluoromethyl-1,4-epoxynaphthalene 72</u>.

Into a 50-mL glass tube containing 2.00 g(7.40 mmoles) diphenylisobenzofuran[91] in 30 mL CCl_4 was condensed 2.60 g (27.1 mmoles) 3,3,3-trifluoropropene. The tube was sealed under vacuum and heated at 83° for 63 hours. The tube was cooled and opened. The pale yellow solution was concentrated by rotary evaporation at reduced pressure to give a viscous oil. Recrystallization from hexane gave 2.42 g (89%) yellowish needles of <u>72</u>: mp 121.0-122.8°; IR(CCl_4) 3070, 3040, 2965, 1606(w), 1461, 1450, 1371, 1314, 1270, 1165(s), 1117, 1095, 983, 694 cm^{-1}; ^1H NMR(100 MHz) δ 8.0-7.0 (complex m, 14H), 3.83(m, 1H), 2.83(t, 1H, J=11 Hz), 2.30(dd, 1H, J_{HH}=12.0 and 4.6 Hz); ^{19}F NMR(100 MHz) ϕ 64.5(d, J_{HF}=8.7 Hz); ^{13}C NMR(300 MHz) δ 148.9, 143.3, 137.8, 135.6, 129.0, 128.5, 128.4, 128.3, 128.2, 127.7, 127.1, 126.2, 121.9, 118.8(aromatics), 125.9(quartet, J_{CF}=278.2 Hz, CF_3), 89.1, 88.3(quat), 48.0(quartet, J_{CF}= 26.3 Hz, CH), 36.7(CH_2); mass spectrum gave M^+ 366.12299 $\pm$ 0.0025(7 ppm), calcd for $C_{23}H_{17}F_3O$ 366.12315 dev -0.00016(0.4 ppm).

The minor isomer <u>73</u> was detected by ^{19}F NMR of the product mixture but was not isolated: ^{19}F NMR ϕ 65.3 (d, J_{HF}=9.1 Hz). Relative yields by ^{19}F NMR integration were 90.4% for <u>72</u> and 9.6% for <u>73</u>.

1,4-Diphenyl-2-trifluoromethylnaphthalene 70 by dehydration of 72.

A 5-mL flask containing 100 mg (0.273 mmole) 72 and 2 mL polyphosphoric acid was heated at 100° for 6 hours. The dark amber mixture was poured into 8 mL water. After extracting with three 5-mL portions ether and washing with 5 mL of 5% $NaHCO_3$ then 5 mL water, the ether solution was dried over anhydrous sodium sulfate. Concentration by rotary evaporation at reduced pressure gave 84 mg amber oil which was revealed by ^1H NMR analysis to be 50% 70 and 50% starting material.

3-Methyl-1,1,1-trifluorobutan-2-ol 75.

A 5-L, 3-necked, round-bottom flask was equipped with magnetic stirrer, nitrogen inlet, addition funnel and Friedrich's condenser. Isopropyl magnesium bromide was prepared in the usual manner from 120 g (4.95 moles) Mg turnings, 609 g (4.95 moles) isopropyl bromide and 2300 mL dry ether.

The flask containing the Grignard solution was cooled in an ice-water bath and 284 g (2.00 moles) ethyltrifluoroacetate in an equal volume of ether was added over a period of 50 minutes. After removal of the ice bath, the dark brown solution was stirred for one hour, as propene was evolved. Stirring was continued overnight for 15 hours.

After careful acidification with 1500 mL dilute HCl, the ether layer was separated. The aqueous layer was extracted with five 300-mL portions of ether. The combined ether layers were dried over anhydrous sodium sulfate. After removal of the ether by fractional distillation, 168.4 g (59%) colorless liquid 75 were collected from 95-100°. Distillation of the product from CaH_2 gave pure material: bp 97-101.5° (lit.[92] bp 99-100°); IR(film) 3420(broad), 2980, 1480, 1280, 1175, 1150, 1080, 1030 cm^{-1}; ^1H NMR(60 MHz) δ 4.0-3.5(complex m, 1H), 3.3(s, 1H), 2.4-1.7(complex m, 1H), 1.1(d, 6H, J_{HH}=7 Hz); ^{19}F NMR(100 MHz) ϕ 76.4(d, J_{HF}=5.1 Hz); mass spectrum gave 141(M^+-1), 125, 122, 80, 79, 73, 55, 43(base), 41.

3-Methyl-1,1,1-trifluorobutan-2-ol acetate 76.

Into a 1000-mL flask equipped with magnetic stirrer and Friedrich condenser was added 224.8 g (1.58 moles) methyltrifluorobutanol 75 and 310 g (3.95 moles) acetyl chloride. The vigorous evolution of HCl was controlled by cooling with an ice-water bath. After 4 hours at reflux, the solution was stirred overnight (15 hours) at room temperature.

The solution was carefully poured into 600 mL water. Separation of the organic layer yielded 270 g (93%) of crude ester which was dried over anhydrous magnesium sulfate and purified by fractional distillation to give 228 g acetate 76: bp 111-121°; IR(film) 2980

1770, 1475, 1375, 1280, 1225, 1170, 1125, 1090, 1050 cm^{-1}; ^1H NMR(60 MHz) δ5.1(d of quartet, 1H, J_{HH}=5.5 and J_{HF}=7.2 Hz, CH), 2.1(m, 4H, CH_3 and CH), 1.0(d, 6H, J_{HH}=6 Hz, CH_3); ^{19}F NMR(100 MHz) ϕ74.1(d, J_{HF}= 7.0 Hz); mass spectrum gave 185(M^++1), 169, 164, 142, 122, 104, 43(base).

2-Methyl-4,4,4-trifluoro-2-butene 77.

A vertical pyrolysis tube (40 cm by 2 cm) packed with glass wool was equipped with pressure-equalizing addition funnel with nitrogen inlet at the top and a flask at the bottom with nitrogen outlet. Over a period of 12 hours, 228 g (1.24 moles) methyltrifluorobutyl acetate 76 were passed through the tube at 500° (2 drops/sec, 12 mL/min N_2).

The brown pyrolysate collected in the ice-chilled flask was poured into 450 mL water. The amber organic layer was separated, treated with K_2CO_3 and purified by fractional distillation to yield 115 g (75%) methyltrifluorobutene 77: bp 48-51°; IR(film) 2990, 2960, 2925, 1685, 1385, 1355, 1275, 1230, 1110 cm^{-1}; ^1H NMR (60 MHz) δ 5.4(complex quartet, 1H, J_{FH}=8.4 Hz), 1.9 (s, 6H, CH_3); mass spectrum gave M^+ 124.0505, calcd for $C_5H_7F_3$ 124.0500 dev 0.0005(4 ppm).

2,3-Dibromo-2-methyl-4,4,4-trifluorobutane 78.

A 500-mL, three-necked flask was equipped with magnetic stirrer, thermometer and addition funnel.

To the flask was added 109 g (0.881 mole) methyltrifluorobutene 77. To the stirred, cooled solution (10°, salt-ice bath) was added 140 g (0.88 mole) Br_2 over a period of 50 minutes. Periodic irradiation with a sun lamp produced a pale yellow solution which was treated with K_2CO_3. Purification by fractional distillation at reduced pressure gave 198 g (79%) dibromomethyltrifluorobutane 78: bp 64-65°/36 mm; IR(film) 2990, 1460(w), 1255, 1185, 1110 cm^{-1}; ^1H NMR(60 MHz) δ 4.5(quartet, 1H, J_{HF}=7.2 Hz), 2.0(s, 6H, CH_3); ^{19}F NMR(100 MHz) ϕ 63.7(d, J_{HF}=7.5 Hz); mass spectrum gave 285, 283, 281, 203(base), 123, 103, 77, 59, 39.

Analysis: Calcd, C 21.15, H 2.48, F 20.07

Found, C 21.39, H 2.61, F 19.82

3-Bromo-2-methyl-4,4,4-trifluoro-2-butene 79.

A 1000-mL, three-necked flask was equipped with magnetic stirrer, pressure-equalizing addition funnel and fractional distillation apparatus with vacuum adapter. To the flask was added 200 g (3.03 moles) KOH pellets and to the funnel was added 197.7 g (0.696 moles) dibromomethyltrifluorobutane 78.

After lowering the pressure to 100 mm, the flask was heated to 90° and the dibromide was added over a period of 2 hours. The product distilled with a head temperature of 48-50°/100 mm. The crude product was dried over molecular sieves and purified by fractional distillation at ambient pressure to give 104.8 g (74%) bromomethyltrifluorobutene 79: bp 106-108°; IR(film) 2940, 1645, 1375, 1355(w), 1285, 1220, 1165, 1130, 920 cm^{-1}; ^1H NMR(60 MHz) δ 2.2-2.0(m, 6H, CH_3); ^{19}F NMR(100 MHz) ϕ 57.4(broad s); mass spectrum gave 204, 202, 189, 183, 123, 103(base), 77, 59, 53, 43, 39.

Analysis: Calcd, C 29.58, H 2.98, F 28.08

Found, C 29.39, H 3.13, F 29.22

1,1-Difluoro-3-methyl-1,2-butadiene 80.

A solution of n-BuLi in heptane was prepared from commercial n-BuLi by flash vacuum distillation of the hexane followed by addition of an equal volume of dry heptane(distilled from P_2O_5 under nitrogen).

A 250-mL, three-necked, round-bottom flask was equipped with magnetic stirrer, pressure-equalizing addition funnel with nitrogen inlet, low-temperature thermometer, and coil reflux condenser with nitrogen outlet. After flame drying the apparatus, the outlet was connected first to a spiral trap in a Dewar of water, then to a trap in a Dewar of dry ice/isopropanol. To the flask was added 16.5 g (0.0813 mole) bromomethyltrifluorobutene 79 and 75 mL dry heptane. To the funnel was added 35 mL (0.0514 mole) of 1.47 M n-BuLi in heptane.

After the flask was cooled with a hexane-slush bath to -85°, the n-BuLi solution was added over a period of 5 minutes. The colorless, slightly cloudy solution was allowed to stir an additional 10 minutes, then warm to room temperature.

After replacement of the funnel and thermometer with glass stoppers, a vacuum system was attached to the dry-ice temperature trap and the pressure was lowered to 180 mm. The flask was heated at reflux (180 mm) for one hour.

The liquid in the dry-ice temperature trap was purified by prep GLPC (10 ft by 1/4 in 10% DNP at 50°, 24 mL/min) to give 1.49 g (28%) pure 80: bp 52-53°; IR(gas) 3000, 2930, 2410(w), 2010, 1500, 1430, 1370(w), 1210, 1085, 905 cm^{-1}; ^{1}H NMR(60 MHz) δ 1.97(t, J_{HF}=5.1 Hz, CH_3); ^{19}F NMR(100 MHz) ϕ 102.2(septet, J_{HF}=5.05 Hz); mass spectrum gave M^+ 104.0367, calcd for $C_5H_6F_2$ 104.04376 dev 0.00009(0.9 ppm).

6-(Difluoromethylene)-5,5-dimethylbicyclo[2.2.1]-hept-2-ene 81.

To a 5-mL flask was added 1.00 g (15 mmoles) cyclopentadiene and 172 mg (1.65 mmoles) 80. After stirring 4 days at room temperature, the product was isolated from unreacted 80, cyclopentadiene and dicyclopentadiene by prep GLPC(10 ft by 1/4 in 10% DNP at 140°, 24 mL/min) to give 106 mg (38%) 81: bp 149-152°; IR(film) 2980, 1765(s), 1460, 1230, 1030, 735 cm^{-1}; ^{1}H NMR(60 MHz) δ6.2(complex m, 2H, olefinic),

3.4(broad s, 1H, di-allylic CH), 2.5(broad s, 1H, allylic CH), 1.8 and 1.5(two d, 2H, CH_2), 1.3(s, 3H, CH_3), 1.03 (s, 3H, CH_3); ^{19}F NMR(100 MHz) ϕ 96.6(d, 1F, J_{FF}=74 Hz), 93.0(d, 1F, J_{FF}=74 Hz); mass spectrum gave M^+ 170.09103, calcd for $C_{16}H_{12}F_2$ 170.09071 dev 0.00032(1 ppm).

<u>1,1-Difluoro-3-(difluoromethylene)-2-methylenecyclobutane 13, 7-(Difluoromethylene)-2-methylene-1,1,5,5-tetrafluorospiro[3.3]heptane 82, and 3-(Difluoromethylene)-7,7,8,8-tetrafluorobicyclo[4.2.0]oct-1(6)-ene 83.</u>

Into a 10-mL glass tube containing 20 mg hydroquinone was condensed 1.10 g (14.5 mmoles) difluoroallene. The tube was sealed under vacuum and kept in the dark at room temperature for 108 hours. The tube was cooled and opened. The volatile components were transferred on the vacuum line to a flask and the products were isolated by prep GLPC (10 ft by 1/4 in 10% TCP at 28°, 60 mL/min) to give 67 mg (6.1%) dimer <u>13</u>: IR(gas) 1765(s), 1430(w), 1320, 1290, 1204, 1133, 1070, 1023 cm^{-1}; ^1H NMR(100 MHz) δ 5.58(m, 1H), 5.35 (m, 1H), 3.20(tt, 2H, J_{HF}=4.3 and 10.0 Hz); ^{19}F NMR (100 MHz) ϕ 80.0(dtt, 1F, J_{FF}=31.0, J_{HF}=4.3 and J_{HF}= 1 Hz), 87.6(dtt, 1F, J_{FF}=31.0, J_{HF}=4.3 and J_{HF}=1 Hz), 98.4(tt, 2F, J_{HF}=10.1 and J_{HF}=2.6 Hz).

From the non-volatile liquid fraction was isolated by prep GLPC (10 ft by 1/4 in 10% TCP at 120°, 60 mL/min) 35 mg (3.2%) trimer <u>82</u>: bp 143.5-144.0°; IR(film)

2955(w), 1792(s), 1430, 1305, 1280, 1262, 1110, 955, 880 cm^{-1}; ^{1}H NMR(100 MHz) δ 5.62(m, 1H), 5.30(m, 1H), 3.3-3.0(complex m, 3H), 2.7(d of m, 1H, J=16.7 Hz); ^{19}F NMR(100 MHz) ϕ 85.6(dd, 1F, J_{FF}=48.7 and J_{HF}=12.3 Hz), 86.9(d, 1F, J_{FF}=48.7 Hz), 98.1(dd, 1F, J_{FF}=220.1 and J_{HF}=11.8 Hz), 101.1(non-first-order m, 2F), 104.9 (dd, 1F, J_{FF}=220.1 and J_{HF}=27.7 Hz); mass spectrum gave M^+ 228.03688 $\pm$ 0.00195(8 ppm), calcd for $C_9H_6F_6$ 228.03737 dev -0.00049(2 ppm).

There were also isolated 100 mg of trimer 83: bp 173.0-174.5°; IR(film) 2930, 1766(s), 1436, 1334(s), 1304, 1254, 1212, 1103(s), 953, 918, 874 cm^{-1}; ^{1}H NMR (60 MHz) δ 2.92(m, 2H), 2.39(m, 4H); ^{19}F NMR(300 MHz) ϕ 90.7(d of pentet, 1F, J_{FF}=50.3 and J_{HF}=1.7 Hz), 94.4 (d of pentet, 1F, J_{FF}=50.3 and J_{HF}=2.0 Hz), 113.6 and 113.7(non-first-order AB pattern, 2F); ^{13}C NMR(100 MHz) δ 153.0(t, J_{CF}=284.7 Hz, $=CF_2$), 119.6(t with non-first-order fine structure, J_{CF}=287, J_{CF}=30.8 and J_{CF}=25.3 Hz, CF_2-CF_2), 81.7(dd, J_{CF}=19.8 and 21.7 Hz, C_3), 20.2(d, J_{CF}= 2.4 Hz, CH_2), 19.5(t, J_{CF}=1.8 Hz, CH_2), 19.1(t, J_{CF}= 2.1 Hz, CH_2); mass spectrum gave M^+ 228.03780 $\pm$ 0.00096 (4 ppm), calcd for $C_9H_6F_6$ 228.03737 dev 0.00043(2 ppm).

The combined yield of isolated products was 18.4%. The reaction was roughly 50% complete judging from the amount of unreacted difluoroallene. Order of elution was 13, 82, then 83. The relative yields were 33% for

13, 17% for 82 and 62% for 83. There were numerous minor products which were not isolated due to their small concentrations.

2,2-Difluoro-3-methylenecyclobutanecarbonitrile 14 and 3-(Difluoromethylene)cyclobutanecarbonitrile 21.

Into a 20-mL glass tube containing 3.00 g (56.5 mmoles) acrylonitrile was condensed 400 mg (5.26 mmoles) difluoroallene. The tube was sealed under vacuum and heated at 110° for 12 hours. The tube was cooled and opened. The excess acrylonitrile was removed by fractional distillation using a 10-cm Vigreux column. After vacuum transfer of the distillation residue, prep GLPC (20 ft by 1/4 in 20% SE-30 at 150°, 40 mL/min) gave 436 mg (64%) of a mixture of 14 and 21 along with 21 mg (1.7%) isolated trimer 83. Further separation by prep GLPC (10 ft by 1/4 in 10% DEGS at 90°, 60 mL/min) gave 262 mg (38.6%) of 14: 174.5-175.0° (lit.[22] bp 104-108°/100 mm); IR(film) 2980, 2260, 1695(w), 1436, 1280, 1163, 1118, 1072, 1000, 932 cm^{-1}; ^1H NMR(100 MHz) δ 5.68(m, 1H), 5.40(m, 1H), 3.69(pentet, 1H, J=9.0 Hz), 3.0(complex m, 2H); ^{13}C NMR(100 MHz) δ 141.4(dd, J_{CF}=19.8 and J_{CF}=21.1 Hz, C_3), 116(=CH_2), 115.9(dd, J_{CF}=282.6 and J_{CF}=289.3 Hz, CF_2), 115.2 (t, J_{CF}=2.4 Hz), 34.3(dd, J_{CF}=21.4 and J_{CF}=26.9 Hz, C_1), 28.3(dd, J_{CF}=4.9 and J_{CF}=10.4 Hz, C_4); ^{19}F NMR(300 MHz) ϕ 91.7(dddd, 1F, J_{FF}=209, J_{HF}=9.9, J_{HF}=3.9, and J_{HF}= 2.3 Hz), 98.8(dd, 1F, J_{FF}=209.0 and J_{HF}=8.2 Hz); mass spectrum gave M^+ 129.0384 ± 0.00051(4 ppm), calcd for $C_6H_5NF_2$ 129.03901 dev -0.00061(5 ppm).

Analysis: Calcd, C 55.82, H 3.90, N 10.85

Found, C 55.85, H 3.92, N 10.84

It also gave 86.6 mg (12.8%) of 21: bp 169.2-169.8°; IR(film) 2990, 2955, 2860, 2243, 1795(s), 1428(w), 1260 (s), 1105 cm^{-1}; ^1H NMR(100 MHz) δ 3.12(complex m); ^{19}F NMR(100 MHz) φ 94.6(complex m); ^{13}C NMR(100 MHz) δ 151.1 (t, J_{CF}=284.4 Hz, =CF_2), 121.4(CN), 82.5(t, J_{CF}=28.4 Hz, C_3), 29.2(t, J_{CF}=2.4 Hz, C_2), 18.8(C_1); mass spectrum gave M^+ 129.03885 ± 0.00057(4 ppm), calcd for $C_6H_5NF_2$ 129.03901 dev -0.00016(1 ppm).

The order of elution with SE-30 was 14, 21, then 83. With DEGS the order was 21 then 14. The GLPC relative yields were 77% for 14 and 23% for 21.

2,2-Difluoro-1-methyl-3-methylenecyclobutanecarbonitrile 86 and 1-Methyl-3-(difluoromethylene)cyclobutanecarbonitrile 87.

Into a 20-mL glass tube containing 4.00 g (59.6 mmoles) methacrylonitrile was condensed 641 mg (8.43 mmoles) difluoroallene. The tube was sealed under vacuum and heated at 110° for 24 hours. After cooling and opening, the excess methacrylonitrile was removed by distillation. After vacuum transfer of the residue, prep GLPC (20 ft by 1/4 in 20% Carbowax 20M at 130°, 40 mL/min gave 402 mg (33%) of 86: bp 171.8-172.0°;

IR(film) 3000, 2950, 2248, 1694(w), 1432, 1279, 1123, 1100, 1053, 1022, 933 cm^{-1}; ^1H NMR(100 MHz) δ 5.71 (m, 1H), 5.43(hextet, 1H, J=2.2 Hz), 3.08(d of m, 1H, J=16.0 Hz), 2.57(d of m, 1H, J=16.0 Hz), 1.6(dd, 3H, J=1.8 and 0.7 Hz); ^{19}F NMR(100 MHz) ϕ 97.9(d of septet, 1F, J_{FF}=206.4 and J_{HF}=2.2 Hz), 105.5(d of m, 1F, J_{FF}= 206.4 Hz); ^{13}C NMR(100 MHz) δ 140.7(dd, J_{CF}=20.1 and 21.4 Hz, C_3), 118.5(d, J_{CF}=4.3 Hz, CN), 116.6(d, J_{CF}= 1.2 Hz, =CH_2), 116.2(dd, J_{CF}=282.9 and 293.9 Hz, CF_2), 41.6(dd, J_{CF}=20.4 and 25.3 Hz, C_1), 37.0(dd, J_{CF}=9.2 and 5.5 Hz, C_4), 18.4(d, J_{CF}=5.5 Hz, CH_3); mass spectrum gave M^+-1 142.04568 ± 0.00084(6 ppm), calcd for $C_7H_6NF_2$ 142.04683 dev -0.00116(8 ppm).

Analysis: Calcd, C 58.74, H 4.93, N 9.79

Found, C 58.62, H 4.93, N 9.74

It also gave 147 mg of a mixture of 87 and trimer 83. Further separation by prep GLPC (20 ft by 1/4 in 20% SE-30 at 110°, 45 mL/min) gave 50.5 mg (7.9%) of trimer 83 and 76 mg (6.3%) of pure 87: bp 160.5-161.0°; IR(film) 2985, 2943, 2880(w), 2853(w), 2240, 1795(s), 1766(w), 1430, 1260(s), 1058 cm^{-1}; ^1H NMR(100 MHz) δ 3.24(d of quartet, 2H, J=15.0 and 3.5 Hz), 2.69(d of quartet, 2H, J=15.0 and 3.2 Hz), 1.58(s, 3H); ^{19}F NMR (100 MHz) ϕ 94.0(pentet, J_{HF}=3.8 Hz); ^{13}C NMR(100 MHz) δ 151.8(t, J_{CF}=284.7 Hz, =CF_2), 123.8(CN), 80.3(t, J_{CF}= 28.4 Hz, C_3), 36.5(t, J_{CF}=2.1 Hz, C_2), 28.0(C_1), 24.9(Me);

mass spectrum gave M^+ 143.05393 $\pm$ 0.0012(8 ppm), calcd for $C_7H_7NF_2$ 143.05466 dev -0.00073(5 ppm).

Analysis: Calcd, C 58.74, H 4.93, N 9.79

Found, C 58.68, H 4.96, N 9.75

The combined yield of isolated products was 39.3%. Order of elution on Carbowax 20M was (83 and 87) then 86. On SE-30 the order was 87 then 83. The GLPC relative yields were 84% for 86 and 16% for 87.

2,2-Dichloro-4-methylene-1,1,3,3-tetrafluorocyclobutane 88 and 1,1-Dichloro-2,2-difluoro-3-(difluoromethylene)cyclobutane 89.

Into a 20-mL glass tube was condensed 2.31 g (17.4 mmoles) 1,1-dichloro-2,2-difluoroethylene and 110 mg (1.45 mmoles) difluoroallene. The tube was sealed under vacuum and heated at 134° for 27 hours. The tube was cooled and opened. The mixture was subjected to prep GLPC (10 ft by 1/4 in 10% TCP at 80°, 60 mL/min) to give 150 mg dichlorodifluoroethylene dimer, 10.6 mg (3.5%) of 88: IR(CCl_4) 1718(w), 1700(w), 1438(w), 1300, 1180, 1091, 924, 890 cm^{-1}; ^1H NMR(60 MHz) δ 6.23(pentet, J_{HF}=2.1 Hz); ^{19}F NMR(100 MHz) ϕ 94.4(t, J_{HF}=2 Hz).

It also gave 49.3 mg (16.3%) of 89: IR(CCl_4) 1787 (s), 1432, 1321(s), 1150, 990, 782 cm^{-1}; ^1H NMR(100 MHz) δ 3.35(t, J=4.6 Hz); ^{19}F NMR(100 MHz) ϕ 65.6(d of m, 1F, J_{FF}=22.5 Hz), 72.3(d of m, 1F, J_{FF}=22.5 Hz),

88.2(dd, 2F, J_{FF}=7.8 and 4.1 Hz); mass spectrum gave M^+ 207.94782 ± 0.00263(13 ppm), calcd for $C_5H_2F_4Cl_2$ 207.94697 dev 0.00085(4 ppm).

The combined yield of isolated products was 20%. Order of elution was 89, dichlorodifluoroethylene dimer, then 88. The GLPC relative yields were 73% for 89 and 27% for 88. At low conversion (134°, 1 hour) the relative yields were 50.6% for 88 and 49.4% for 89.

(E)-1,1-Dichloro-2,2-difluoro-3-(fluoromethylene)-cyclobutane 90, 1,1-Dichloro-2,2,4-trifluoro-3-methyl-enecyclobutane 91, and (Z)-1,1-Dichloro-2,2-difluoro-3-(fluoromethylene)cyclobutane 92.

Into a 5-mL glass tube was condensed 2.50 g (18.8 mmoles) 1,1-dichloro-2,2-difluoroethylene and 105 mg (1.8 mmoles) fluoroallene. The tube was sealed under vacuum and heated at 134° for 8 hours. The tube was cooled and opened. The products were isolated by prep GLPC (10 ft by 1/4 in 10% TCP at 130°, 60 mL/min) to give 220 mg dichlorodifluoroethylene dimer, 40.3 mg (11.7%) of 90: IR($CDCl_3$) 1732(s), 1425, 1290, 1237, 1205, 1150(s), 1100, 1030, 1000, 978 cm^{-1}; ^1H NMR(100 MHz) δ7.17(d of pentet, 1H, J_{HF}=77.4 and J=3.1 Hz), 3.39(m, 2H); ^{19}F NMR(100 MHz) φ96.4(dd, 2F, J_{HF}=3.2 and J_{FF}=5.2 Hz), 124.8(d of pentet, 1F, J_{HF}=77.4 and J=4.9 Hz); mass spectrum gave M^+ 189.95566 ± 0.00168 (9 ppm), calcd for $C_5H_3Cl_2F_3$ 189.95639 dev -.00073 (4 ppm).

There were also isolated 65.4 mg (19.0%) of <u>91</u>: IR(CCl_4),1300, 1149, 1084, 876, 735(s) cm^{-1}; ^1H NMR (100 MHz) δ 6.09(m, 1H), 5.97(m, 1H), 5.39(d of pentet, 1H, J_{HF}=56.2 and J=2.5 Hz); ^{19}F NMR(100 MHz) ϕ 99.1 (dd of quartet, 1F, J_{FF}=203.9, J_{FF}=9.2 and J_{HF}=2.5 Hz), 102.3(dd of quartet, 1F, J_{FF}=203.9, J_{FF}=8.4, and J_{HF}= 2.5 Hz), 174.1(dtt, 1F, J_{HF}= 56.2, J_{FF}=8.8, and J_{HF}= 3.7 Hz); mass spectrum gave M^+ 189.95678 $\pm$ 0.00327 (17 ppm), calcd for $C_5H_3Cl_2F_3$ 189.95639 dev 0.00039 (2 ppm).

Finally there were isolated 47.1 mg (14%) of <u>92</u>: IR(film) 1723(s), 1429, 1361, 1257, 1147, 1118, 990, 977, 872 cm^{-1}; ^1H NMR(60 MHz) δ 6.65(d of m, 1H, J_{HF}=77.4 Hz), 3.32(dd, 2H, J_{HF}=4.9 and J_{HH}=2.5 Hz); ^{19}F NMR(100 MHz) ϕ 98.1(d, 2F, J_{FF}=9.1 Hz), 116.6(d of septet, 1F, J_{HF}= 77.4 and J=4.8 Hz; mass spectrum gave 190(M^+, 7), 155(100), 132(20), 96(55).

The combined yield of isolated products was 45%. Order of elution was dichlorodifluoroethylene dimer, <u>90</u>, then <u>91</u>, followed by <u>92</u>. The GLPC relative yields were 25.3% for <u>90</u>, 49.3% for <u>91</u> and 25.4% for <u>92</u>. In a separate experiment at 80° for 30 hours, the relative yields were 23.4%, 51.1%, and 25.5%.

<u>4-(Difluoromethylene)-4,5-dihydro-3H-pyrazole 101</u>.

An ether solution of diazomethane (75 mL) was prepared from 7.00 g (32.7 mmoles) N-methyl-N-nitroso-p-toluenesulfonamide. The solution was vacuum transferred to a glass tube. Into the tube was condensed 2.50 g (32.8 mmoles) difluoroallene. The tube was sealed under vacuum and allowed to warm to room temperature. The yellow solution turned rapidly colorless.

After concentration by rotary evaporation at reduced pressure, the residue was distilled at reduced pressure using a short-path distillation apparatus. The clear, colorless distillate was collected in a flask cooled with a salt-ice bath. The unstable product was stored on dry ice under nitrogen. A total of 2.0 g (52%) 101 was obtained: bp 41.0-41.5^{o}/31 mm and 48.5-49.0^{o}/42 mm; UV_{max} 320 nm; IR(film) 2930, 1797(s), 1554, 1428, 1300, 1275, 1080, 930 cm^{-1}; ^{1}H NMR(60 MHz) δ 5.1(t, J_{HF}=3.7 Hz); ^{19}F NMR(100 MHz) ϕ 84.8(pentet, J_{HF}=3.8 Hz); mass spectrum gave M^{+} 118.03375 $\pm$ 0.00125 (11 ppm), calcd for $C_4H_4N_2F_2$ 118.03425 dev -0.00050 (4 ppm).

NMR yield: Into an NMR tube containing 53.5 mg benzene(cooled in an ice bath) was added 0.350 mL diazomethane solution in ether. Into the tube was condensed 0.227 mmole difluoroallene and the tube was sealed under vacuum. After 15 minutes at room temperature the ^{1}H NMR of the yellow solution was integrated. The average of 5 integrations gave an NMR yield of 94.6%.

<u>4-(Difluoromethylene)-4,5-dihydro-3,3-dimethyl-3H-pyrazole 104 and 5-(Difluoromethylene)-4,5-dihydro-3,3-dimethyl-3H-pyrazole 105</u>.

An ether solution of diazopropane[93] (20 mL) was prepared from 2.50 g (34.7 mmoles) acetone hydrazone. The solution was vacuum transferred to a 50-mL glass tube and 2.71 g (35.7 mmoles) difluoroallene were condensed into the tube, which was sealed under vacuum. After 3.5 hours at -78°, the orange diazopropane color was completely gone. The tube was opened and the solution was concentrated by rotary evaporation at reduced pressure to give 0.81 g clear amber liquid. Analysis by ^{1}H NMR indicated relative yields of 48.7% for <u>104</u>, 31.4% for <u>105</u> and 19.9% acetone azine. The unstable products were not able to be isolated from the reaction mixture.

Analysis of the reaction mixture by IR, ^{1}H NMR, and ^{19}F NMR gave for <u>105</u>: IR(CCl_4) 1756(S); ^{1}H NMR (60 MHz) δ2.1(t, 2H, J_{HF}=3.7 Hz), 1.3(s, 6H, CH_3); ^{19}F NMR(100 MHz) ϕ83.2(d of t, 1F, J_{FF}=25 and J_{HF}= 3.8 Hz), 94.4(d of t, 1F, J_{FF}=25 and J_{HF}=3.9 Hz).

Spectra of <u>104</u> from the mixture were identical to spectra of the pure material prepared by the alternate route. The relative yields obtained by ^{1}H NMR integration were 61% for <u>104</u> and 39% for <u>105</u>.

NMR yield: Into an NMR tube containing 0.400 mL diazopropane solution in ether was condensed 0.1053 mmole difluoroallene. The tube was sealed under vacuum and warmed to room temperature. The ^{1}H NMR spectrum indicated no difluoroallene remained. The tube was cooled and opened. To the very pale orange solution was added 46.6 mg (0.637 mmole) dimethylformamide as an internal standard. The broad singlet 7.8 ppm for DMF and the triplet at 5.0 ppm for 104 were integrated. After correcting for the minor isomeric product 105 (which was obscured by the ether solvent) by using the ^{19}F NMR ratio of the two products, an ^{1}H NMR yield of 99% was calculated.based on difluoroallene.

4-(Difluoromethylene)-4,5-dihydro-3,3-dimethyl-3H-pyrazole 104.

To an ether solution of diazomethane prepared from 3.5 g (16.3 mmoles) N-methyl-N-nitroso-p-toluenesulfonamide in 30 mL ether was added 858 mg (8.25 mmoles) difluorodimethylallene 80. After 3 hours at room temperature, the pale yellow solution was concentrated at reduced pressure by rotary evaporation and the residue was distilled at reduced pressure using a short-path distillation apparatus. The clear, colorless distillate was collected in a dry-ice-chilled flask. The unstable liquid was stored on dry ice under nitrogen.

A total isolated yield of 0.891 g (74%) 104 was obtained: bp 37-38°/19 mm and 48-50°/36 mm; IR(CCl_4) 2985, 2940, 1780(s), 1556, 1463, 1285, 1245, 1135, 1050, 920 cm^{-1}; ^{1}H NMR(60 MHz) δ 5.19(t, 2H, J_{HF}= 3.8 Hz), 1.5(s, 6H); ^{19}F NMR(100 MHz) ϕ 81.3(d of t, 1F, J_{FF}=59 and J_{HF}=3.6 Hz), 89.7(d of t, 1F, J_{FF}=59 and J_{HF}=3.7 Hz); ^{13}C NMR(100 MHz) δ 149.9(t, J_{CF}=284 Hz, =CF_2), 89.0(t, J_{CF}=22 Hz, C_4), 88.5(d, J_{CF}=5 Hz, C_3), 75.9(d, J_{CF}=5 Hz, CH_2), 23.9(CH_3); mass spectrum gave M^+ 146.0655 ± 0.00134(9 ppm), calcd for $C_6H_8N_2F_2$ 146.06555 dev -0.00005(0.4 ppm).

Analysis: Calcd, C 49.31, H 5.52, N 19.17

Found, C 49.08, H 5.56, N 18.99

NMR yield: To an ice-chilled NMR tube containing 35.5 mg difluorodimethylallene 80 and 26.9 mg benzene as internal standard was added a slight molar excess of diazomethane solution in ether. Integration of the ^{1}H NMR spectrum indicated a yield of 90% based on difluorodimethylallene.

5-(Difluoromethylene)-4,5-dihydro-3,3-diphenyl-3H-pyrazole 108 and 4-(Difluoromethylene)-4,5-dihydro-3,3-diphenyl-3H-pyrazole 107.

Into a 60-mL glass tube containing 2.4 g (12.4 mmoles diphenyldiazomethane[94] and 20 mL ether was condensed 1.03 g (13.6 mmoles) difluoroallene. The tube was sealed under vacuum. After 5 hours at room temperature, the deep red color had faded to pale yellow. The tube was opened and the ether was removed by rotary evaporation at reduced pressure to give 3.18 g (95%) pale yellow solid. Recrystallization from hexane gave a pale yellow solid <u>108</u> mp 60-65°; IR(CCl_4) 3070, 3040, 1757(s), 1604, 1499, 1452, 1307, 1180, 1096, 699 cm^{-1}; ^1H NMR(60 MHz) δ 7.22(s, 10H), 3.02(t, 2H, J_{HF}=3.6 Hz); ^{19}F NMR(100 MHz) ϕ 79.8(d of t, 1F, J_{FF}=17 and J_{HF}=3.6 Hz), 91.5(d of t, 1F, J_{FF}= 17 and J_{HF}=3.8 Hz); mass spectrum (12 eV) gave 242 (M^+-28, 100), 241(69), 222(85), 204(10), 191(16), 165 (12), 127(26), 109(15).

Analysis: Calcd, C 71.10, H 4.47, F 14.06

Found, C 71.33, H 4.65, F 13.84

Compound <u>108</u> was very unstable in chloroform solution and unstable to silica gel. The mother liquor from recrystallization was concentrated by rotary evaporation at reduced pressure to give an oil which was subjected to silica gel chromatography using $CHCl_3$ solvent. A yellow oil was obtained which was <u>107</u>: IR(CCl_4) 3070, 3040, 1780(s), 1603, 1560, 1497, 1450, 1284, 1105, 1036, 697(s) cm^{-1}; ^1H NMR(60 MHz) δ 7.3(m, 10H), 5.34(t, 2H, J_{HF}=3.5 Hz); ^{19}F NMR(100 MHz) ϕ 78.6(d of t, 1F, J_{FF}=44 and J_{HF}=3.5 Hz), 82.0

(d of t, J_{FF}=44 and J_{HF}=3.5 Hz); mass spectrum gave 270(M^+, 3), 242(96), 221(100), 213(67), 193(60), 165 (66), 105(91), 77(94).

The relative yields of products by ^1H NMR integration were 14% for <u>107</u> and 86% for <u>108</u>.

<u>4'-(Difluoromethylene)-4',5'-dihydrospiro[9H-fluorene-9,3'-[3H]pyrazole] 110, 5'-(Difluoromethylene)-4',5'-dihydrospiro[9H-fluorene-9,3'-[3H]pyrazole] 111, and 2-(Difluoromethylene)spiro[cyclopropane-1,9'-[9H]-fluorene] 112.</u>

Into a 20-mL glass tube containing 0.485 g (2.52 mmoles) diazofluorene[95] and 7 mL ether was condensed 0.574 g (7.56 mmoles) difluoroallene. The tube was sealed under vacuum. After 3 hours at room temperature, the deep red color had faded to pale amber. The tube was opened and the solvent was removed by rotary evaporation at reduced pressure to give 0.655 g (98%) thick amber oil. Analysis by ^1H NMR indicated relative yields of 72% for <u>111</u> and 28% for <u>112</u>. Flash chromatography using silica gel and eluting with 95% hexane/5% EtOAc gave 132 mg (20%) of yellow solid <u>112</u>: R_f=0.54; mp 78-82°(from hexane); IR(CCl_4) 3075, 3050, 3020, 2985, 1844(s), 1453, 1320, 1240(s), 1175, 680 cm^{-1}; ^1H NMR(100 MHz) δ 7.85(m, 2H), 7.6-7.0(m, 6H), 2.42(t, 2H, J_{HF}=5.2 Hz); ^{19}F NMR(300 MHz) ϕ 82.3(d of t, 1F, J_{FF}=56.5 and J_{HF}=5.2 Hz), 86.2(d of t, 1F, J_{FF}=56.5

and J_{HF}=4.8 Hz); ^{13}C NMR(100 MHz) δ 149.9(dd, J_{CF}=281.4 and 283.4 Hz, =CF_2), 144.6, 140.4(substituted aromatic), 127.4, 127.3, 120.2(aromatic), 78.2(t, J_{CF}=36.0 Hz, C_2), 34.4(d, J_{CF}=6.1 Hz, C_1), 20.0(d, J_{CF}=4.9 Hz, CH_2); mass spectrum gave M^+ 240.07408 ± 0.00245(10 ppm), calcd for $C_{16}H_{10}F_2$ 240.07506 dev -0.00098(4 ppm).

Analysis: Calcd, C 79.99, H 4.20

Found, C 79.89, H 4.23

Compound 112 reacted slowly with oxygen in solution, but was indefinately stable in the absence of oxygen and the purified solid was stored under nitrogen. The major product 111 was very unstable in $CHCl_3$ solution and did not elute from the silica gel column. Spectroscopic analysis of the product mixture containing 111 and 112 gave for pyrazoline 111: IR(CCl_4) 1756(s) cm^{-1}; ^1H NMR (60 MHz) δ 2.97(t, J_{HF}=3.8 Hz, CH_2); ^{19}F NMR(100 MHz) ϕ 79.2(d of t, 1F, J_{FF}=15.2 and J_{HF}=3.6 Hz), 89.4(d of t, 1F, J_{FF}=15.2 and J_{HF}=3.8 Hz).

When the reaction was monitored by ^1H NMR(100 MHz), the CH_2 protons of the unstable intermediate 110 were observed: ^1H NMR(100 MHz) δ 5.65(t, J_{HF}=3.7 Hz). By the time the reaction was complete, 110 was no longer present.

4,5-Dihydro-4-(fluoromethylene)-3H-pyrazole 113.

An ether solution of diazomethane (70 mL) was prepared from 7.00 g (32.7 mmoles) N-methyl-N-nitroso-p-toluenesulfonamide. The solution was vacuum transferred to a 150-mL glass tube and 1.9 g (32.7 mmoles) fluoroallene were condensed into the tube, which was sealed under vacuum. After 10 minutes at room temperature, the clear, colorless solution was concentrated by rotary evaporation at 200 mm pressure. The residue was distilled at reduced pressure. The pale yellow distillate was collected in an ice-chilled flask. A total of 2.1 g (64%) 113 was obtained: bp 59-60°/23 mm; IR(CCl_4) 2920, 2822, 1710(s), 1545, 1420, 1340, 1263, 1183, 1092(s), 926 cm^{-1}; ^1H NMR(300 MHz) δ 6.67(d of pentet, 1H, J_{HF}=83.9 and J_{HH}=2.7 Hz), 5.16(t, 2H, J=2.8 Hz), 5.06(t, 2H, J=2.8 Hz); ^{19}F NMR(100 MHz) ϕ 121.7(d of pentet, J_{HF}=84.0 and 3 Hz); mass spectrum gave M^+ 100.04442 $\pm$ 0.00195(20 ppm), calcd for $C_4H_5N_2F$ 100.04368 dev -0.00075(8 ppm).

The unstable product 113 was stored on dry ice under nitrogen.

4-(Difluoromethylene)-2,3-diphenylisoxazolidine 116.

Into a 30-mL glass tube containing 1.00 g (5.07 mmoles) diphenylnitrone[96] and 10 mL $CHCl_3$ was condensed 1.11 g (14.6 mmoles) difluoroallene. The tube was kept at room temperature for 3 hours. The tube was opened and the solution was concentrated by rotary evaporation at reduced pressure to give 1.35 g (97%) amber oil. Purification by silica gel chromatography using benzene as solvent gave 1.29 g (93%) yellow oil which was 116: IR(CCl_4) 3074, 3040, 2872, 1787(s), 1600, 1492, 1453, 1278(s), 1075, 1030, 691 cm^{-1}; ^1H NMR(100 MHz) δ 7.5-6.9(complex m, 10H), 5.31(broad s, 1H), 4.61(non-first-order AB pattern, 2H); ^{19}F NMR ϕ 87.0(non-first-order AB pattern, J_{AB}=54.8 Hz, downfield F is d with t fine structure, J_{HF}=1.1 and 3.3 Hz, upfield F is d with t fine structure, J_{HF}=2.5 and 3.3 Hz); ^{13}C NMR(100 MHz) δ 149.9(N-subst aromatic), 149.7 (t, J_{CF}=285.0 Hz, =CF_2), 139.0(t, J_{CF}=2.4 Hz, C-subst aromatic), 129.1, 128.8, 128.1, 127.1(aromatic), 94.6 (t, J_{CF}=22.0 Hz, olefinic), 122.9(para-N-subst aromatic), 115.5(ortho-N-subst aromatic), 69.3(CH), 65.2(CH_2); mass spectrum gave M^+ 273.09670 $\pm$ 0.00129(5 ppm), calcd for $C_{16}H_{13}F_2NO$ 273.09652 dev 0.00017(0.6 ppm).

Analysis: Calcd, C 70.32, H 4.79, N 5.13

Found, C 70.22, H 4.81, N 5.12

4-(Difluoromethylene)-2-methyl-3-phenylisoxazolidine 117.

Into an NMR tube containing 100 mg (0.740 mmole) phenylmethylnitrone[97] and 0.40 mL $CDCl_3$ was condensed 113 mg (1.48 mmoles) difluoroallene. The tube was sealed under vacuum. After 3 hours at room temperature, the tube was cooled and opened. Concentration of the solution by rotary evaporation at reduced pressure gave 149 mg (95%) colorless liquid 117: IR(film) 3068, 3035, 3000, 2965, 2916, 2860, 1785(s), 1604(w), 1457, 1265(s), 1103, 1022 cm^{-1}; ^1H NMR(100 MHz) δ 7.35(m, 5H),

4.61(m, 2H), 4.56(m, 1H), 2.67(s, 3H); ^{19}F NMR(100 MHz) ϕ 96.8(d, J_{FF}=56.1 Hz), 94.5(d, J_{FF}=56.1 Hz); ^{13}C NMR (100 MHz) δ 149.6(t, J_{CF}=285.0 Hz, =CF_2), 137.4(subst aromatic), 128.6, 128.3, 128.1(aromatic), 95.8(t, J_{CF}= 20.8 Hz, olefinic), 71.8(CH), 65.1(CH_2), 43.2(CH_3); mass spectrum gave M^+ 211.08181 ± 0.00273(13 ppm), calcd for $C_{11}H_{11}NOF_2$ 211.08087 dev 0.00094(4 ppm).

Analysis: Calcd, C 62.55, H 5.25, N 6.63

Found, C 62.61, H 5.27, N 6.61

<u>4-(Difluoromethylene)-2,3,3-triphenylisoxazolidine 119</u>.

Into an NMR tube containing 0.50 mL $CDCl_3$ and 100 mg (0.366 mmole) triphenylnitrone[98] was condensed 56 mg (0.737 mmole) difluoroallene. The tube was sealed under vacuum and kept at room temperature. After 5 days the tube was cooled and opened. The solvent was removed by rotary evaporation at reduced pressure to give an oil. Flash chromatography using silica gel and 95% hexane/5% EtOAc gave 34.3 mg (27%) colorless oil R_f=0.43 which was pure <u>119</u>: ^1H NMR (60 MHz) δ7.5-7.2(m, 10H), 7.0-6.6(m, 5H), 4.91(t, 2H, J_{HF}=3.6 Hz); ^{19}F NMR(100 MHz) ϕ86.4(d, 1F, J_{FF}=51.5 Hz), 87.2(d of t, 1F, J_{FF}=51.5 and J_{HF}=3.2 Hz).

<u>4-(Difluoromethylene)-5,5-dimethyl-2,3-diphenylisoxazolidine 120</u>.

Into an NMR tube was added 100 mg(0.507 mmole) diphenylnitrone,[96] 0.40 mL $CDCl_3$, and 100 mg (0.961 mmole) difluorodimethylallene 80. After 40 hours at room temperature, the solvent was removed by rotary evaporation at reduced pressure to give 148 mg (97%) clear, colorless liquid 120: IR(CCl_4) 3074, 3040, 2993, 2941, 1774(s), 1600, 1493, 1456, 1300, 1270, 1250, 1133, 1062, 1038, 883 cm^{-1}; ^1H NMR(60 MHz) δ 7.5-6.9 (m, 10H), 5.26(t, 1H, J_{HF}=3.0 Hz), 1.67(s, 3H), 1.61 (s, 3H); ^{19}F NMR(100 MHz) ϕ 87.7(dd, 1F, J_{FF}=58.6 and J_{HF}=3.2 Hz), 91.5(d, 1F, J_{FF}=58.6 Hz); mass spectrum gave M^+ 301.12825 $\pm$ 0.00498(16 ppm), calcd for $C_{18}H_{17}NOF_2$ 301.12782 dev 0.00043(1 ppm).

4-(Difluoromethylene)-3-phenyl-2,5,5-trimethylisoxazolidine 121.

Into an NMR tube was added 100 mg (0.740 mmole) phenylmethylnitrone,[97] 0.40 mL $CDCl_3$ and 100 mg (0.96 mmole) difluorodimethylallene 80. After 9 days at room temperature, the solvent was removed by rotary evaporation at reduced pressure to give 70 mg (40%) oil which was purified by flash chromatography using silica gel and 90% hexane/10% EtOAc. A total of 40 mg (23%) clear, colorless liquid 121 was obtained: R_f=0.4; IR(film) 3075, 3040, 2994, 2940, 2880, 1777(s), 1680(w), 1457, 1292, 1273, 1249, 1122, 1104, 1050 cm^{-1}; ^1H NMR(60 MHz) δ7.36(s, 5H), 4.15(t, 1H, J_{HF}=4.3 Hz), 2.60(s, 3H), 1.63(s, 3H), 1.53(s, 3H); ^{19}F NMR(100 MHz) ϕ 89.6(dd, 1F, J_{FF}=59.8 and J_{HF}=2.4 Hz), 91.4(dd, 1F, J_{FF}=59.8 and J_{HF}=4.4 Hz).

<u>(E)- and (Z)-2,3-Diphenyl-4-(fluoromethylene)isoxazolidines 122 and 123.</u>

Into an NMR tube containing 100 mg (0.507 mmole) diphenylnitrone[96] and 0.40 mL $CDCl_3$ was condensed 96 mg (1.65 mmoles) fluoroallene. The tube was sealed under vacuum and kept at room temperature for 36 hours. The tube was cooled and opened. Removal of the solvent by rotary evaporation at reduced pressure gave 129 mg (99.7%) dark amber oil which was subjected to flash chromatography using silica gel and 95% hexane/5% EtOAc. A total of 84 mg (65%) colorless liquid (E)-isomer <u>122</u> was isolated: R_f=0.33; IR(film) 3074, 3040, 2930, 2870, 1713(s), 1603(s), 1493(s), 1456, 1095(s), 1030 cm^{-1}; ^1H NMR(300 MHz) δ 7.56(d, 2H, J=7.5 Hz), 7.35 (t, 2H, J=7.3 Hz), 7.3-7.2(m, 3H), 7.06(d, 2H, J= 7.7 Hz), 6.97(t, 1H, J=7.3 Hz), 6.55(d,d,t, 1H, J_{HF}= 81.9, J_{HH}=1.7 and J_{HH}=1.8 Hz), 5.52(broad s, 1H), 4.58(m, 2H, non-first-order AB pattern); ^{19}F NMR(300 MHz) ϕ 128.9(d, J_{HF}=82.0 Hz); ^{13}C NMR(100 MHz) δ 150.2 (N-subst aromatic), 141.2(d, J_{CF}=255.1 Hz, =CHF), 139.3(d, J_{CF}=2.5 Hz, C-subst aromatic), 129.0, 128.6, 127.8, 127.2(aromatic), 125.6(d, J_{CF}=11.0 Hz, olefinic), 122.6(para-N-subst aromatic), 115.3(ortho-N-subst aromatic), 69.9(CH), 65.7(d, J_{CF}=6.1 Hz, CH_2); mass spectrum gave M^+ 255.10581 $\pm$ 0.00141(5 ppm), calcd for $C_{16}H_{14}FNO$ 255.10594 dev -0.00013(0.5 ppm).

There were also isolated 13 mg (10%) of (Z)-isomer 123: R_f=0.43; IR(CCl_4) 3074, 3036, 2918, 2862, 1714(s), 1601, 1490(s), 1456, 1102(s), 1030, 698 cm^{-1}; ^{1}H NMR (300 MHz) δ 7.49(d, 1H, J=7.0 Hz), 7.39(t, 2H, J=7.1 Hz), 7.35-7.2(m, 3H), 7.06(d, 2H, J=7.7 Hz), 7.00 (t, 1H, J=7.3 Hz), 6.46(d,d,t, 1H, J_{HF}=81.3, J_{HH}= 1.6, and J_{HH}=2.2 Hz), 5.10(broad s, 1H), 4.87(d, 1H, J_{HH}=12.2 Hz), 4.77(d, 1H, J_{HH}=12.2 Hz); ^{19}F NMR(300 MHz) φ 129.6(d of m, J_{HF}=81.3 Hz); ^{13}C NMR(300 MHz) δ 149.9(N-subst aromatic), 142.9(d, J_{CF}=255.5 Hz, =CHF), 139.6(d, J_{CF}=2.1 Hz, C-subst aromatic), 128.9, 128.8, 128.1, 127.1(aromatic), 126.8(d, J_{CF}=10.2 Hz, olefinic), 123.0(para-N-subst aromatic), 116.0(ortho-N-subst aromatic), 70.1(d, J_{CF}=5.5 Hz, CH), 66.4(CH_2); mass spectrum gave M^+ 255.10588 ± 0.00158(6 ppm), calcd for $C_{16}H_{14}ONF$ 255.10594 dev -0.00006(0.2 ppm).

The relative yields determined by ^{1}H NMR and ^{19}F NMR integration of the reaction mixture were 86% for 122 and 14% for 123.

4-(Difluoromethylene)-4,5-dihydro-3-(2,4,6-timethylphenyl)isoxazole 125 and 3,8-Bis-(2,4,6-trimethylphenyl)-4,4-difluoro-1,6-dioxa-2,7-diazaspiro[4.4]-nona-2,7-diene 126.

Into an NMR tube containing 100 mg (0.620 mmole) mesitylnitrile oxide[67] and 0.50 mL $CDCl_3$ was condensed 94 mg (1.24 mmoles) difluoroallene. The tube was sealed under vacuum. After 60 minutes at room temperature, the tube was cooled and opened. The solvent was removed by rotary evaporation at reduced pressure to give 144 mg (98%) crude liquid product. Flash chromatography using silica gel and 96% hexane/4% EtOAc gave 91.5 mg (62%) colorless liquid <u>125</u>: R_f=0.29 ^{1}H NMR(60 MHz) δ 6.96(s, 2H), 5.24(t, 2H, J_{HF}=6.1 Hz), 2.32(s, 3H), 2.22(s, 6H); ^{19}F NMR(300 MHz) φ 77.5(d of t, 1F, J_{HF}=5.5 and J_{FF}=26.7 Hz), 79.2(d of t, 1F, J_{HF}= 6.3 and J_{FF}=26.7 Hz).

It also gave 17 mg (7%) white solid <u>126</u>: R_f=0.18; mp 104-107°; ^{1}H NMR(100 MHz) δ 6.93(s, 2H), 6.90(s, 2H), 3.41(t, 2H, J_{HF}=1.2 Hz), 2.38(s, 6H), 2.30(s, 6H), 2.23(s, 6H); ^{19}F NMR(300 MHz) φ 94.2(d, 1F, J_{FF}=213 Hz), 95.4(d, 1F, J_{FF}=213 Hz); ^{13}C NMR(300 MHz) δ 186.2(t, J_{CF}=14 Hz, C=N), 158.1(s, C=N), 141.5, 139.7, 138.9, 134.2, 130.6, 128.74, 128.72, 117.8(aromatic) 113.2 (t, J_{CF}=29 Hz, quaternary), 46.4(CH_2), 21.3, 21.1, 19.9, 19.2(CH_3).

The relative yields by ^{1}H NMR integration of the reaction mixture were 77% for <u>125</u> and 23% for <u>126</u>. The neat <u>125</u> decomposes slowly at room temperature.

<u>Tetrahydro-3-(difluoromethylene)-2,2,5,5-furantetracarbonitrile 130</u>.

Into a 30-mL glass tube containing 0.500 g (3.47 mmoles)tetracyanoethylene oxide[99] and 20 mL of

1,2-dichloroethane was condensed 1.03 g (13.6 mmoles) difluoroallene. The tube was sealed under vacuum and heated at 130° for 10 hours. The tube was cooled and opened. The amber mixture was filtered and then concentrated by rotary evaporation at reduced pressure to give 1.00 g of oily brown solid which was recrystallized from 1.5 mL benzene to give 144 mg (19%) tan solid 130: mp 100-105°; IR(CH_2Cl_2) 2345, 2275, 1786(s), 1336, 1150, 948 cm^{-1}; ^1H NMR(acetone-d_6, 60 MHz) δ 4.25 (t, J_{HF}=3.5 Hz); ^{19}F NMR(100 MHz) ϕ 72.5(d of t, 1F, J_{FF}=21.6 and J_{HF}=3.6 Hz), 73.3(d of t, 1F, J_{FF}=21.6 and J_{HF}=3.4 Hz); mass spectrum gave M^+ 220.02195 ± 0.00286(13 ppm), calcd for $C_9H_2F_2N_4O$ 220.01967 dev -0.00228(10 ppm).

Compound 130 was very sensitive to water and its spectra were obtained in dry solvents.

1-Fluoro-2-methylenecyclopropane 136 and (Fluoromethylene)-cyclopropane 137.

Through a 15 cm by 2 cm Vigreux column heated at 380° was vacuum transferred 1.3 g (13 mmoles) pyrazoline 113 over a period of 2.5 hours at a pressure of 0.6-1.0 mm. A total of 650 mg (69%) crude liquid pyrolysate was collected in a vacuum trap at dry-ice temperature. The products were isolated by prep GLPC (20 ft by 1/4 in 20% DNP at 50°, 60 mL/min) to give 72 mg (7.7%) of 136: IR(CCl_4) 3080, 3040, 3005, 2985, 1828(w), 1755(w), 1438,

1322, 1143(s), 1110(s), 910(s), 835 cm^{-1}; ^1H NMR(100 MHz) δ 5.86(m, 1H), 5.61(m, 1H), 4.92(d of m, 1H, J_{HF}= 68.6 Hz, non-first-order), 1.65-1.45(m, 2H); ^{19}F NMR (100 MHz) ϕ 169.6(d of m, J_{HF}=68.0 Hz); mass spectrum gave M^+ 72.03732 $\pm$ 0.00107(15 ppm), calcd for C_4H_5F 72.03753 dev -0.00021(3 ppm).

There were also isolated 110 mg (11.7%) of <u>137</u>: IR (CCl_4) 3075, 3000, 1790(s), 1435, 1326, 1125(s), 1038, 977 cm^{-1}; ^1H NMR(100 MHz) δ 6.88(d of pentet, 1H, J_{HF}= 91.5 and J_{HH}=2.2 Hz), 1.27(dd, 4H, J_{HF}=4.6 and J_{HH}=2.2 Hz), ^{19}F NMR(100 MHz) ϕ 130.6(d of pentet, J_{HF}=91.5 and J_{HF}=4.6 Hz); mass spectrum gave M^+ 72.03751 $\pm$ 0.00083 (12 ppm), calcd for C_4H_5F 72.03753 dev -0.00002(0.2 ppm).

The order of elution was <u>136</u> then <u>137</u>.

<u>1,1-Difluoro-2-(1-methylethylidene)cyclopropane 138, 1,1-Difluoro-2,2-dimethyl-3-methylenecyclopropane 139, and 2-(Difluoromethylene)-1,1-dimethylcyclopropane 140.</u>

Photolysis of 200 mg (1.37 mmoles) pyrazoline <u>104</u> in a 500-mL gas sample bulb for 16 hours using a Rayonet Photoreactor (350 nm) gave 98 mg (61%) crude photolysis mixture which was subjected to prep GLPC (20 ft by 1/4 in 15% ODPN, ambient, 30 mL/min) to give 17 mg (11%) of <u>138</u>: IR(gas) 2990, 2955, 2930, 1854(w), 1784(m), 1458, 1414, 1330, 1202(s), 1130, 1061 cm^{-1}; ^1H NMR(60 MHz) δ 1.9(m); ^{19}F NMR(100 MHz) ϕ 130.6(complex m); mass spectrum gave M^+ 118.0590 $\pm$ 0.00063(5 ppm), calcd for $C_6H_8F_2$ 118.0594 dev -0.0004(3 ppm).

It also gave 4 mg (2%) of 139, identical to 139 prepared by addition of difluorocarbene to dimethylallene, and 22 mg (14%) of 140: IR(gas) 3062, 2980, 2940, 2885, 1840(s), 1440, 1316, 1230(s), 1140 cm^{-1}; ^1H NMR(60 MHz) δ 1.43(m); ^{19}F NMR(100 MHz) ϕ 86.6(d of t, 1F, J_{FF}=73.2 and J_{HF}=4.1 Hz), 90.9(d, 1F, J_{FF}=73.2 Hz); mass spectrum gave M^+ 118.05948 $\pm$ 0.0007(6 ppm), calcd for $C_6H_8F_2$ 118.05941 dev 0.00007(0.6 ppm).

The combined yield of isolated products was 27%. Order of elution was 140, 139, then 138.

1,1-Difluoro-2,2-dimethyl-1-methylenecyclopropane 139.

Into a 10 mL glass tube containing 3.8 g (11.0 mmoles) phenyltrifluoromethylmercury,[100] 4.00 g (26.7 mmoles) anhydrous NaI and a catalytic amount (30 mg) tetra-n-butylammonium iodide was condensed 1.4 g (20.6 mmoles) dimethylallene 141. The tube was sealed under vacuum and heated at 80° for 16.5 hours. The tube was cooled and opened. Vacuum transfer gave 1.7 g of liquid which was subjected to prep GLPC (20 ft by 1/4 in 15% ODPN, ambient, 30 mL/min) to give 0.815 g recovered dimethylallene and 0.368 g (28%) 139: bp 63.5-63.8^o; IR(gas) 3100, 3008, 2983, 2955, 2895, 1853(w), 1762(m), 1450, 1332, 1243, 1172(s), 990, 924, 881 cm^{-1}; ^1H NMR(60 MHz) δ 5.91(t, 1H, J_{HF}=1.6 Hz), 5.61(t, 1H, J_{HF}=2.3 Hz), 1.24(t, 6H, J_{HF}=2.2 Hz); ^{19}F NMR(100 MHz) ϕ 139.9(complex m); mass spectrum gave

M^+ 118.05908 ± 0.001427(12 ppm) calcd for $C_6H_8F_2$ 118.05941 dev -0.00033(3 ppm).

Order of elution was dimethylallene then 139.

1,1-Diphenyl-2-(difluoromethylene)cyclopropane 142.

A 60-mL glass tube containing 100 mg (0.370 mmole) pyrazoline 108 and 40 mL pentane was sealed under vacuum and irradiated in a Rayonet Photoreactor (350 nm) for 15 hours. The cloudy mixture was concentrated by rotary evaporation at reduced pressure to give 89 mg (99%) yellow oil which was 142 (purification by flash chromatography using silica gel and hexane): R_f=0.39; IR(CCl_4) 3095, 3070, 3036, 1845(s), 1690(w), 1602(w), 1498, 1231, 1166 cm^{-1}; ^1H NMR(100 MHz) δ 7.3-7.1(complex m, 10H), 1.98(t, 2H, J_{HF}=4.5 Hz); ^{19}F NMR(100 MHz) ϕ 84.3(d of t, 1F, J_{FF}=64.7 and J_{HF}=4.5 Hz), 90.0(d of t, 1F, J_{FF}=64.7 and J_{HF}=4.5 Hz); ^{13}C NMR(100 MHz) δ 151.6(d of d, J_{CF}=281.4 and 276.5 Hz, =CF_2), 141.8 (d, J_{CF}=2.4 Hz, subst aromatic), 128.6, 128.1, 127.1 (aromatic), 78.1(t, J_{CF}=35.4 Hz, C_2), 36.1(d, J_{CF}= 6.1 Hz, C_1), 20.8(d, J_{CF}=3.7 Hz, CH_2); mass spectrum gave M^+ 242.0895 ± 0.0021(9 ppm), calcd for $C_{16}H_{12}F_2$ 242.0907 dev -0.00116(5 ppm).

Analysis: Calcd, C 79.32, H 4.99

Found, C 79.17, H 5.05

In a similar experiment, pyrazoline 107 was

irradiated for 5 hours to give a mixture of 15% 107 and 85% 142 as indicated by integration of the ^1H NMR spectrum.

2-(Difluoromethylene)spiro[cyclopropane-1,9'-[9H]-fluorene] 112.

Into a 700-mL gas sample bulb with teflon Rotoflo stopcock containing 5.00 g (26.0 mmoles) diazofluorene[95] and 200 mL ether was condensed 2.37 g (31.2 mmoles) difluoroallene. The tube was kept at room temperature for 5.5 hours, then connected to a vacuum line and 50 mL ether were transferred off in order to remove excess difluoroallene.

The tube was irradiated in a Rayonet Photoreactor (350 nm) for 8 hours. The nitrogen evolved was removed on the vacuum line and irradiation was continued for a total of 28 additional hours.

After 36 hours irradiation, the solvent was rapidly removed by rotary evaporation at reduced pressure to give a yellow solid, which was immediately and rapidly purified by flash chromatography using silica gel and 95% hexane/5% EtOAc to give 3.4 g (54%) yellow solid 112 identical to that obtained as a direct product of the cycloaddition reaction without irradiation. Solutions of 112 must be kept cold and the procedure followed rapidly in order to minimize reaction with oxygen.

5,5-Difluoro-4-methylenespiro[1,2-dioxolane-3,9'-[9H]fluorene] 145, 4-(Difluoromethylene)spiro[1,2-dioxolane-3,9'-[9H]fluorene] 146, and 3,3-Difluoro-4-fluorenylidene-1,2-dioxolane 147.

A solution of 338 mg (1.41 mmoles) 112 in 10 mL $CHCl_3$ was exposed to a slow flow of oxygen which was bubbled through the solution. Analysis by TLC after 32 hours indicated the starting material was gone. The solution was concentrated by rotary evaporation at reduced pressure to give an oily yellow solid. Flash chromatography using silica gel and hexane gave 172 mg (45%) pale yellow solid 145: R_f=0.18; mp 75-78.5^o; IR(CCl_4) 3074, 1831(w), 1690(vw), 1612(w), 1453, 1334, 1134, 1110(s), 1089 cm^{-1}; ^1H NMR(300 MHz) δ 7.58(d, 2H, J=7.5 Hz), 7.51(d, 2H, J=7.5 Hz), 7.39 (t, 2H, J=7.5 Hz), 7.27(t, 2H, J=7.5 Hz), 5.72(d of t, 1H, J_{HH}=1.5 and J_{HF}=2.4 Hz), 5.10(d of t, 1H, J_{HH}= 1.8 and J_{HF}=2.4 Hz); ^{19}F NMR(100 MHz) ϕ 77.7(t, J_{HF}= 2.4 Hz); ^{13}C NMR(100 MHz) δ 146.4(t, J_{CF}=27.2 Hz, C_4), 142.2, 140.7(subst aromatic), 130.9, 128.7, 125.6, 120.4(aromatic), 127.6(t, J_{CF}=266.5 Hz, CF_2), 115.0 (=CH_2), 94.8(t, J_{CF}=3 Hz, quat C); mass spectrum gave M^+ 272.06549 $\pm$ 0.00531(20 ppm), calcd for $C_{10}H_{16}O_2F_2$ 272.06489 dev 0.00060(2 ppm).

Analysis: Calcd, C 70.59, H 3.70

Found, C 70.76, H 3.75

It also gave 70 mg (18%) of <u>146</u>: R_f=0.07; mp 136-138°; IR(CCl_4) 3075, 2925, 2870, 1789(s), 1453, 1305, 1291, 1120 cm^{-1}; ^1H NMR(100 MHz) δ 7.64-7.20(complex m, 8H), 5.16(t, 2H, J_{HF}=3.4 Hz); ^{19}F NMR(300 MHz) φ 85.4(d of t, 1F, J_{FF}=48.6 and J_{HF}=3.4 Hz), 87.8(d of t, 1F, J_{FF}=48.6 and J_{HF}=3.4 Hz); ^{13}C NMR(100 MHz) δ 148,1(d of d, J_{CF}=285.6 and 290.5 Hz, =CF_2), 142.1, 140.6(subst aromatic), 130.4, 128.4, 125.0, 120.3(aromatic), 99.6(d of d, J_{CF}=23.2 and 19.5 Hz, C_4), 90.3(quat C), 70.5(d, J_{CF}=2.4 Hz, CH_2); mass spectrum gave M^+ 272.06604 ± 0.00196(7 ppm), calcd for $C_{16}H_{10}O_2F_2$ 272.06489 dev 0.00116(4 ppm).

Finally it gave 15 mg (4%) of <u>147</u>: R_f=0.09; mp 130.5-131.5°; IR(CCl_4) 3070, 1840(w), 1670, 1607, 1454, 1278, 1140, 1117, 1096, 1083 cm^{-1}; ^1H NMR(100 MHz) δ 7.97(m, 1H), 7.64(m, 2H), 7.48-7.08(m, 5H), 5.55(t, 2H, J_{HF}=1.9 Hz); ^{19}F NMR(300 MHz) φ 78.0(quartet, J_{HF}=1.9 Hz); ^{13}C NMR(300 MHz) δ 141.5, 141.2, 136.7, 136.0, 133.8(subst aromatic and olefinic), 130.3, 130.2, 128.1, 127.8, 124.3, 120.3, 119.8, 126.2(t, J_{CF}=8 Hz) (aromatic), 75.5(t, J_{CF}=2.7 Hz, CH_2); mass spectrum gave M^+ 272.06561 ± 0.00161(6 ppm), calcd for $C_{16}H_{10}O_2F_2$ 272.06489 dev 0.00072(3 ppm).

The relative yields determined by ^1H NMR integration were 58.6% for <u>145</u>, 36.8% for <u>146</u>, and 4.6% for <u>147</u>. The combined yield of isolated products was 67%.

4,4-Difluoro-5-methylenespiro[cyclopentane-1,9'-[9H]-fluorene]-2-carbonitrile 152, 5-(Difluoromethylene)-spiro[cyclopentane-1,9'-[9H]-fluorene]-2-carbonitrile 153, and 3,3-Difluoro-4-fluorenylidenecyclopentane-carbonitrile 154.

Into a 30-mL glass tube containing 1.00 g (4.16 mmoles) 112 under nitrogen, was condensed 6.0 mL (4.8 g, 91 mmoles) acrylonitrile. The tube was sealed under vacuum and heated at 70° for 6 hours. The tube was opened and the cloudy, amber mixture was concentrated by rotary evaporation at reduced pressure to give an amber oil. Flash chromatography using silica gel and 90% hexane/10% EtOAc gave 491 mg (49%) recovered starting material, R_f=0.70, 390 mg (32%) white solid 152: R_f=0.43; mp 124.5-125.5°; IR(CCl_4) 3075, 2248(w), 1670(w), 1453, 1124, 933 cm^{-1}; ^1H NMR (100 MHz) δ 7.7(m, 2H), 7.3(m, 4H), 5.68(d of d of d, 1H, J_{HF}=2.0 and 4.3, J_{gem}=0.85 Hz), 4.90(d of d of d, 1H, J_{HF}=2.0 and 3.9, J_{gem}=0.85 Hz), 3.63(d of d of d, 1H, J_{HF}=1.5, J_{HH}=8.3 and 11.4 Hz), 3.1-2.8(m, 2H); ^{19}F NMR(100 MHz) ϕ 88.3 and 98.5(AB pattern, J_{FF}=249.8, downfield F has t of t fine structure with J_{HF}=17.0 and 4.1 Hz, upfield F has m fine structure); ^{13}C NMR (300 MHz) δ 147.5, 147.3, 140.7, 140.1(subst aromatics), 129.1, 129.0, 128.4, 128.2, 124.0, 123.4, 120.5, 120.3 (aromatics), 147.1(t, J_{CF}=21.9 Hz, subs olefinic),

123.9(t, J_{CF}=247.7 Hz, CF_2), 117.0(CN), 116.9(d, J_{CF}= 5 Hz, =CH_2), 60.6(d, J_{CF}=2.2 Hz, quat C), 38.8(t, J_{CF}= 26.3 Hz, CH_2), 36.9(d, J_{CF}=6.6 Hz, CH); mass spectrum gave M^+ 293.10121 ± 0.00151(5 ppm), calcd for $C_{19}H_{13}NF_2$ 293.10161 dev -0.00040(1 ppm).

Analysis: Calcd, C 77.80, H 4.47, N 4.78

Found, C 77.60, H 4.49, N 4.75

It also gave 162 mg of <u>153</u> and <u>154</u>, R_f=0.18-0.23 which was further separated by flash chromatography using silica gel and 80% hexane/20% ether to give 64 mg (5%) white solid <u>153</u>: R_f=0.28; mp 178.2-180.3°; IR(CCl_4) 3075, 2975, 2246(vw), 1765(s), 1453, 1274 cm^{-1}; ^1H NMR(100 MHz) δ 7.68(m, 2H), 7.3(m, 4H), 3.33(d of d, 1H, J=7.4 and 10.5 Hz), 2.8(m, 2H), 2.47(m, 2H); ^{19}F NMR(100 MHz) ϕ 85.7(d of t, 1F, J_{FF}=42.3 and J_{HF}=3.7 Hz), 86.2(d of m, 1F, J_{FF}=42.3 Hz); ^{13}C NMR(100 MHz) δ 146.9, 146.2(t, J_{CF}=2.4 Hz), 140.3, 140.1(subst aromatics), 128.9, 128.6, 128.0, 127.9, 124.0, 122.7, 120.3(aromatic), 150.5(dd, J_{CF}=287.5 and 291.1 Hz, =CF_2), 117.9(CN), 93.5 (dd, J_{CF}=23.2 and 17.1 Hz, olefinic), 59.2(quat C), 43.2(d, J_{CF}=2.4 Hz, CH), 29.9(CH_2), 26.2(CH_2); mass spectrum gave M^+ 293.10189 ± 0.00069(2 ppm), calcd for $C_{19}H_{13}NF_2$ 293.10160 dev 0.00029(1 ppm).

Finally it gave 67 mg (5%) of <u>154</u>: R_f=0.20; mp 164.5-167°; ^1H NMR(100 MHz) δ 8.1(m, 1H), 7.6(m, 2H), 7.3(m, 5H), 3.4-2.3(complex m, 5H); ^{19}F NMR(300 MHz)

ϕ 89.5 and 92.4(AB pattern, J_{FF}=251.2 Hz); ^{13}C NMR (acetone-d_6, 300 MHz) δ 142.0, 141.8, 138.7, 138.5, 135.8(subst aromatics and olefinic), 130.4, 130.2, 128.6, 128.3, 127.0, 120.7, 120.4(aromatics), 132.6 (t, J_{CF}=21.9 Hz, C_4 olefinic), 127.5(t, J_{CF}=9.9 Hz, aromatic), 128.4(t, J_{CF}=248.2 Hz, CF_2), 121.0(CN), 41.3(t, J_{CF}=27.4 Hz, C_2), 37.6(t, J_{CF}=4.2 Hz, C_5), 25.0(dd, J_{CF}=3.4 and 5.5 Hz, CH); mass spectrum gave M^+ 293.10129 ± 0.00104(4 ppm), calcd for $C_{19}H_{13}NF_2$ 293.10160 dev -0.00032(1 ppm).

The total combined yield of isolated products was 42% with 49% recovered starting material. On a smaller scale, an isolated yield of 75% products and 15% starting material was obtained. The relative yields of products determined by integration of the ^{19}F NMR spectrum of the reaction mixture were 68.2% for 152, 15.6% for 153, and 16.1% for 154.

2-Methylenespiro[cyclopropane-1,9'-[9H]fluorene] 157.

Into a 150-mL glass tube containing 2.00 g (10.4 mmoles) diazofluorene[95] and 50 mL CCl_4 was condensed 2.00 g (52 mmoles) allene. The tube was sealed under vacuum and irradiated in a Rayonet Photoreactor for 6 hours. The tube was cooled and opened. The dark red-brown solution was concentrated to give an oil which was purified by flash chromatography using silica gel and hexane. A total of 0.764 g crude solid, R_f=0.32, were obtained. Further purification by fractional crystallization from hexane gave white needles (mp 149.5-150.5) of an undesired by-product and colorless rhombohedral crystals of 157 which were separated by hand to give 90 mg (4.2%) pure 157: mp 104-111°; IR(CCl_4) 3075, 3022, 2975, 1450, 1120, 893 cm^{-1}; ^1H NMR(60 MHz) δ 7.9(m, 2H), 7.4-7.1(m, 4H), 5.75(t, 1H, J=2.3 Hz), 5.52(t, 1H, J=3.0 Hz), 2.30 (t, 2H, J=2.8 Hz); ^{13}C NMR(100 MHz) δ 147.4, 140.1, 138.6(subst aromatic and olefinic), 126.9, 126.2, 119.9(aromatics), 104.7(=CH_2), 34.1(quat), 21.2(CH_2); mass spectrum gave M^+ 204.09302 ± 0.00188(9 ppm), calcd for $C_{16}H_{12}$ 204.0939 dev -0.00088(4 ppm).

Analysis: Calcd, C 94.08, H 5.92

Found, C 93.94, H 5.97

4-Methylenespiro[1,2-dioxolane-3,9'-[9H]fluorene] 158 and 4-Fluorenylidene-1,2-dioxolane 159.

A solution of 63 mg (0.308 mmole) 157 in 5 mL $CHCl_3$ was exposed to a slow flow of oxygen which was bubbled through the solution. Analysis by ^1H NMR after 43 hours indicated 57% completion by integration. The reaction was continued. After 7 days, the solution was concentrated by rotary evaporation at reduced pressure to give a yellow solid. Flash chromatography using silica gel and 96% hexane/4% EtOAc gave 7 mg (11%) recovered starting material and 55 mg (75%) crude product mixture. Recrystallization from hexane/$CHCl_3$ gave 24 mg (33%) pure major isomer 158: R_f=0.24 mp 148.0-149.6°; IR(CCl_4) 3072, 2908, 2853, 1678(vw), 1452(s), 900 cm^{-1}; ^1H NMR(100 MHz) δ 7.6-7.2(complex m, 8H), 5.08(t, 2H, J=2.0 Hz), 5.03(d of t, 1 H, J= 1.9 and 1.0 Hz), 4.64(d of t, 1H, J=2.2 and 1.0 Hz); ^{13}C NMR(100 MHz) δ 155.5(subst olefinic), 144.6, 140.7 (subst aromatics), 129.8, 128.3, 125.3, 120.1(aromatics), 105.6(=CH_2), 93.0(quat), 74.3(CH_2); mass spectrum gave M^+ 236.08347 ± 0.00193(8 ppm), calcd for $C_{16}H_{12}O_2$ 236.08373 dev -0.00026(1 ppm).

Analysis: Calcd, C 81.34, H 5.12

Found, C 81.12, H 5.13

Characterization of the minor product 159 relies on the ^1H NMR spectrum of the reaction mixture which showed a singlet at 5.17 ppm for the CH_2 protons of 159. Integration of the ^1H NMR spectrum of the product mixture indicated relative yields of 90.8% for 158 and 9.2% for 159.

Determination of Product Ratios
Kinetics and Thermodynamics

Thermolysis of Pyrazoline 101 in the Gas Phase.

After carefully purifying pyrazoline 101 by distillation and vacuum transfer, 5 mm of pure material was expanded into a well-conditioned thermolysis vessel at 162.2°. Samples were taken periodically and the product ratios were determined by GLPC analysis (7 ft by 1/8 in 10% ODPN, ambient, 20 mL/min). The two products obtained were the difluoromethylene-cyclopropanes 131 and 133 which were identified by comparison of their GLPC retention times with authentic samples[73] and by ^{1}H NMR analysis of the thermolysis mixture. Order of elution was 133 then 131. The product ratios as a function of time and pressure are presented in Table XII. Each ratio is an average of three GLPC analyses.

Photolysis of Pyrazoline 101 in the Gas Phase.

Into a 25-mL pyrex gas sample bulb with teflon Rotoflo stopcock was expanded 2 mm of pyrazoline 101. Photolysis in a Rayonet Photoreactor gave three products which were identified by comparison of their GLPC retention times with authentic samples and by ^{1}H NMR analysis of the photolysis mixture from a larger scale reaction using a 500-mL gas sample bulb. The product ratios of allene, 131 and 133 are averages of three GLPC analyses per sample (ODPN). The effect of time of photolysis and of adding the indicated pressure of argon is presented in Table XIII.

Table XII. Product Ratios from Thermolysis of 101 in the Gas Phase at 162.2°.

Time (min)	Pressure (mm)	131	133
5	5	55.36	44.64
15	5	54.20	45.80
15	700	54.41	45.59
35	5	54.01	45.99
35	700	54.30	45.70

Table XIII. Product Ratios from Photolysis of 101 in the Gas Phase at 38°.

Time (min)	Pressure (mm)	131	133	allene
30	2	30.1	31.6	38.3
30	653	55.3	43.5	1.2
30	657	56.6	42.2	1.2
30	760	55.6	43.1	1.3
180	657	55.7	42.7	1.5

Thermolysis of Pyrazoline 101 in Solution.

Into a 25-mL glass tube containing 5 mL degassed CCl_4 was condensed 12 mg (0.10 mmole) of pyrazoline 101. The tube was sealed under vacuum and heated at 150° for the indicated time. Results for five identical samples are presented in Table XIV. Each product ratio is an average of three GLPC analyses (ODPN). Products 131 and 133 eluted well before the solvent. Due to poor signal to noise ratio for the sample heated for 10 minutes, it was not included in the average.

Table XIV. Product Ratios from Thermolysis of 101 in Solution at 150°.

Time (min)	131	133
10	57.90	42.10
20	59.69	40.31
30	59.54	40.46
40	59.61	40.39
60	60.03	39.97
average	59.72	40.28

Photolysis of Pyrazoline 101 in Solution.

Into a 25-mL pyrex gas sample bulb with teflon

Rotoflo stopcock containing 5 mL of degassed CCl_4 was condensed 12 mg (0.10 mmole) pyrazoline 101. Photolysis of four identical samples in a Rayonet Photoreactor (350 nm) for 1, 2, 3, and 8 hours gave the product ratios indicated in Table XV. Each ratio is an average of three GLPC analyses (ODPN).

Table XV. Product Ratios from Photolysis of 101 in solution at 38°.

Time (h)	131	133
1	61.67	38.33
2	60.87	39.13
3	62.26	37.74
8	61.44	38.56
average	61.6	38.4

Thermolysis of Pyrazoline 113 in the Gas Phase.

Into a well-conditioned thermolysis vessel at 168.0^{o} was expanded 1 mm of pyrazoline 113. The product ratio was monitored with respect to time by GLPC analysis (20 ft by 1/8 in 20% DNP, ambient, 86 mL/min). Also the effect of added argon was determined at 10 minutes. Order of elution was 136 then 137. The results are presented in Table XVI.

Table XVI. Product Ratios from Thermolysis of 113 in the Gas Phase at 168.0°.

Time (min)	Pressure (mm)	136	137
5	1	81.80	18.20
10	1	82.04	17.96
10	1	81.50	18.50
10	745	86.80	13.20
15	1	79.39	20.61
20	1	83.88	16.12

Photolysis of Pyrazoline 113 in the Gas Phase.

Into a 25-mL pyrex gas sample bulb with teflon Rotoflo stopcock was expanded 2 mm of pyrazoline 113. The bulb was then pressurized with the indicated amount of argon and the sample was irradiated in a Rayonet Photoreactor (350 nm). Analysis by GLPC (20 ft by 1/8 in 20% ODPN and 10 ft by 1/8 in 10% DNP in series, ambient, 25 mL/min) gave the product ratios presented in Table XVII. The order of elution was 136 then 137. Each ratio is an average of three GLPC analyses.

Thermolysis of Pyrazoline 113 in Solution.

Into a 30-mL glass tube containing 5 mL distilled, degassed CH_2ClCH_2Cl was condensed 21 mg (0.210 mmole) pyrazoline 113. The tube was sealed under vacuum and

Table XVII. Product Ratios from Photolysis of 113 in the Gas Phase at 38°.

Time (h)	Pressure (mm)	136	137
1	2	47.96	52.04
1	2	47.65	52.35
1	6	47.29	52.71
1	20	46.78	53.22
1	50	46.24	53.76
1	108	46.87	53.13
1	419	48.46	51.54
1	760	49.22	50.78
1	773	49.83	50.17
2	760	49.26	50.74

heated at 140° for 1 hour. An identical sample was heated for 2 hours at 140°. The product ratios for 136 and 137 were determined by GLPC analysis (DNP). Each ratio is the average of three analyses. The results are presented in Table XVIII.

Photolysis of Pyrazoline 113 in Solution.

Into a 30-mL pyrex gas sample bulb equipped with teflon Rotoflo stopcock containing 5 mL distilled, degassed CH_2ClCH_2Cl was condensed 21 mg (0.210 mmole) of pyrazoline 113. Two identical samples were irradiated at 38° in a Rayonet Photoreactor (350 nm) for 1 and 2 hours. Analysis by GLPC (DNP) gave the product ratios for 136 and 137 indicated in Table XIX. Each ratio is the average of three analyses.

Table XVIII. Product Ratios from Thermolysis of 113 in Solution at 140°.

Time (h)	136	137
1	89.40	10.60
2	91.55	8.44
average	90.48	9.52

Table XIX. Product Ratios from Photolysis of 113 in Solution at 38°.

Time (h)	136	137
1	64.83	35.17
2	64.37	35.63
average	64.60	35.40

Equilibrium of 136 and 137 in the Gas Phase.

Into a well-conditioned thermolysis vessel at the indicated temperature was expanded 10 mm of fluoromethylenecyclopropane 136. The isomerization to 137 was monitored by GLPC (DNP) and equilibrium ratios were obtained after 20 half lives. The results for the equilibrium constant K = 137/136 are presented in Table XX. Each ratio is the average of three analyses.

Table XX. Equilibrium Constants for 136 and 137 in the Gas Phase.

Temperature (°C)	K
229.0	7.7009
243.0	7.1900
257.2	6.7760
277.0	6.1906
289.6	5.8283

The equilibrium was verified from the other isomer 137 at 243.0°. Starting with the GLPC purified 137 after 90 min at 243.0° a value for the equilibrium constant of K = 7.217 was obtained.

Thermolysis of Pyrazoline 104 in the Gas Phase.

Into a well-conditioned thermolysis vessel at 161.8° was expanded 3 mm of pyrazoline 104. Samples were taken periodically and the product ratios were determined by GLPC analysis (10 ft by 1/8 in 10% ODPN, ambient, 4.6 mL/min). The effect of added argon was also determined after 20 minutes thermolysis. The order of elution for the products was 140, 139, then 138. Averages of three analyses per sample gave the product ratios presented in Table XXI.

Table XXI. Product Ratios from Thermolysis of 104 in the Gas Phase at 161.8°.

Time (min)	Pressure (mm)	138	139	140
5	3	41.27	17.11	41.62
10	3	41.50	17.44	41.06
20	3	41.96	16.71	41.33
20	279	42.32	16.34	41.35
30	3	40.64	16.72	42.65
60	3	43.12	15.52	41.36

Thermolysis of Pyrazoline 104 in the Gas Phase at 82°.

Into a 30-mL gas sample bulb with teflon Rotoflo stopcock were expanded 3 mm of pyrazoline 104. Six identical samples were prepared and heated at 82-83° for the times indicated in Table XXII. The product ratios were determined by GLPC analysis as before.

Table XXII. Product Ratios from Thermolysis of 104 in the Gas Phase at 82-83°.

Time (h)	138	139	140
2	33.54	18.54	47.92
5	33.28	18.68	48.04
8	32.51	18.69	48.80
11	33.52	18.57	47.91
29	35.15	18.17	46.68
49	36.94	17.59	45.47

Photolysis of Pyrazoline 104 in the Gas Phase.

Into a 25-mL pyrex gas sample bulb with teflon Rotoflo stopcock was expanded 5 mm of pyrazoline 104. The bulb was pressurized with the indicated pressure of argon and irradiated in a Rayonet Photoreactor (350 nm). Analysis by GLPC (ODPN) gave the product ratios indicated in Table XXIII. Each ratio is the average of three analyses. In addition to the three methylenecyclopropanes 138, 139, and 140, dimethylallene 141 was observed.as indicated by its retention time and by ^{1}H NMR of the photolysis mixture.

Table XXIII. Product Ratios from Photolysis of 104 in the Gas Phase at 38°.

Time (h)	Pressure (mm)	138	139	140	141
1	0.05	41.43	7.22	24.33	27.02
1	1	43.45	11.84	37.65	7.06
1	5	39.48	15.59	43.12	1.81
1	103	34.21	18.95	46.27	0.57
1	220	31.65	20.57	47.37	0.41
1	400	30.95	21.42	47.30	0.32
1	577	30.06	22.49	47.20	0.25
1	740	29.35	22.73	47.72	0.20
2	749	29.36	22.80	47.85	——

Thermolysis of Pyrazoline 104 in Solution.

Into a 25-mL glass tube containing 5 mL distilled, degassed CH_2ClCH_2Cl was condensed 7.4 mg (0.0506 mmole) pyrazoline 104. The tube was sealed under vacuum and heated at 155°. The product ratios obtained for four identical samples heated for 10, 20, 30, and 50 minutes are presented in Table XXIV. Ratios are the average of three GLPC analyses per sample (ODPN). The sample at 10 minutes is excluded from the average due to poor signal to noise ratio.

Table XXIV. Product Ratios from Thermolysis of 104 in Solution at 155°.

Time (min)	138	139	140
10	48.35	15.03	36.62
20	52.34	14.83	32.82
30	52.83	15.16	32.01
50	55.16	15.16	29.68
average	53.44	15.05	31.50

Photolysis of Pyrazoline 104 in Solution.

Into a 25-mL pyrex gas sample bulb with teflon Rotoflo stopcock containing 5 mL distilled, degassed CH_2ClCH_2Cl was condensed 7.4 mg (0.0506 mmole) of pyrazoline 104. Photolysis in a Rayonet Photoreactor of three identical samples for 1, 2, and 4 hours gave the product ratios indicated in Table XXV. Dilution of a sample by a factor of two did not change the ratios obtained by GLPC analysis (ODPN).

Equilibrium of 138, 139, and 140 in the Gas Phase.

Into a well-conditioned thermolysis vessel at the indicated temperature was expanded 10 mm of the methylenecyclopropane 139. The isomerization to 138 and 140 was monitored and equilibrium ratios were determined

Table XXV. Product Ratios from Photolysis of 104 in Solution at 38°.

Time (h)	138	139	140
1	55.24	9.68	35.08
2	55.60	9.40	35.00
4	55.54	9.33	35.13
average	55.46	9.47	35.07

by GLPC analysis (ODPN) after 20 half lives. The equilibrium ratios in Table XXVI are the average of three analyses per sample.

Table XXVI. Equilbrium Constants for 138, 139 and 140 in the Gas Phase.

Temperature (°C)	K 140/139	K 140/138	K 139/138
278.7	13.4852	0.37892	0.028099
256.7	15.1170	0.37035	0.024499
243.7	15.8198	0.36583	0.023125
221.5	17.9709	0.35742	0.019889
200.2	20.2694	0.34686	0.017113

The equilibrium was verified by thermolysis of GLPC-purified 138 at 244.0°. After 60 minutes, the equilibrium ratio of 138 to 139 to 140 was 72.093 to 1.640 to 26.267. At 243.7° starting with 139 , the equilibrium ratio was 72.000 to 1.665 to 26.340.

Themolysis of 139; Kinetic Run.

Into a well-conditioned thermolysis vessel at 200.8° was expanded 10 mm of 139. The reaction was monitored by GLPC (ODPN) and the product ratios in Table XXVII were observed.

Table XXVII. Kinetic Run for Thermolysis of 139 at 200.8°.

Time (min)	138	139	140
2	8.83	83.27	7.90
4	16.10	69.40	14.50
6	22.08	58.23	19.69
8	27.31	48.69	23.99
10	31.44	40.87	27.69
15	39.59	26.49	33.92
30	51.16	7.98	40.86

APPENDIX: IR AND ^1H NMR SPECTRA

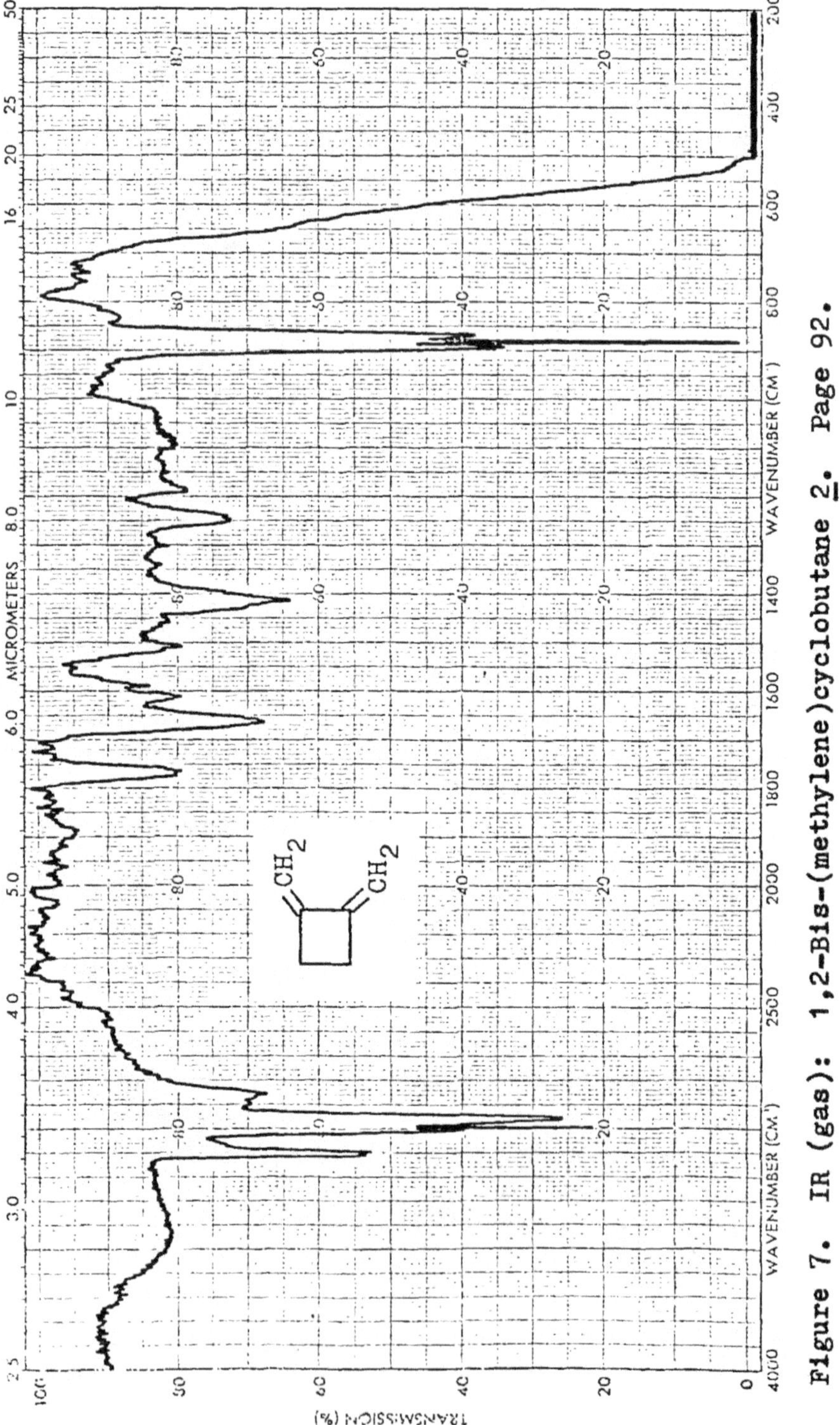

Figure 7. IR (gas): 1,2-Bis-(methylene)cyclobutane 2. Page 92.

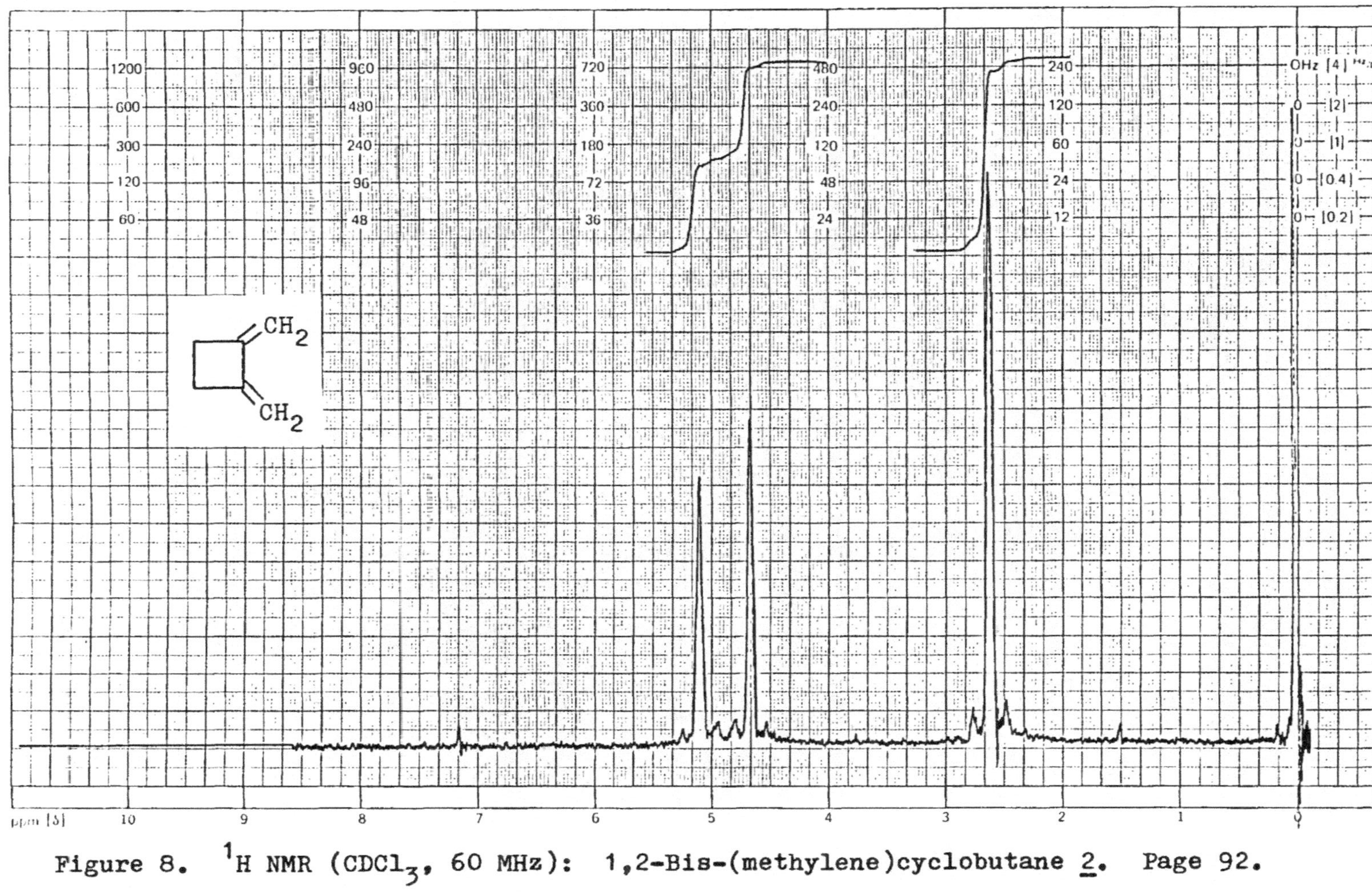

Figure 8. ^{1}H NMR ($CDCl_3$, 60 MHz): 1,2-Bis-(methylene)cyclobutane 2. Page 92.

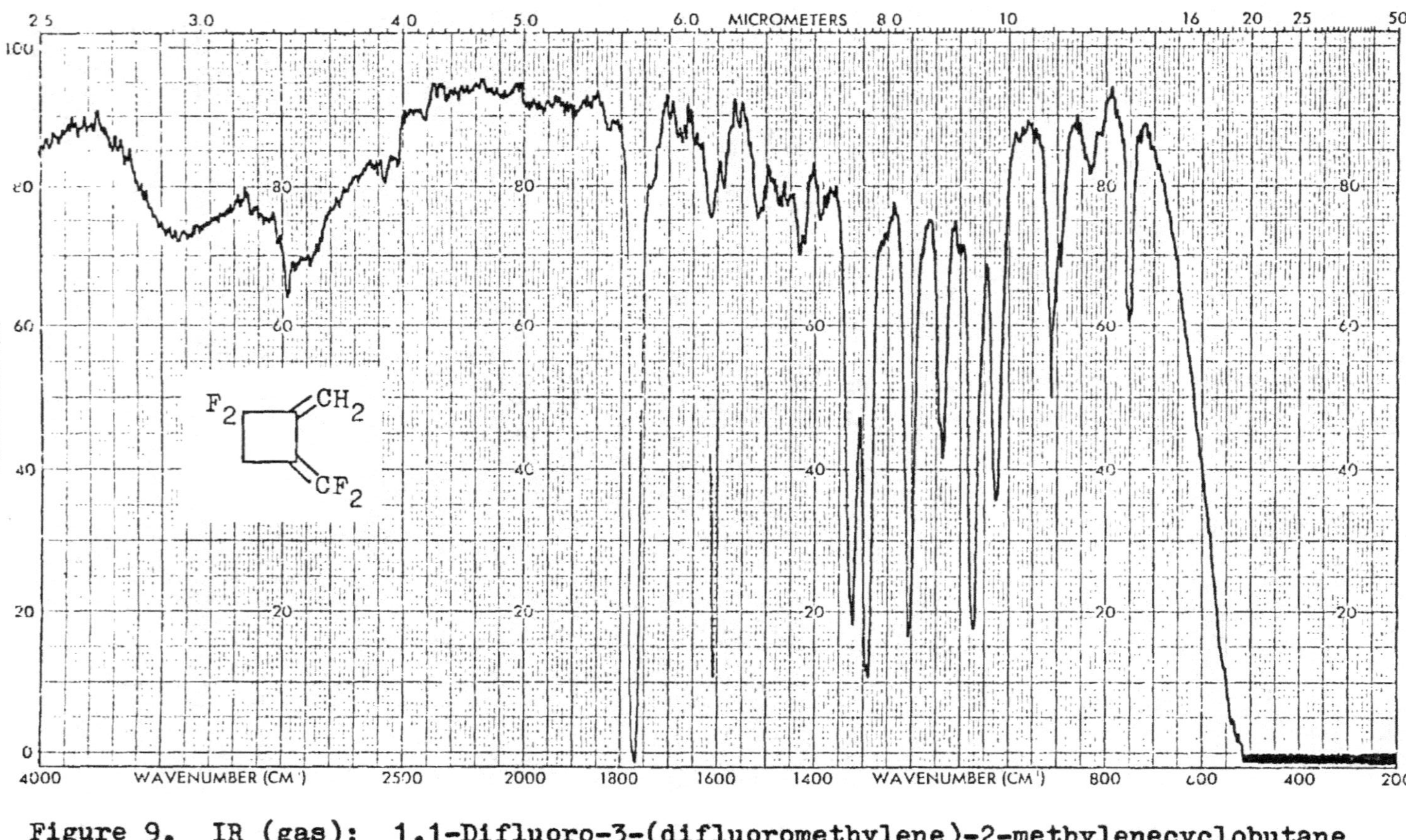

Figure 9. IR (gas): 1,1-Difluoro-3-(difluoromethylene)-2-methylenecyclobutane **13**. Page 107.

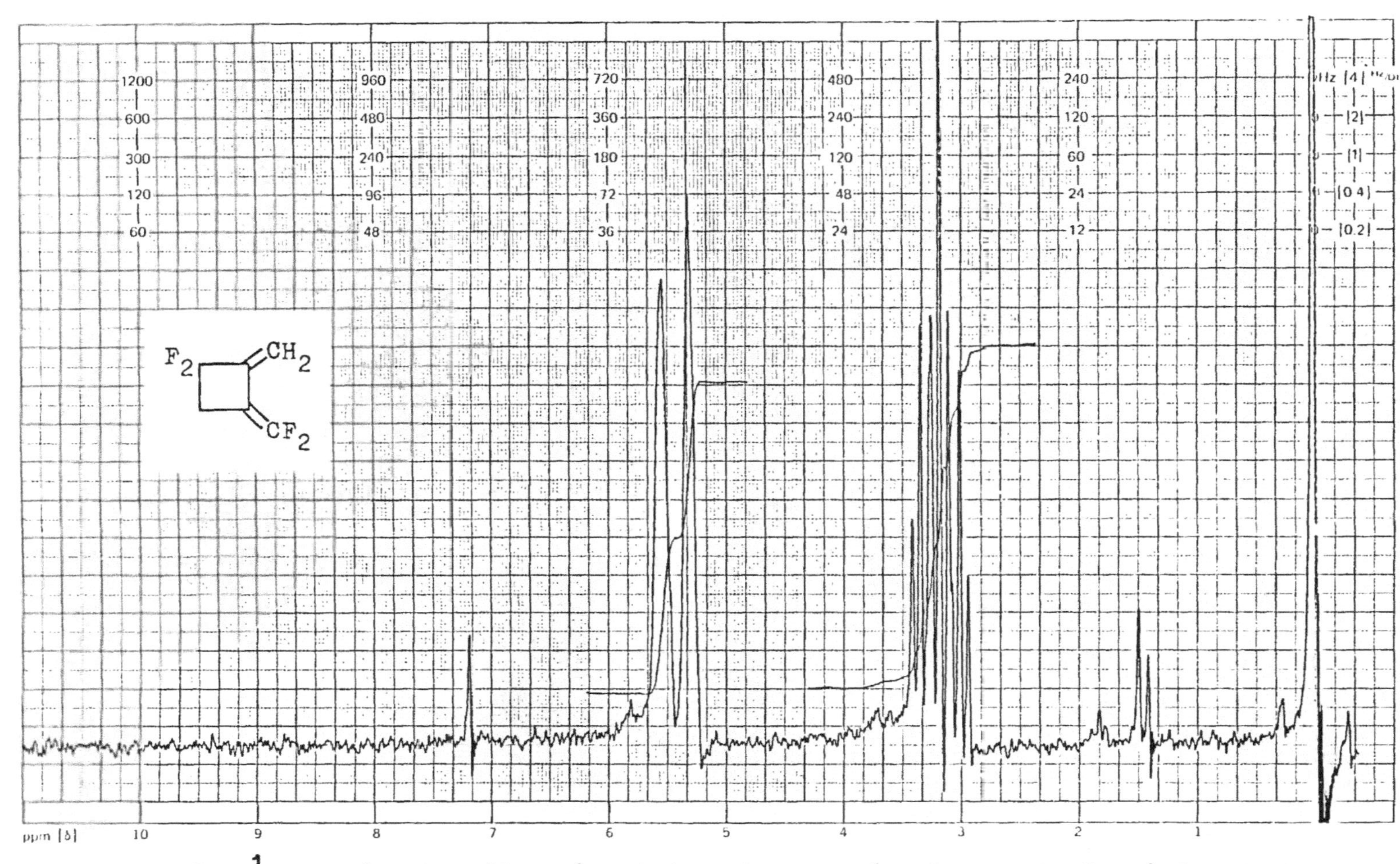

Figure 10. ^{1}H NMR ($CDCl_3$, 60 MHz): 1,1-Difluoro-3-(difluoromethylene)-2-methylene-cyclobutane **13**. Page 107.

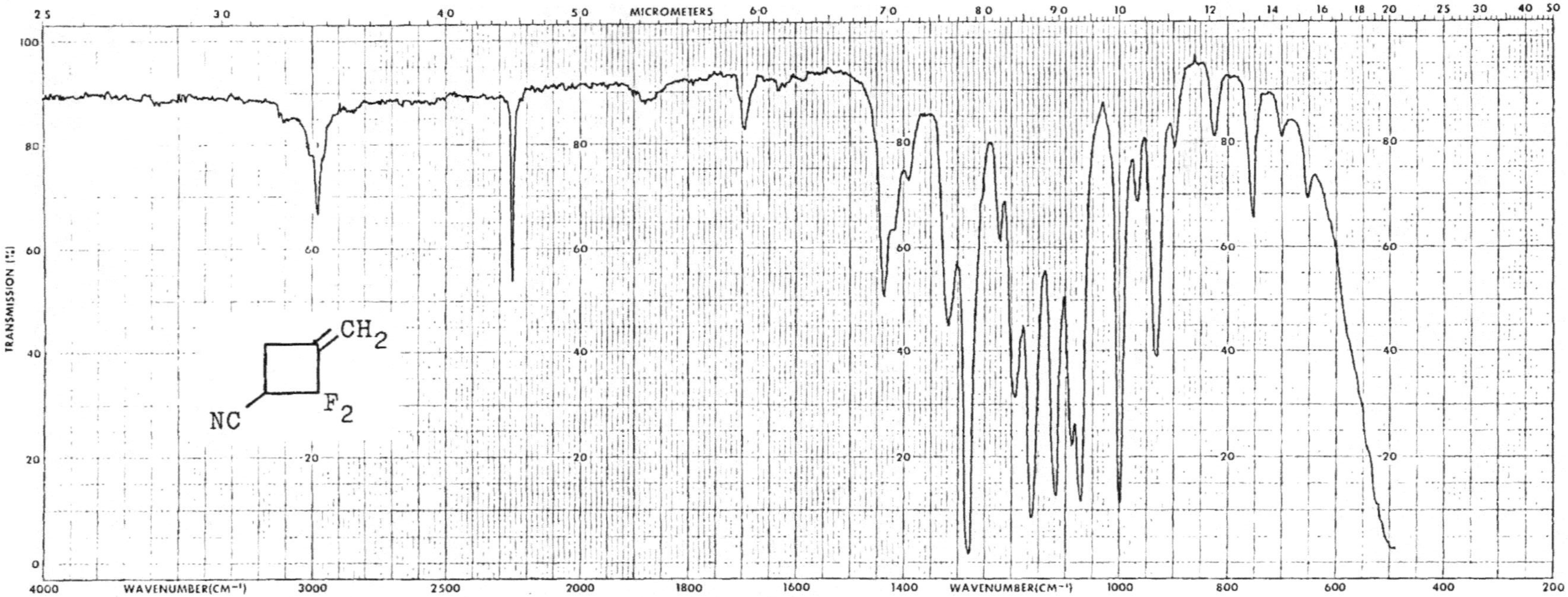

Figure 11. IR (film): 2,2-Difluoro-3-methylenecyclobutanecarbonitrile <u>14</u>. Page 109.

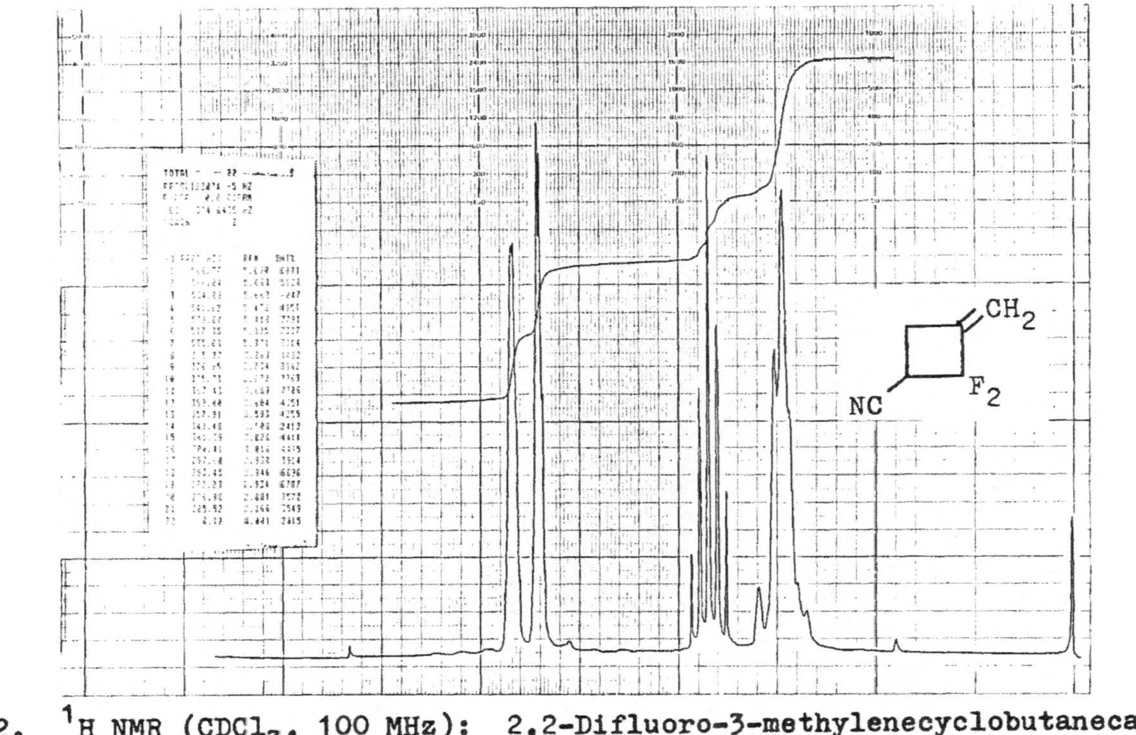

Figure 12. ^{1}H NMR ($CDCl_3$, 100 MHz): 2,2-Difluoro-3-methylenecyclobutanecarbonitrile <u>14</u>. Page 109.

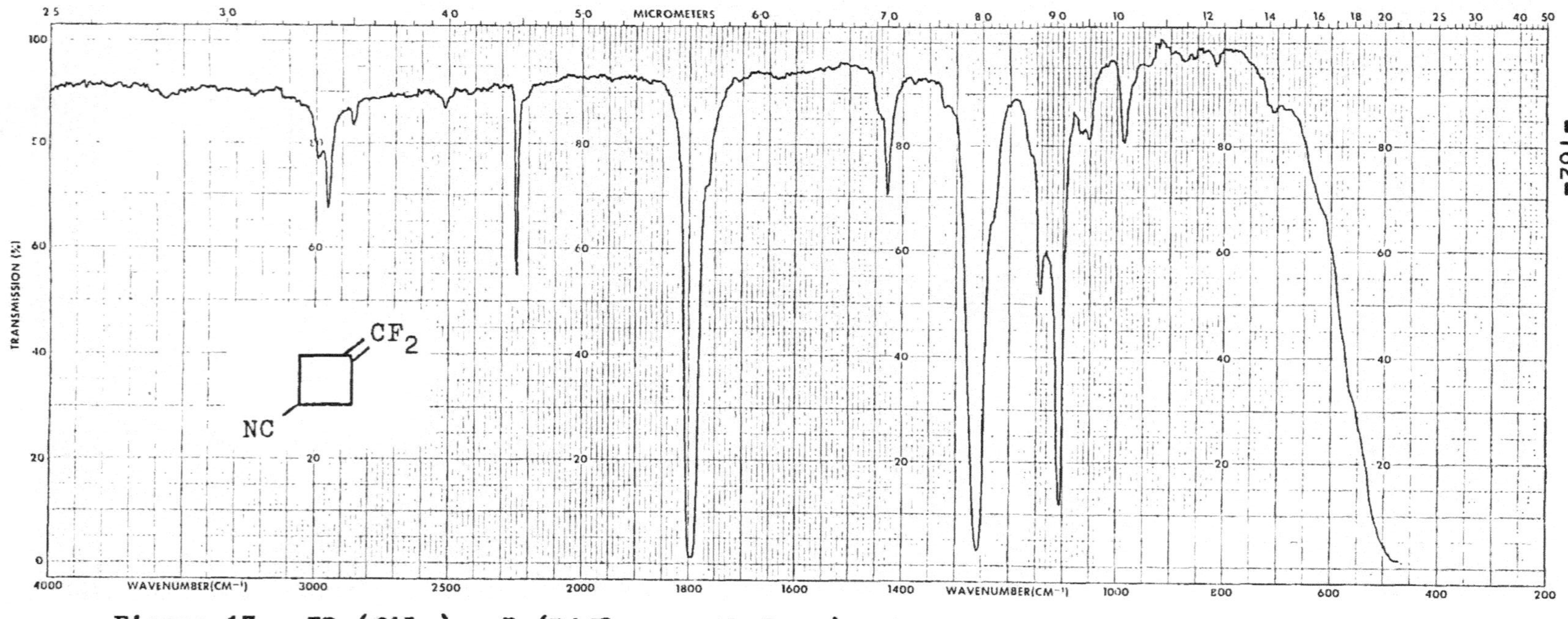

Figure 13. IR (film): 3-(Difluoromethylene)cyclobutanecarbonitrile 21. Page 109.

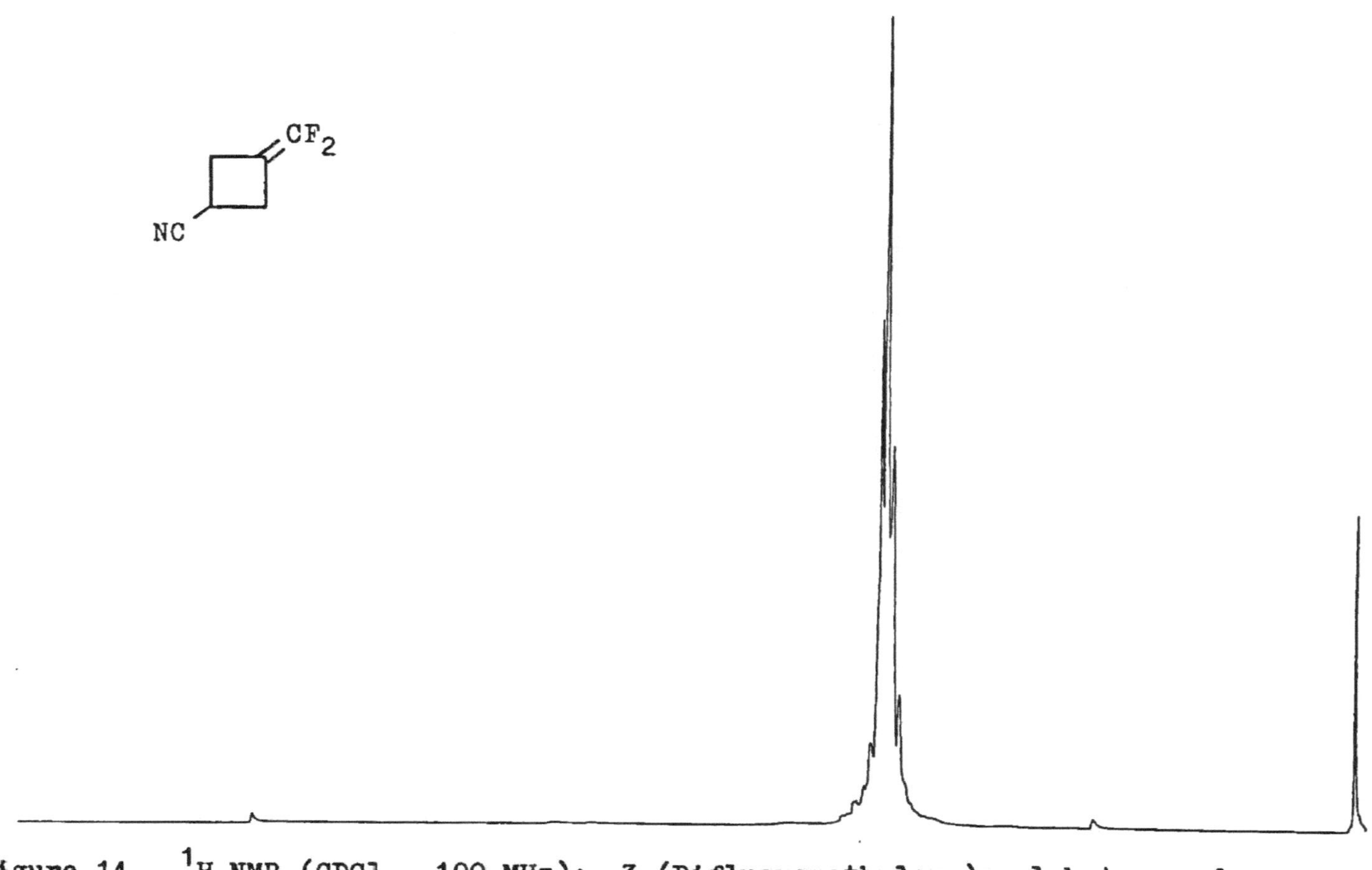

Figure 14. ^{1}H NMR ($CDCl_3$, 100 MHz): 3-(Difluoromethylene)cyclobutanecarbonitrile 21. Page 109.

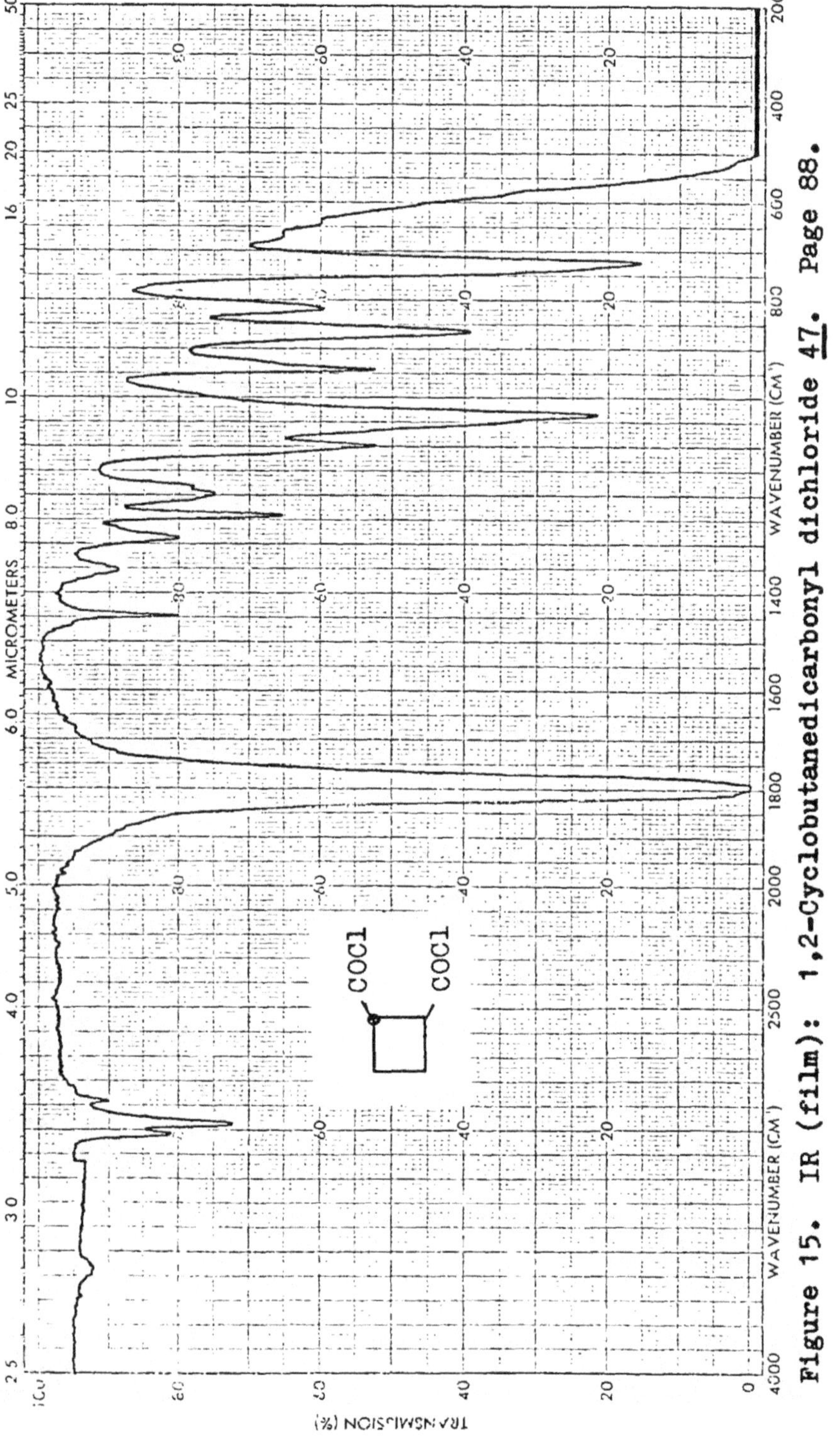

Figure 15. IR (film): 1,2-Cyclobutanedicarbonyl dichloride 47. Page 88.

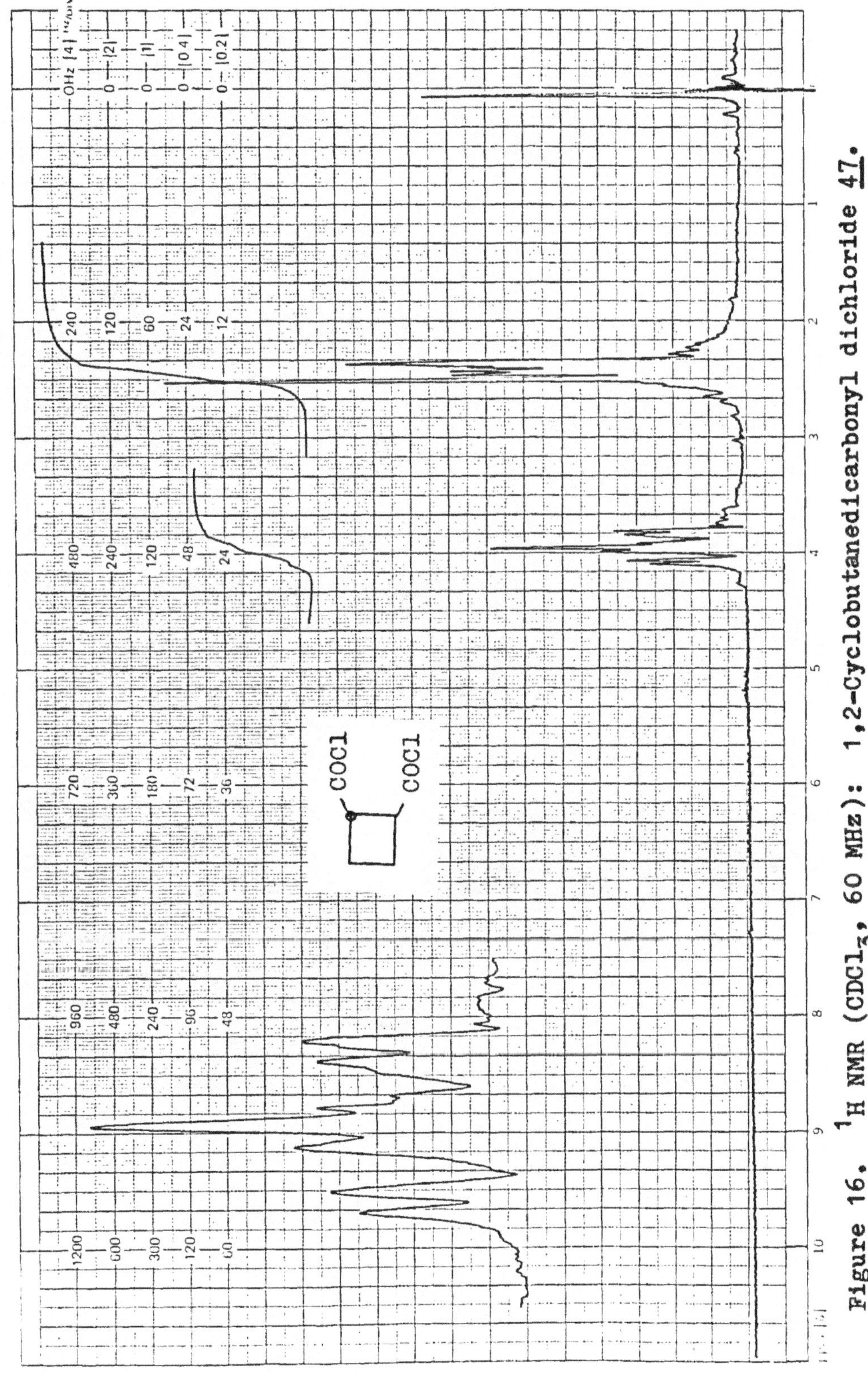

Figure 16. ^{1}**H NMR (CDCl$_3$, 60 MHz): 1,2-Cyclobutanedicarbonyl dichloride** **47.**
Page 88.

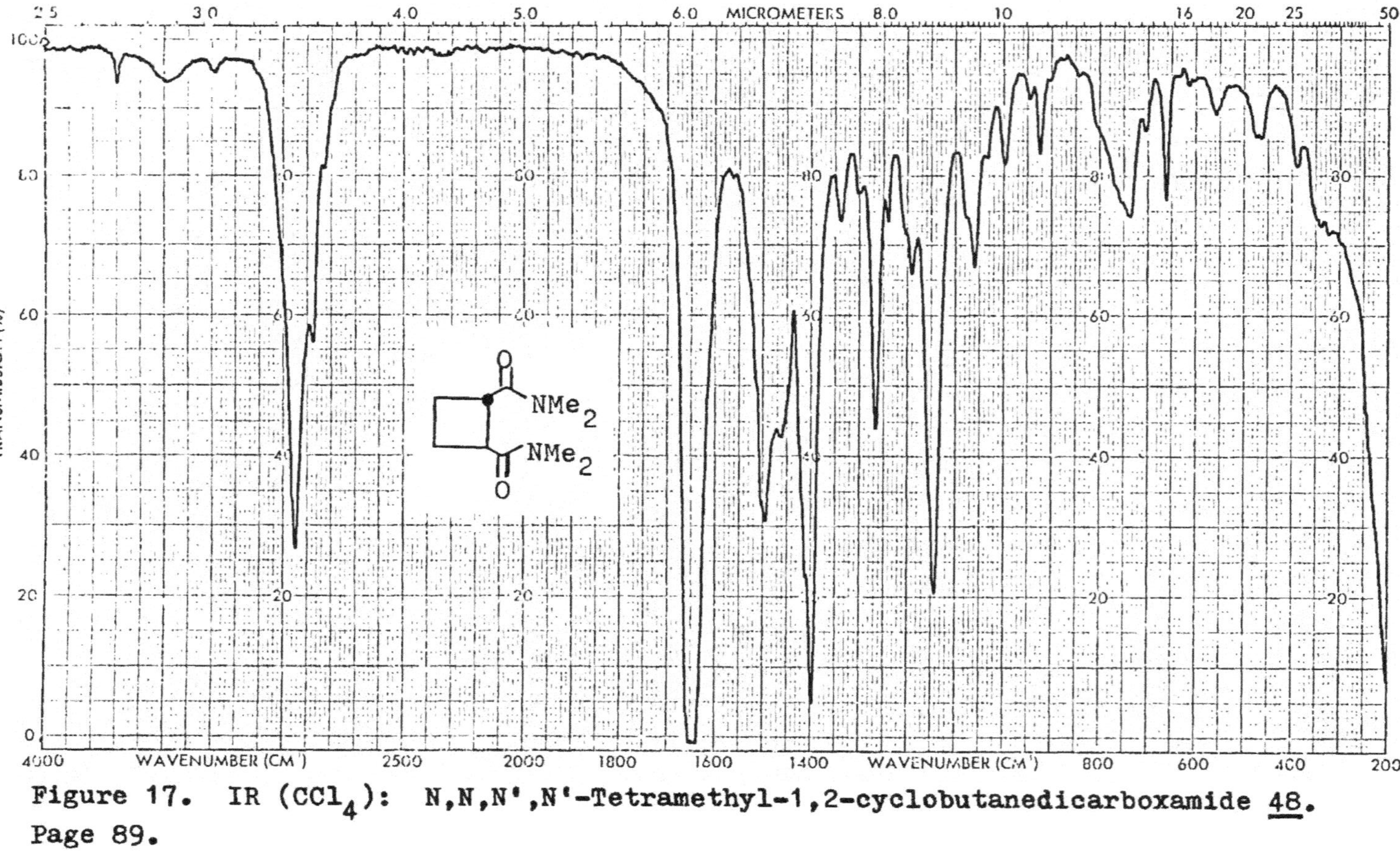

Figure 17. IR (CCl_4): N,N,N',N'-Tetramethyl-1,2-cyclobutanedicarboxamide 48. Page 89.

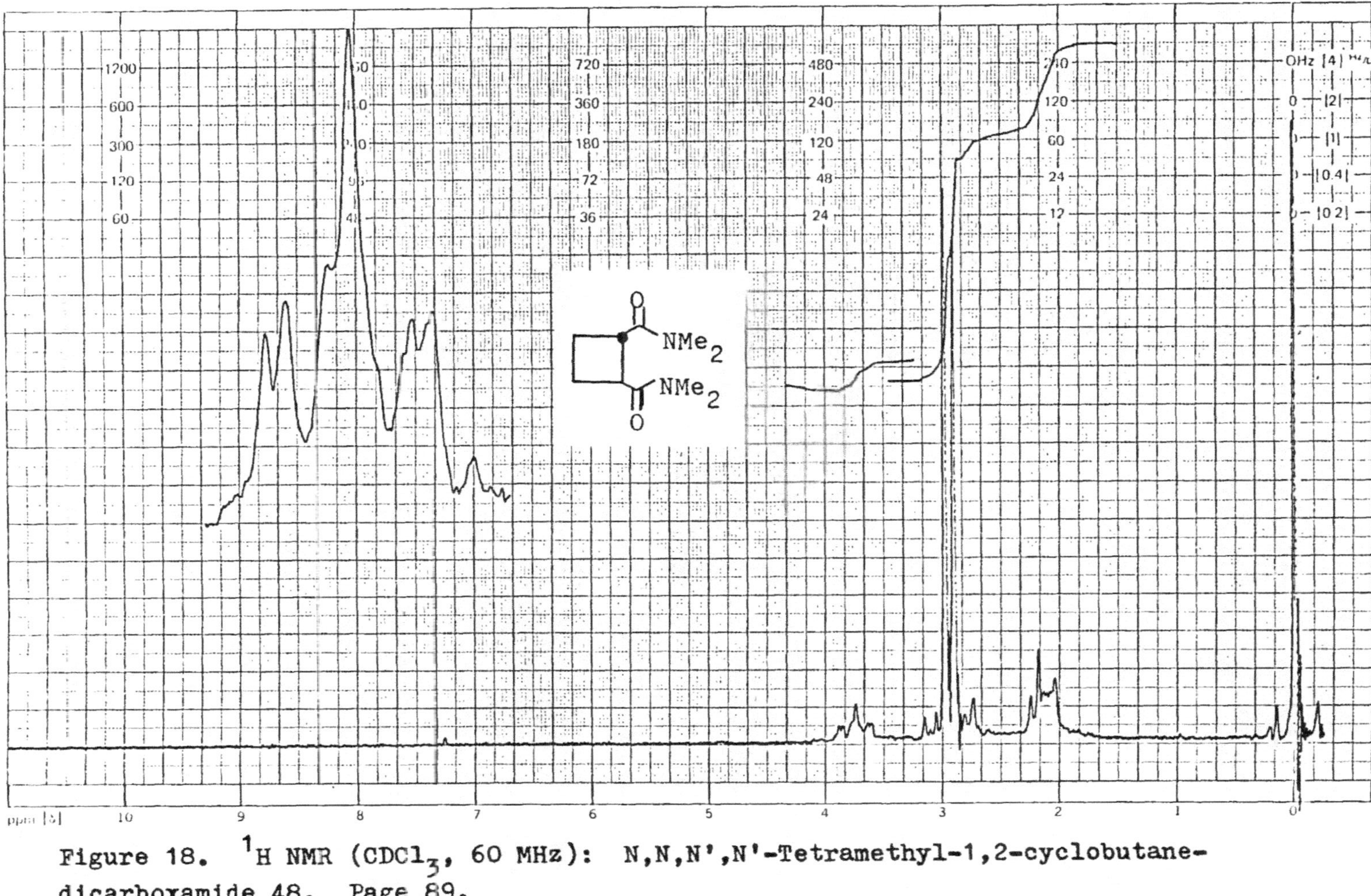

Figure 18. ^{1}H NMR ($CDCl_3$, 60 MHz): N,N,N',N'-Tetramethyl-1,2-cyclobutanedicarboxamide <u>48</u>. Page 89.

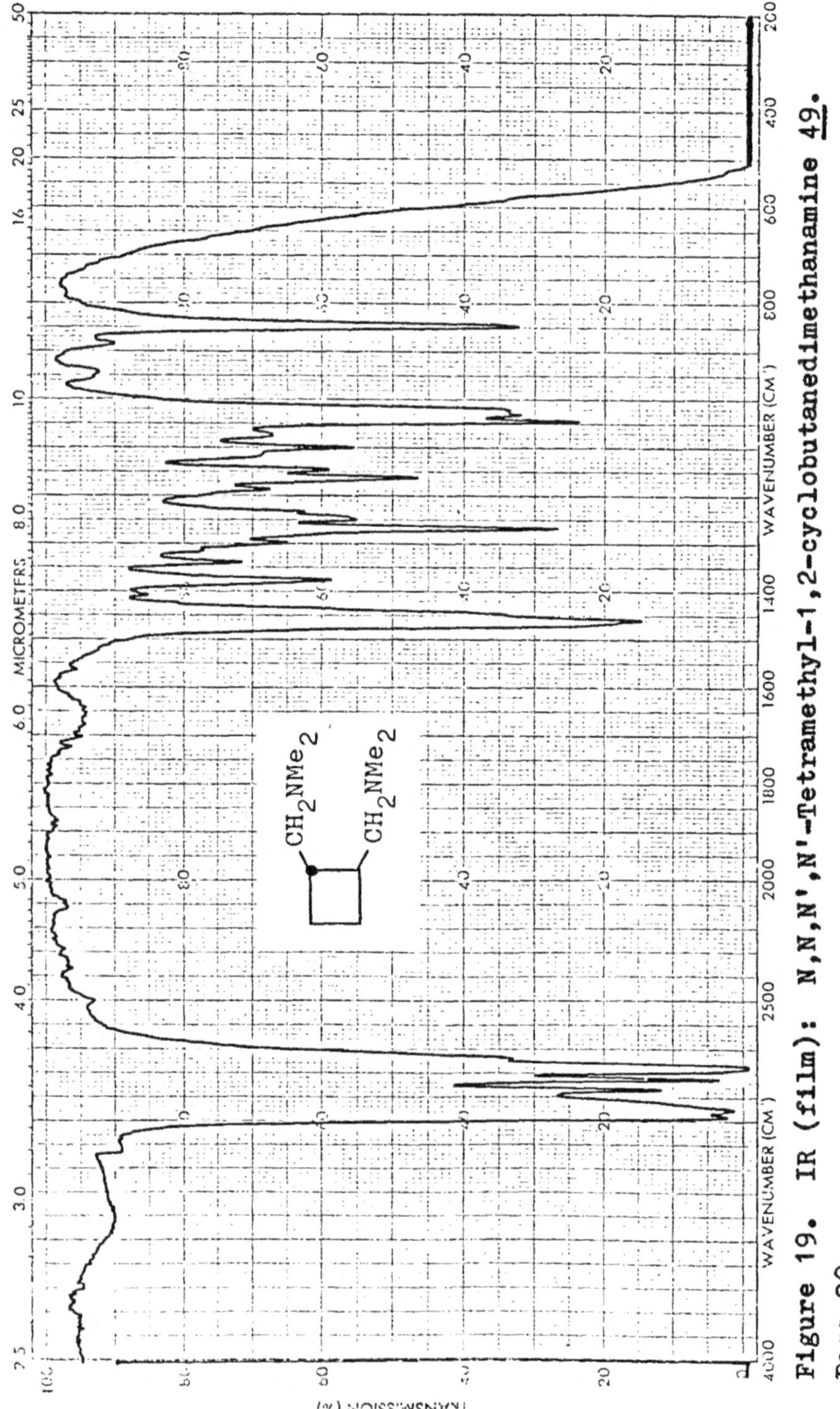

Figure 19. IR (film): N,N,N',N'-Tetramethyl-1,2-cyclobutanedimethanamine 49.
Page 90.

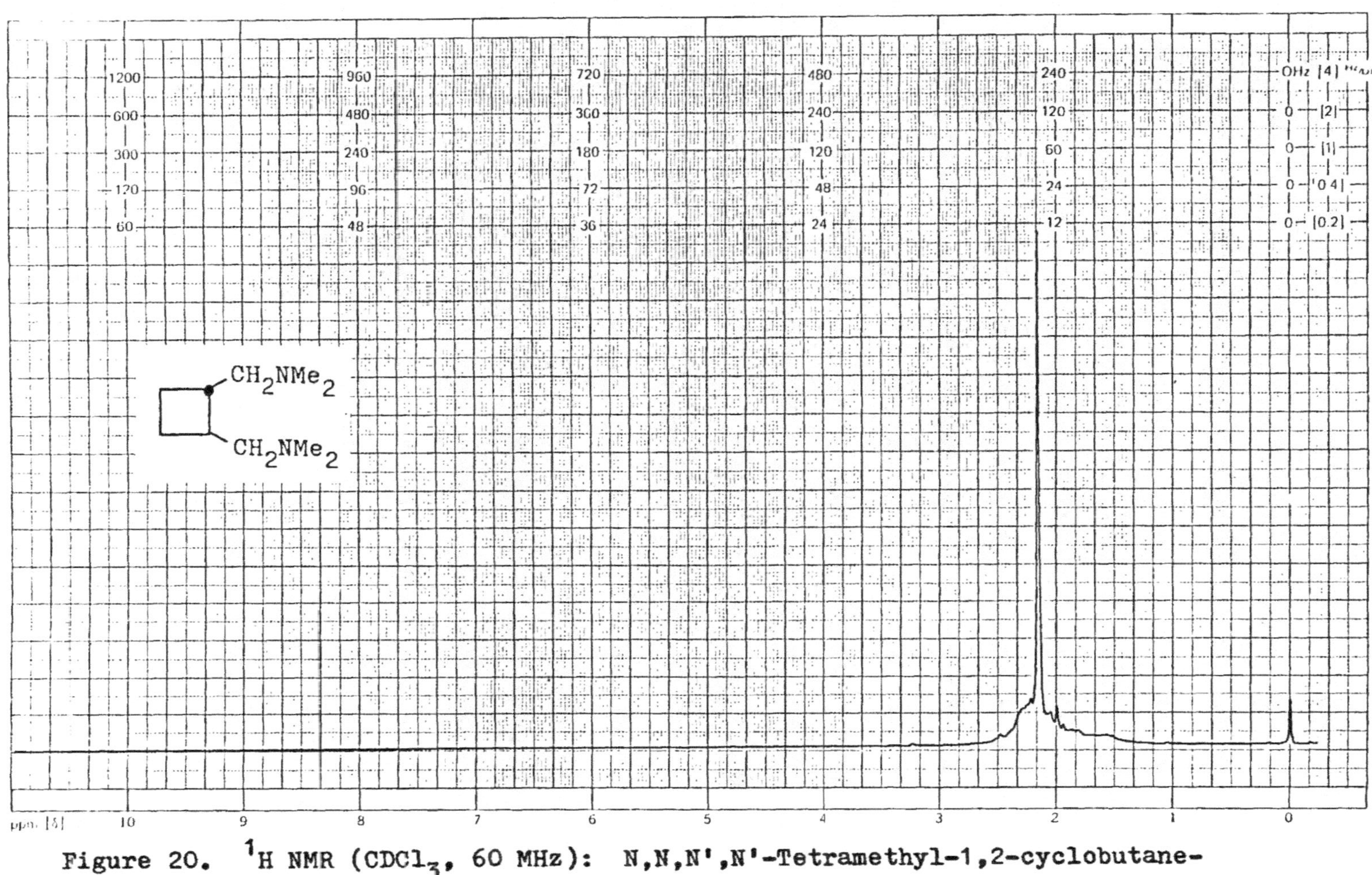

Figure 20. ^{1}H NMR ($CDCl_3$, 60 MHz): N,N,N',N'-Tetramethyl-1,2-cyclobutane-dimethanamine <u>49</u>. Page 90.

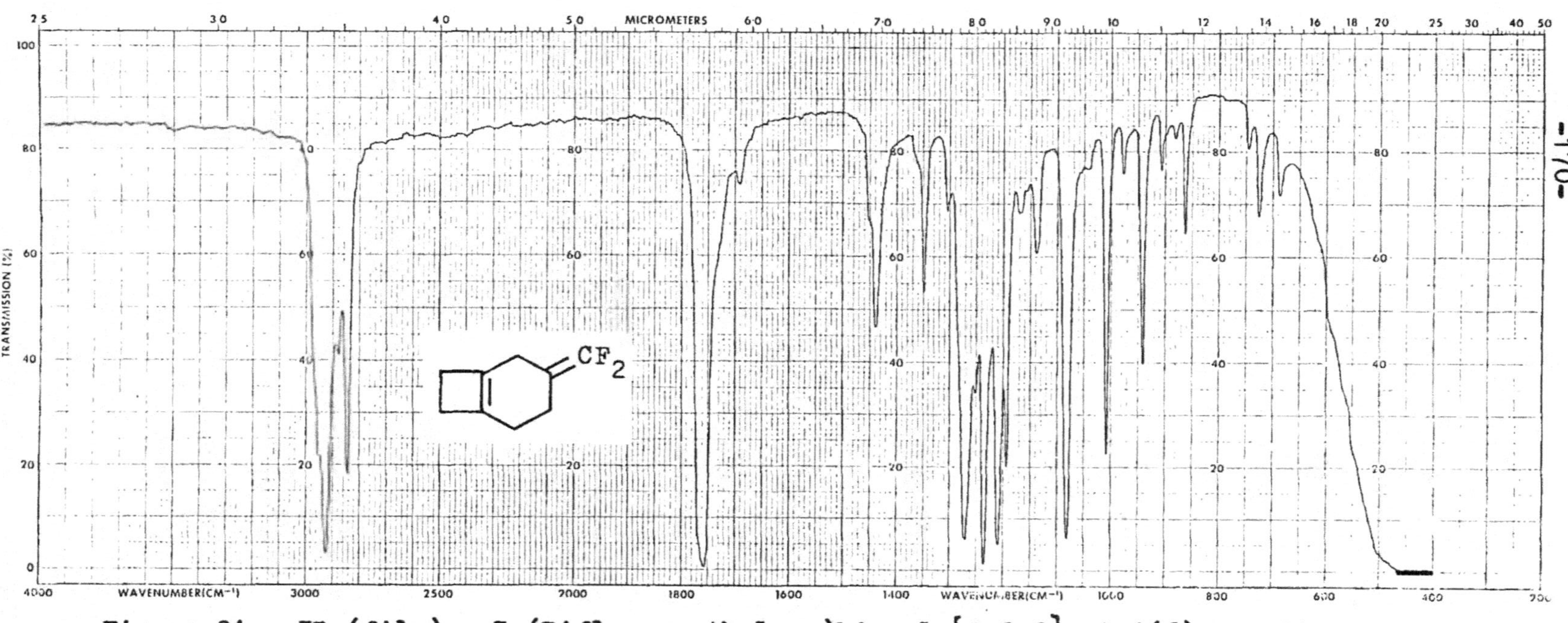

Figure 21. IR (film): 3-(Difluoromethylene)bicyclo[4.2.0]oct-1(6)-ene <u>51</u>. Page 93.

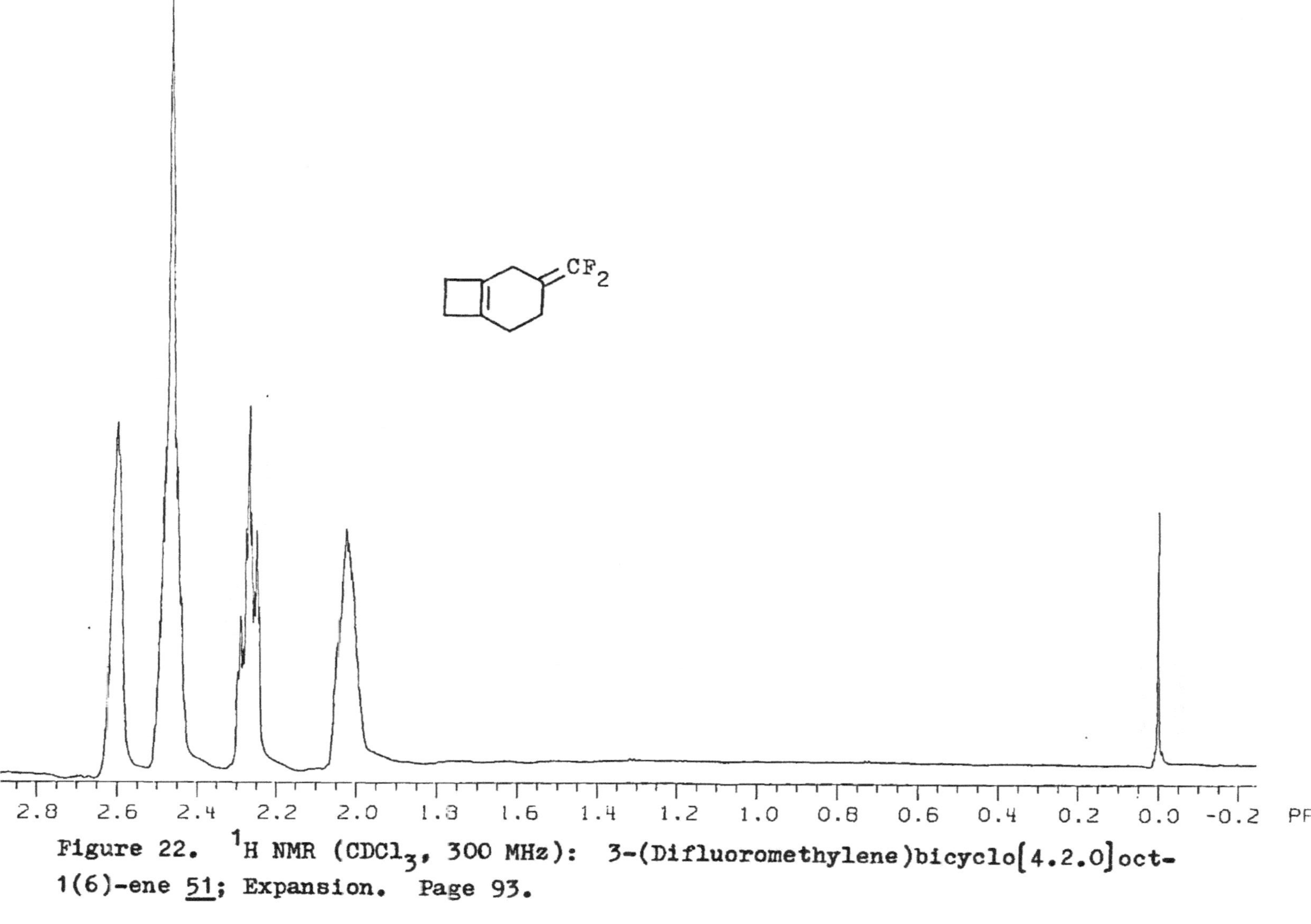

Figure 22. ^{1}H NMR ($CDCl_3$, 300 MHz): 3-(Difluoromethylene)bicyclo[4.2.0]oct-1(6)-ene <u>51</u>; Expansion. Page 93.

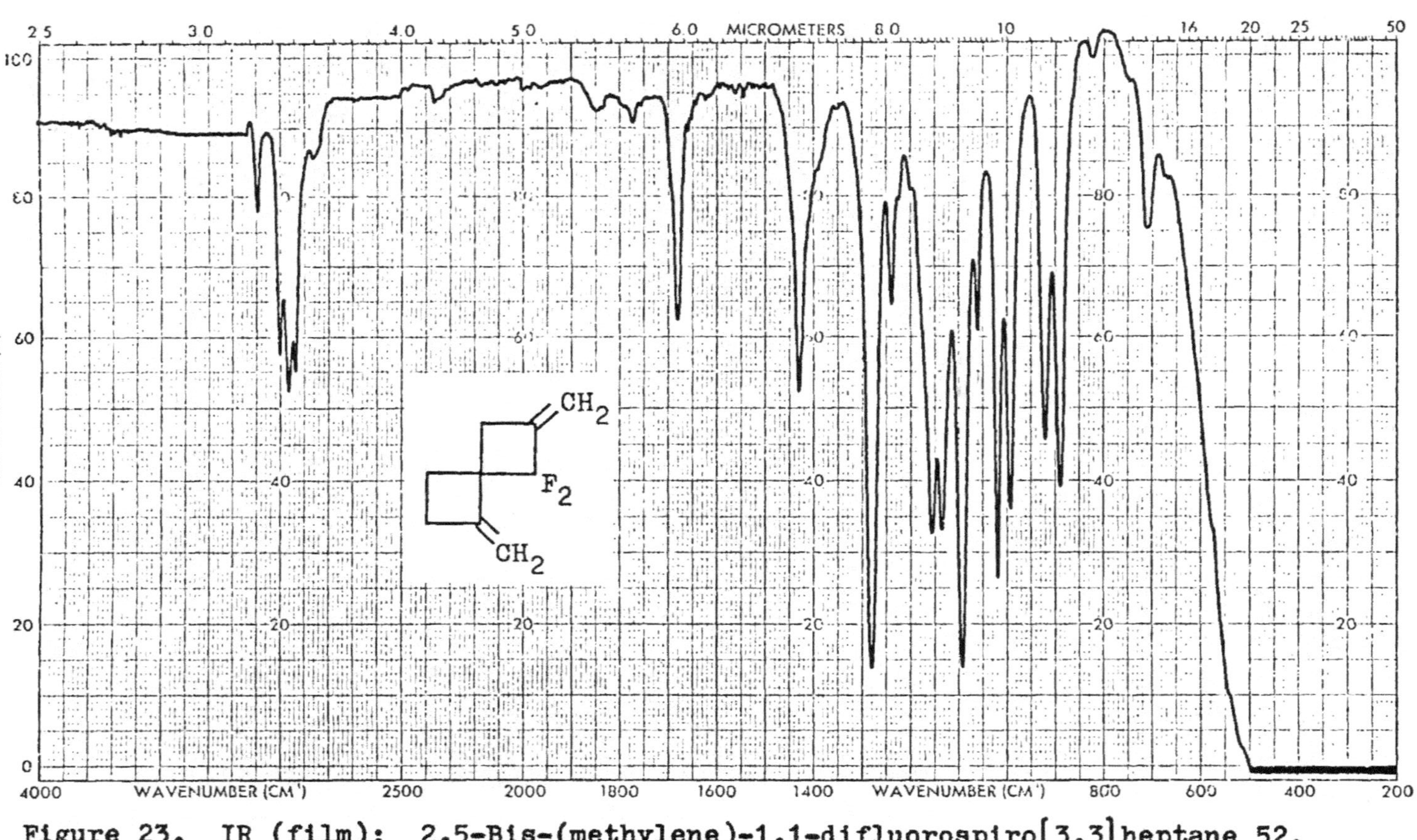

Figure 23. IR (film): 2,5-Bis-(methylene)-1,1-difluorospiro[3.3]heptane 52. Page 93.

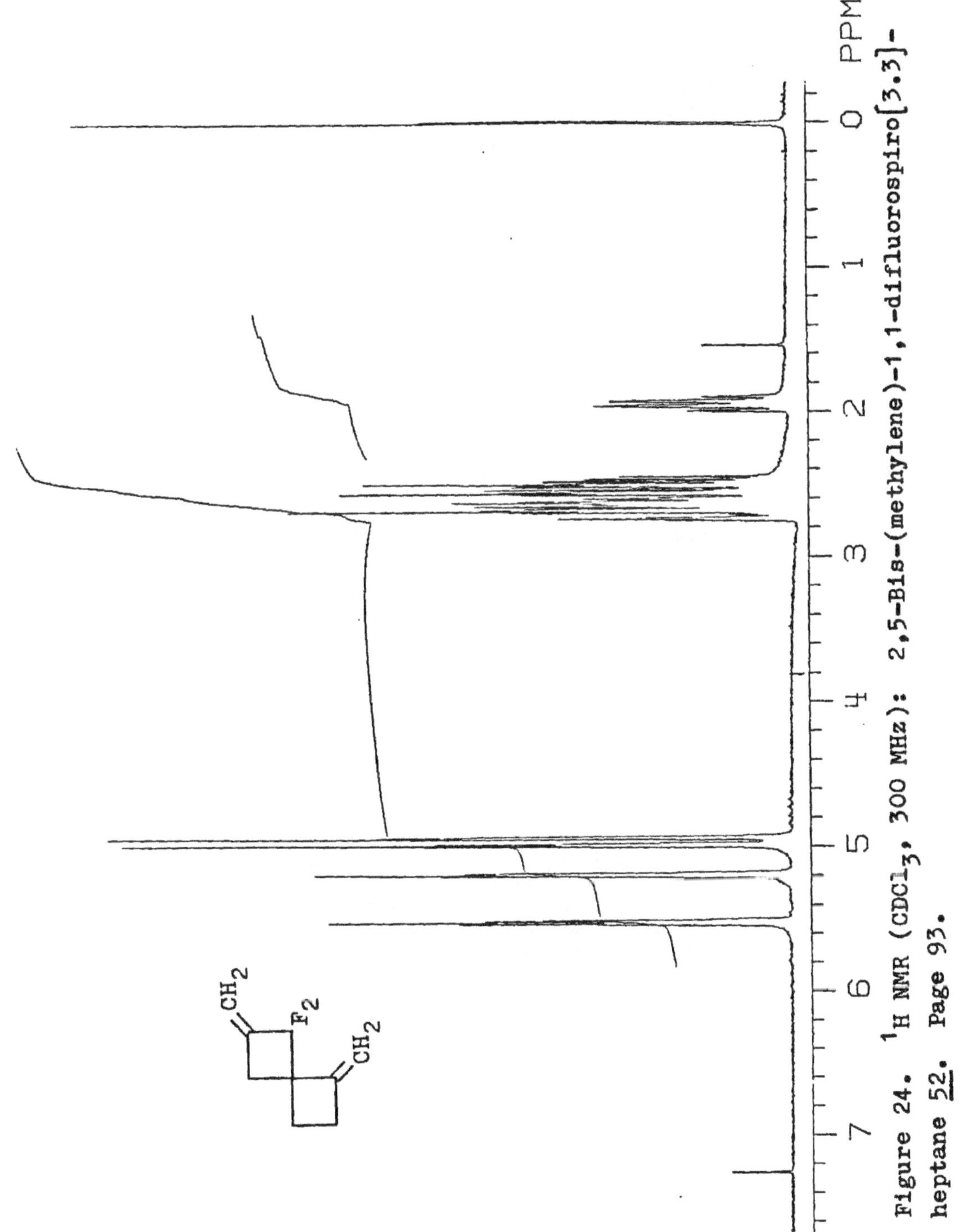

Figure 24. ^{1}H NMR ($CDCl_3$, 300 MHz): 2,5-Bis-(methylene)-1,1-difluorospiro[3.3]-heptane 52. Page 93.

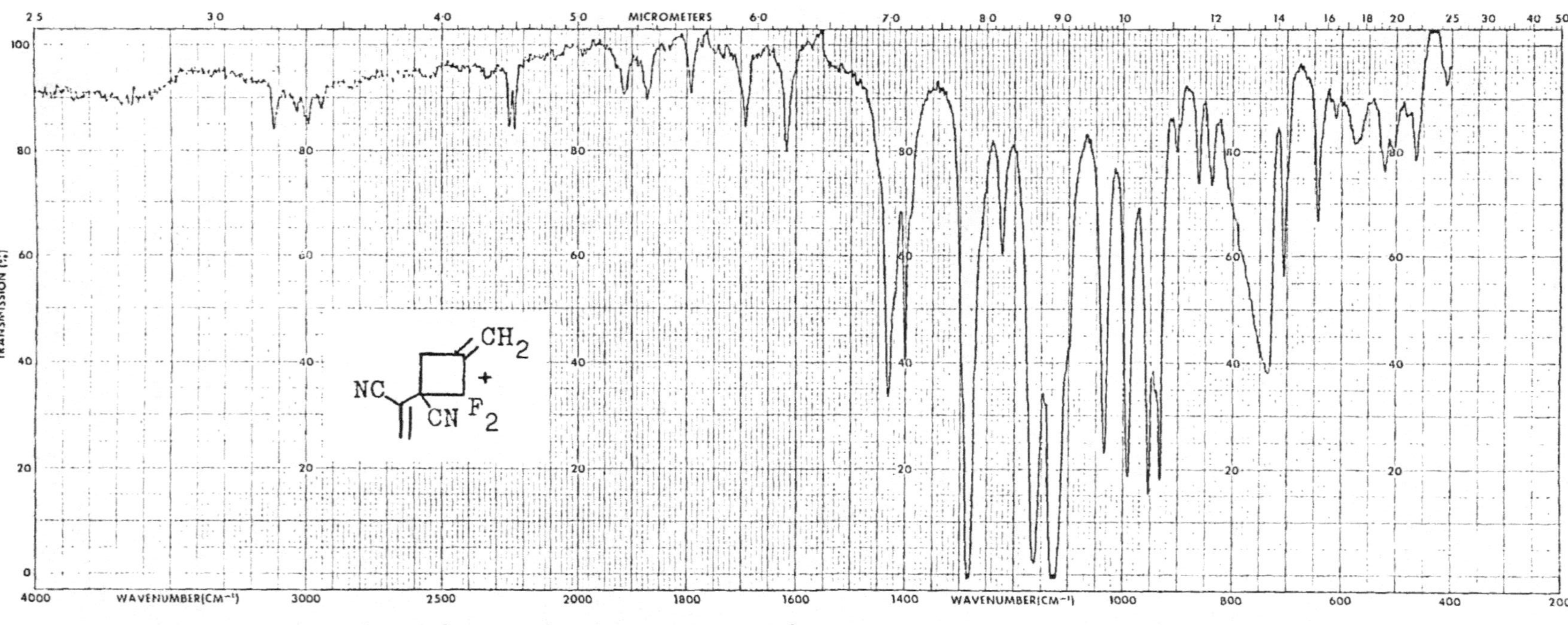

Figure 25. IR (CCl_4): 1-(1-Cyanoethenyl)-2,2-difluoro-3-methylenecyclobutane-carbonitrile 61. Page 95.

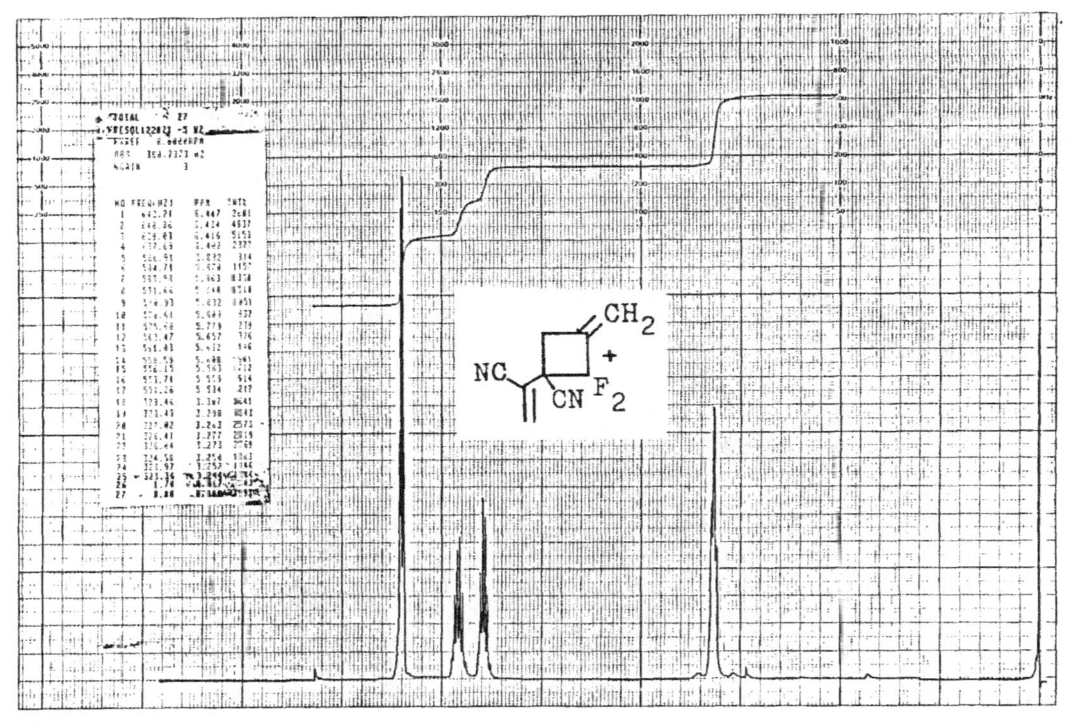

Figure 26. ^{1}H NMR ($CDCl_3$, 100 MHz): 1-(1-Cyanoethenyl)-2,2-difluoro-3-methylenecyclobutanecarbonitrile <u>61</u>. Page 95.

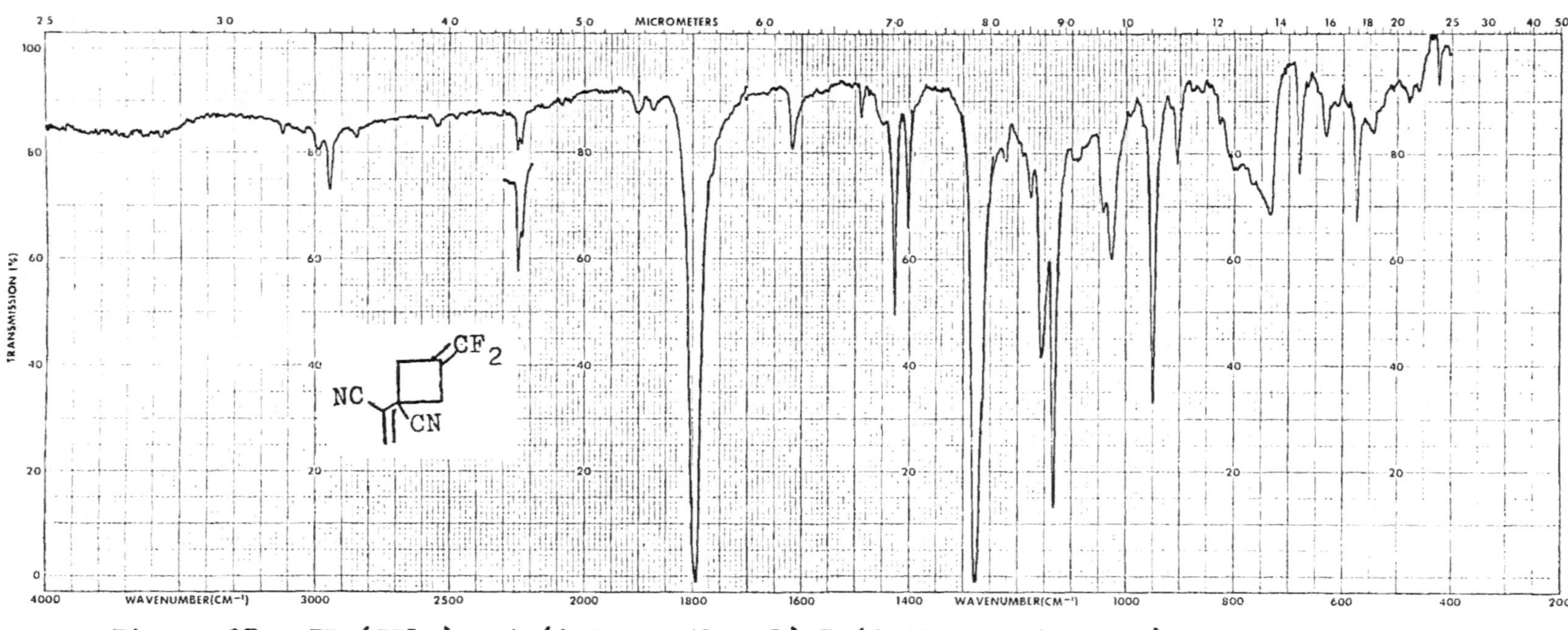

Figure 27. IR (CCl_4): 1-(1-Cyanoethenyl)-3-(difluoromethylene)cyclobutanecarbonitrile <u>62</u>. Page 95.

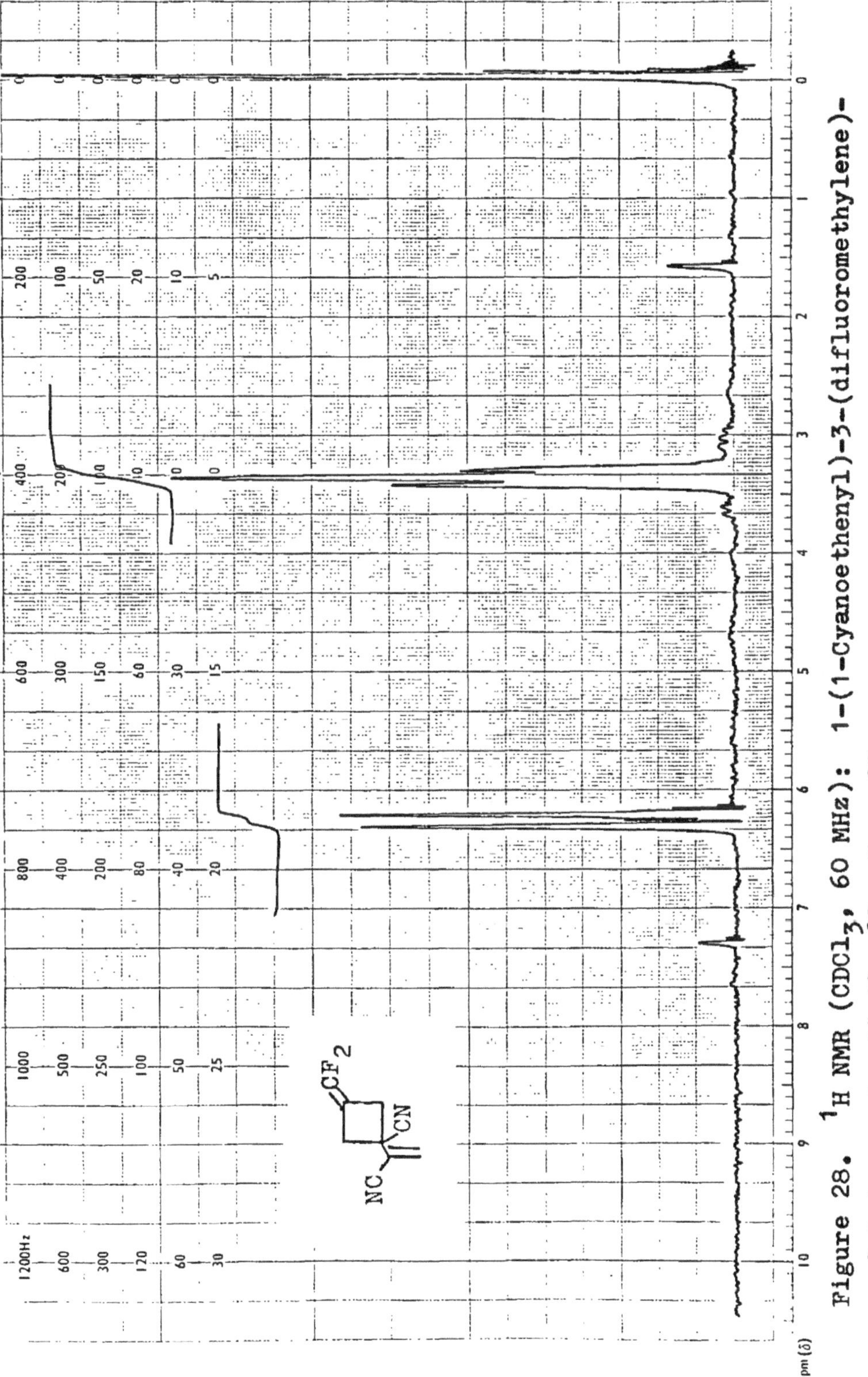

Figure 28. ^{1}H NMR ($CDCl_3$, 60 MHz): 1-(1-Cyanoethenyl)-3-(difluoromethylene)-cyclobutanecarbonitrile <u>62</u>. Page 95.

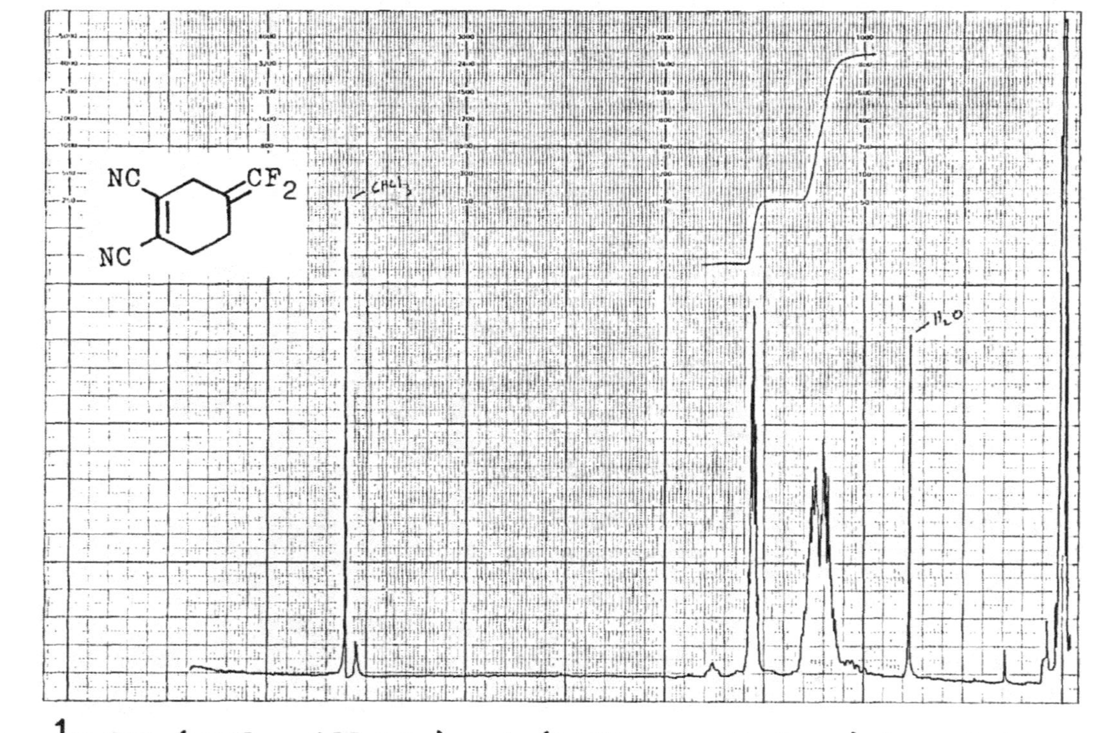

Figure 29. ^{1}H NMR ($CDCl_3$, 100 MHz): 4-(Difluoromethylene)-1-cyclohexene-1,2-dicarbonitrile <u>63</u>. Page 95.

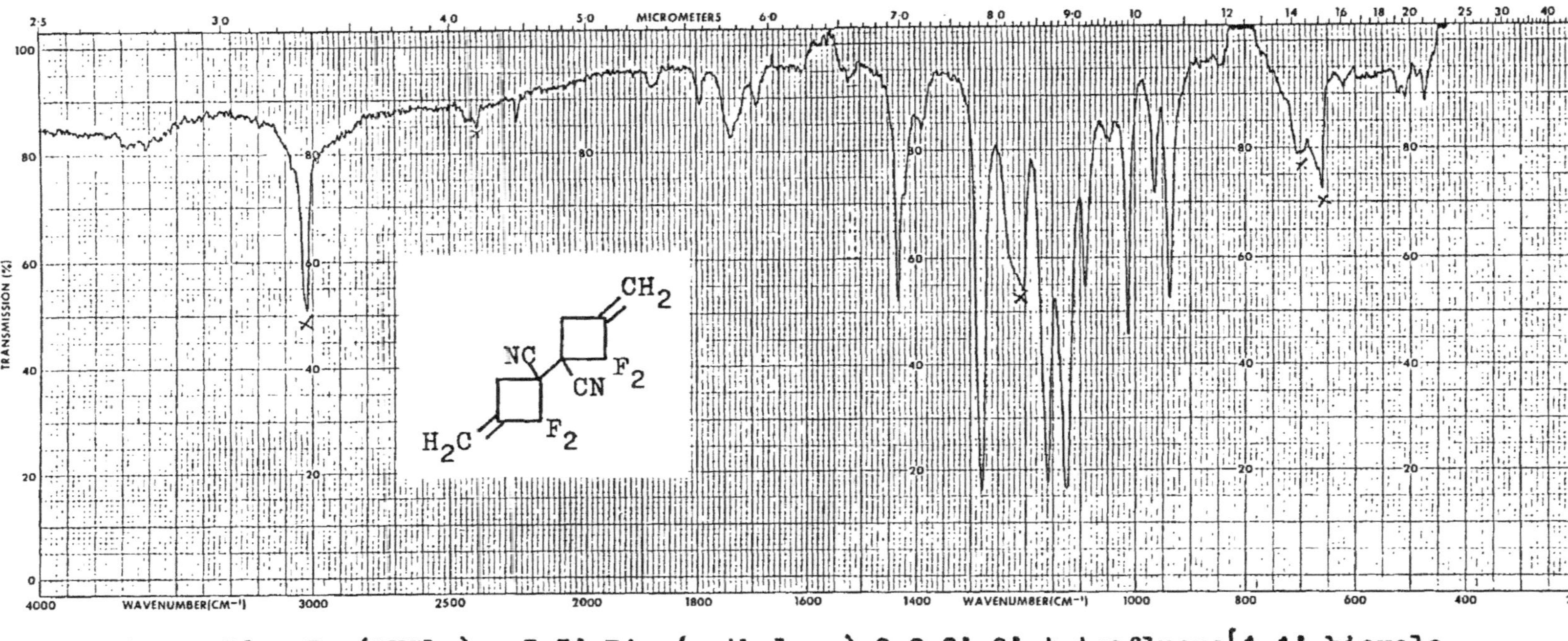

Figure 30. IR ($CHCl_3$): 3,3'-Bis-(methylene)-2,2,2',2'-tetrafluoro[1,1'-bicyclobutyl]-1,1'-dicarbonitrile 64. Page 95.

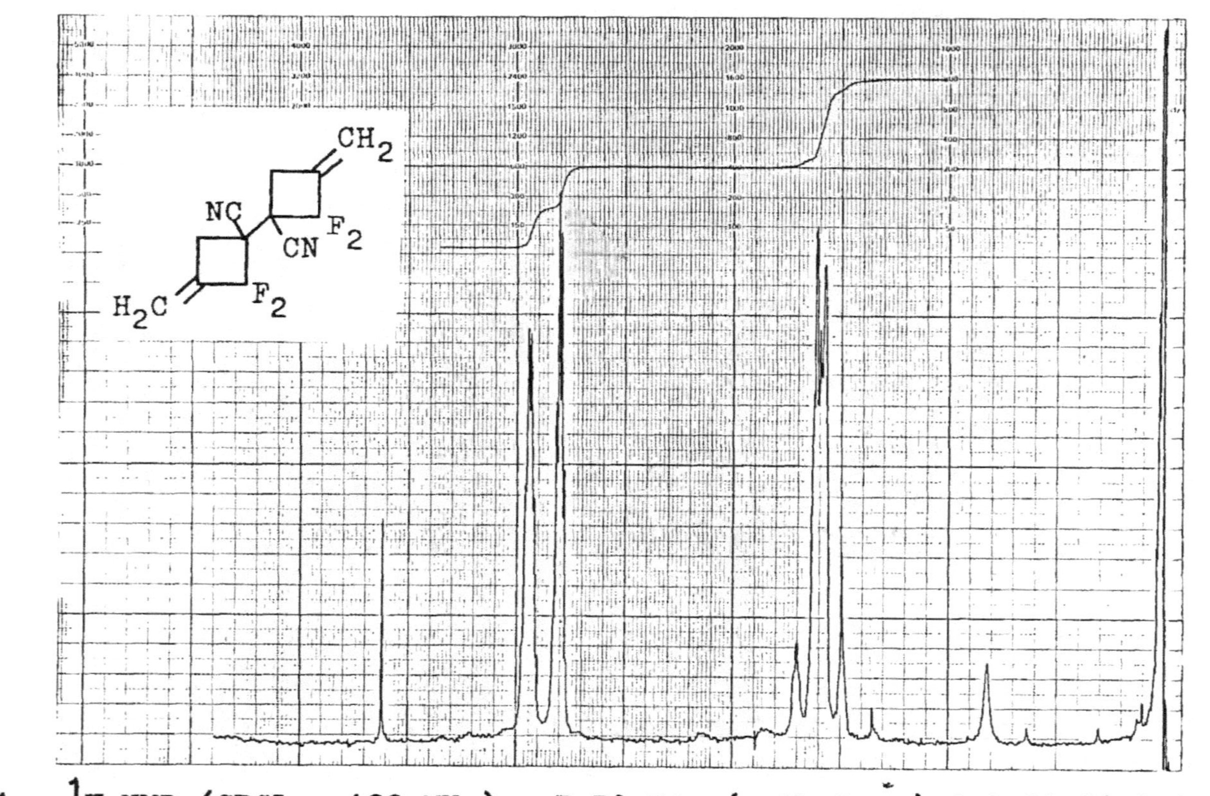

Figure 31. ^1H NMR ($CDCl_3$, 100 MHz): 3,3'-Bis-(methylene)-2,2,2',2'-tetrafluoro-[1,1'-bicyclobutyl]-1,1'-dicarbonitrile <u>64</u>. Page 95.

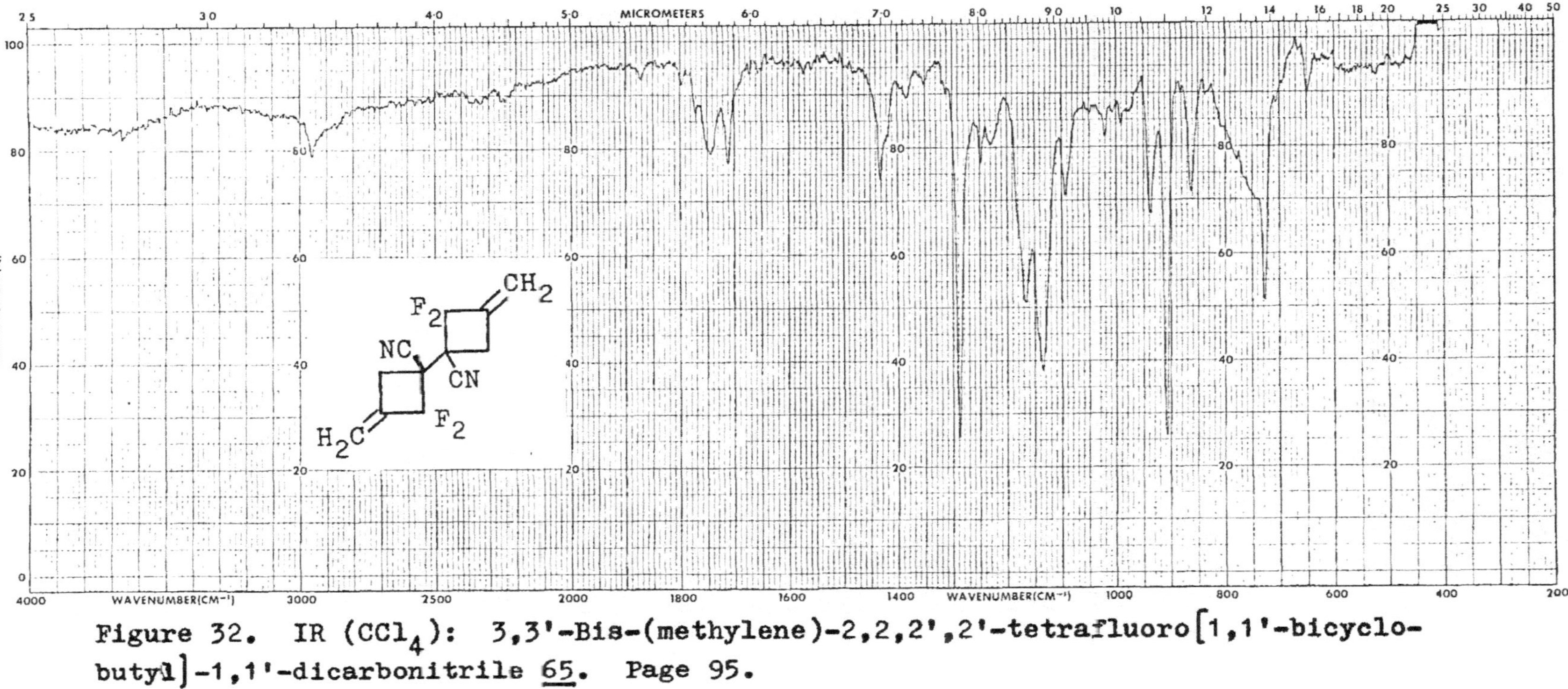

Figure 32. IR (CCl_4): 3,3'-Bis-(methylene)-2,2,2',2'-tetrafluoro[1,1'-bicyclobutyl]-1,1'-dicarbonitrile <u>65</u>. Page 95.

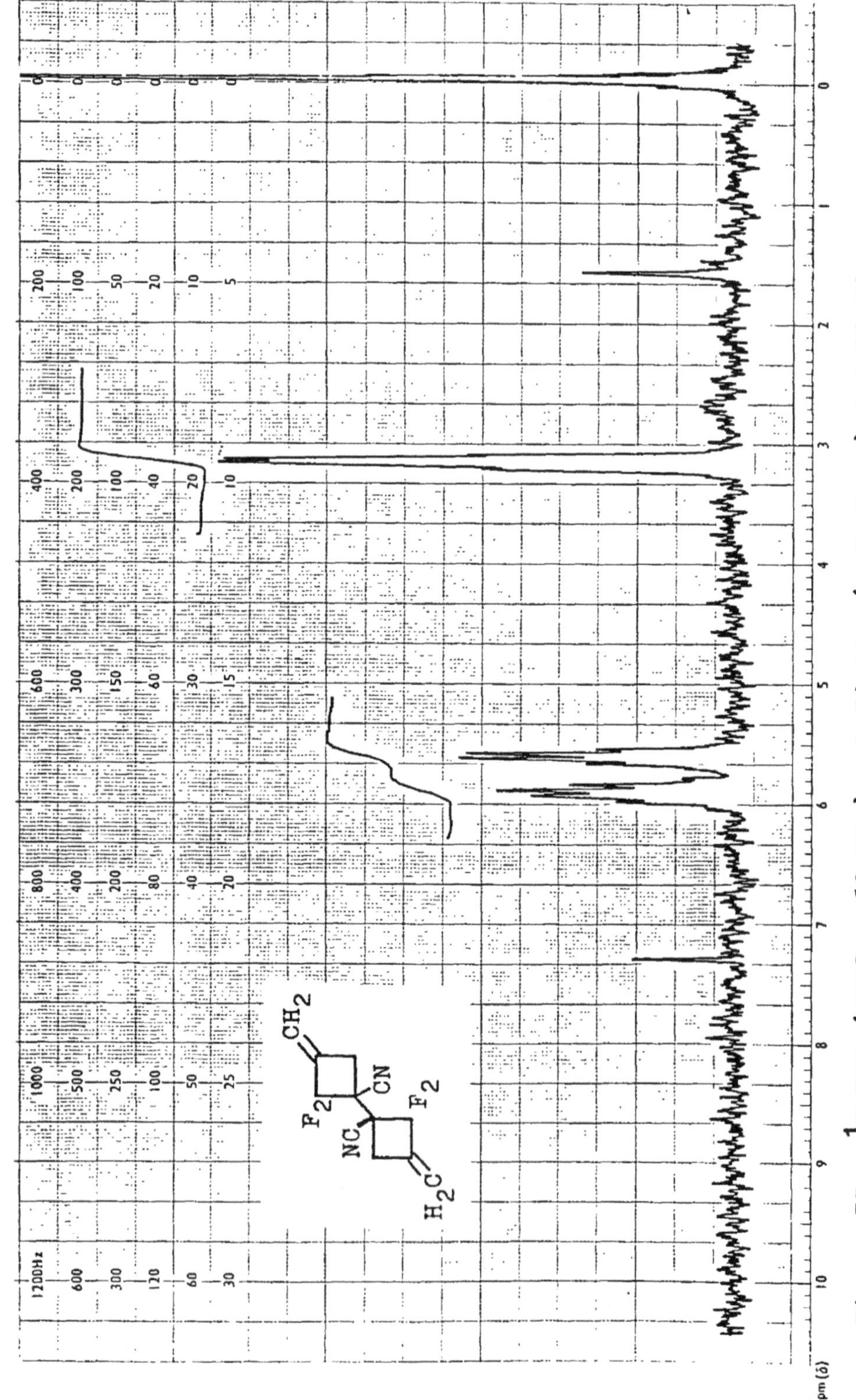

Figure 33. ^{1}H NMR ($CDCl_3$, 60 MHz): 3,3'-Bis-(methylene)-2,2,2',2'-tetrafluoro-[1,1'-bicyclobutyl]-1,1'-dicarbonitrile <u>65</u>. Page 95.

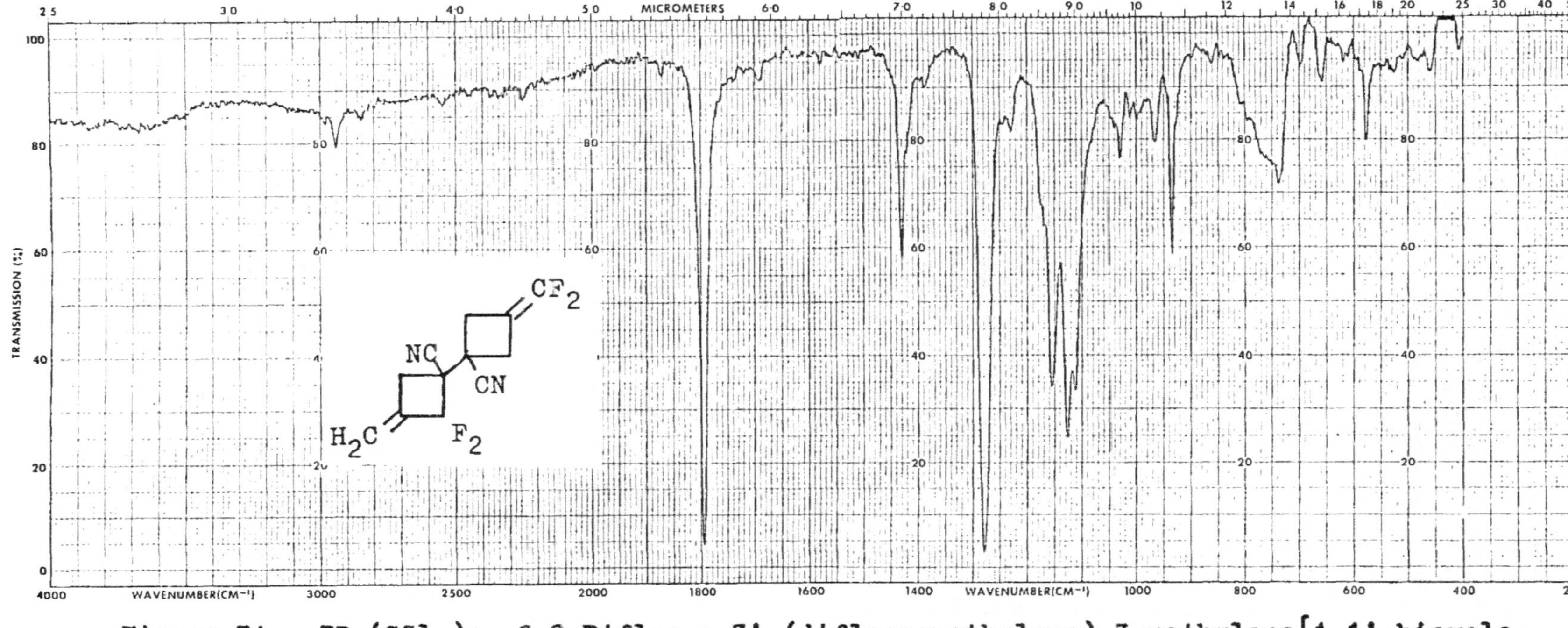

Figure 34. IR (CCl_4): 2,2-Difluoro-3'-(difluoromethylene)-3-methylene[1,1'-bicyclobutyl]-1,1'-dicarbonitrile 66. Page 95.

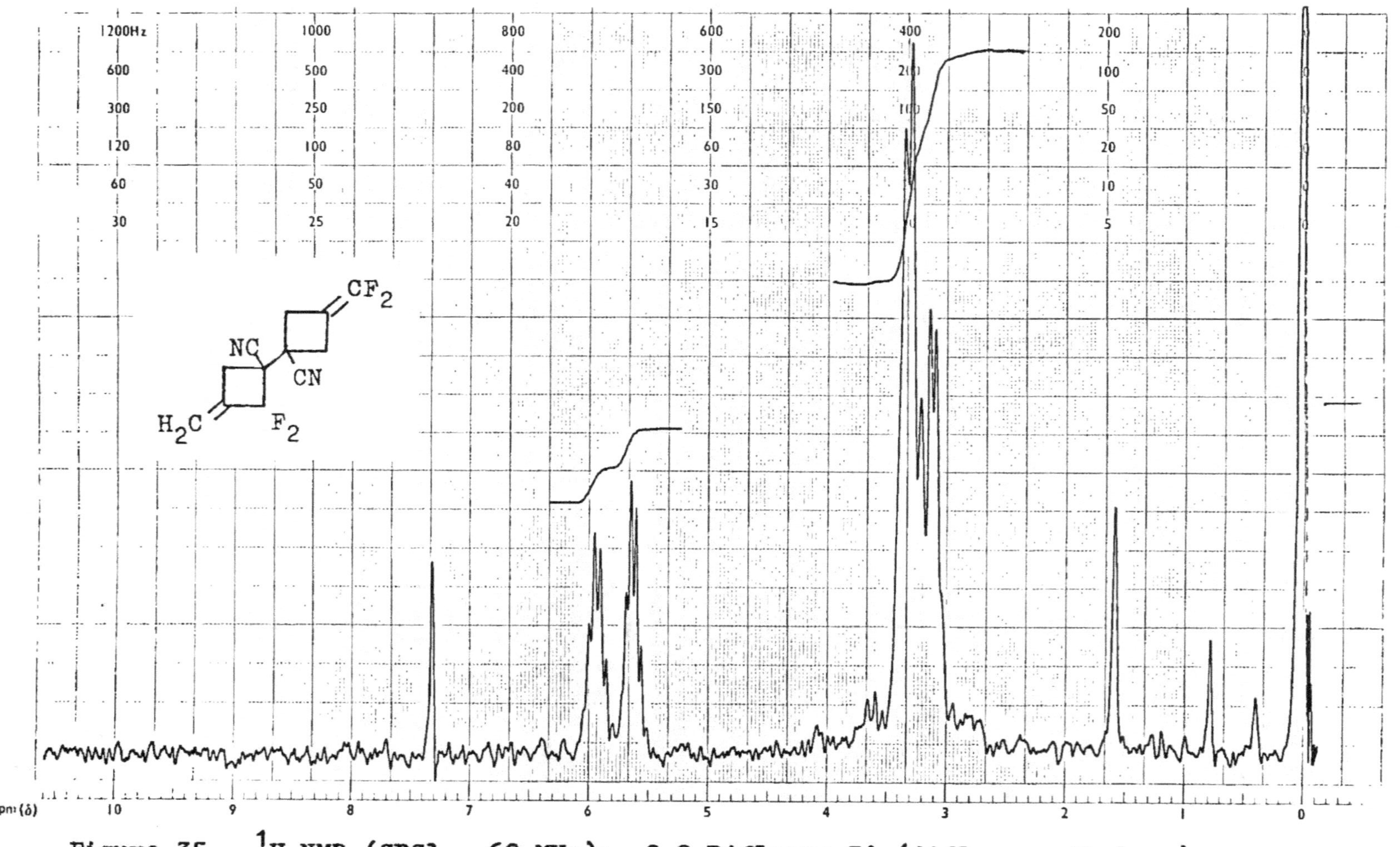

Figure 35. ^{1}H NMR ($CDCl_3$, 60 MHz): 2,2-Difluoro-3'-(difluoromethylene)-3-methylene[1,1'-bicyclobutyl]-1,1'-dicarbonitrile <u>66</u>. Page 95.

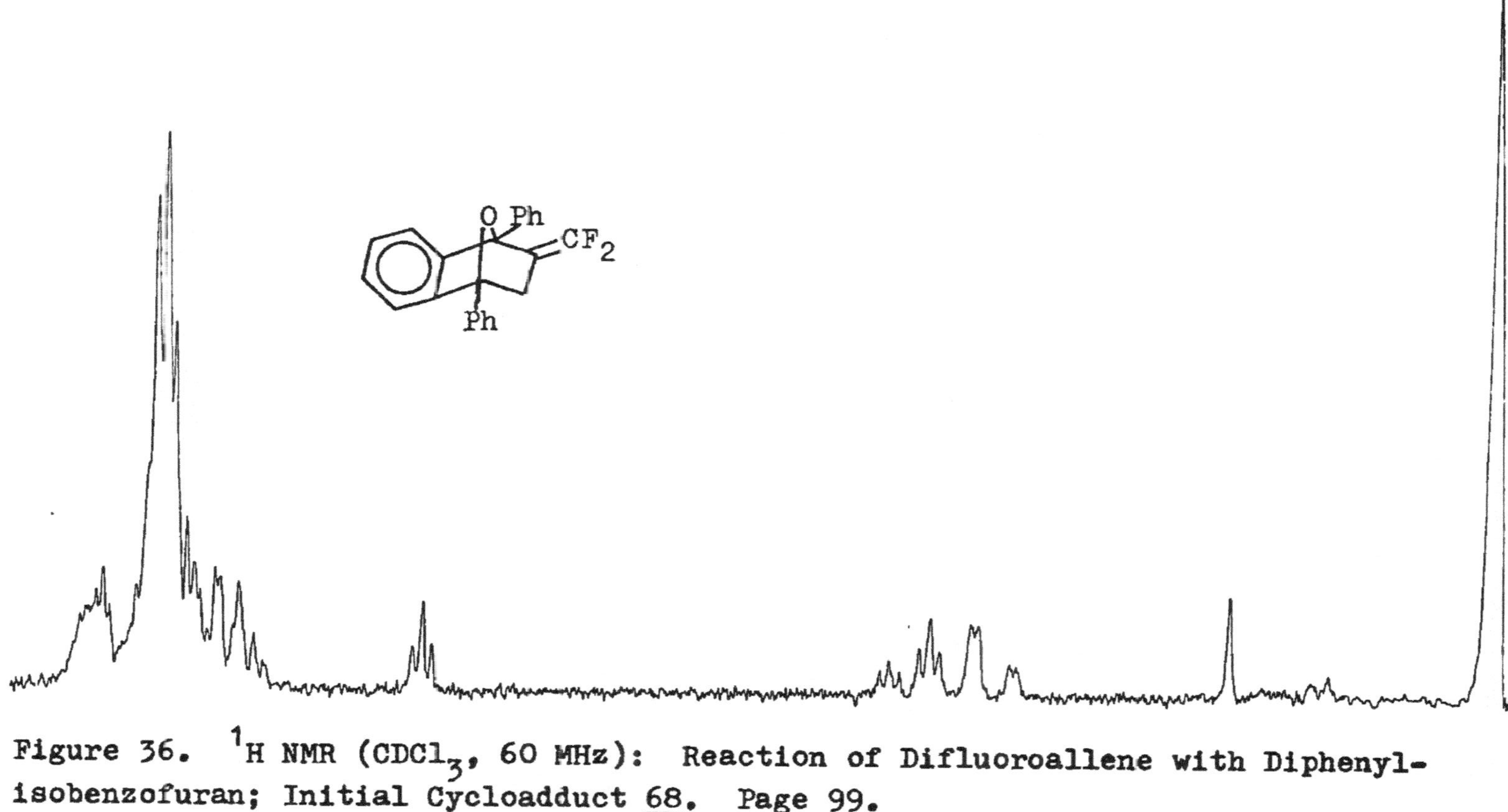

Figure 36. ^{1}H NMR ($CDCl_3$, 60 MHz): Reaction of Difluoroallene with Diphenylisobenzofuran; Initial Cycloadduct 68. Page 99.

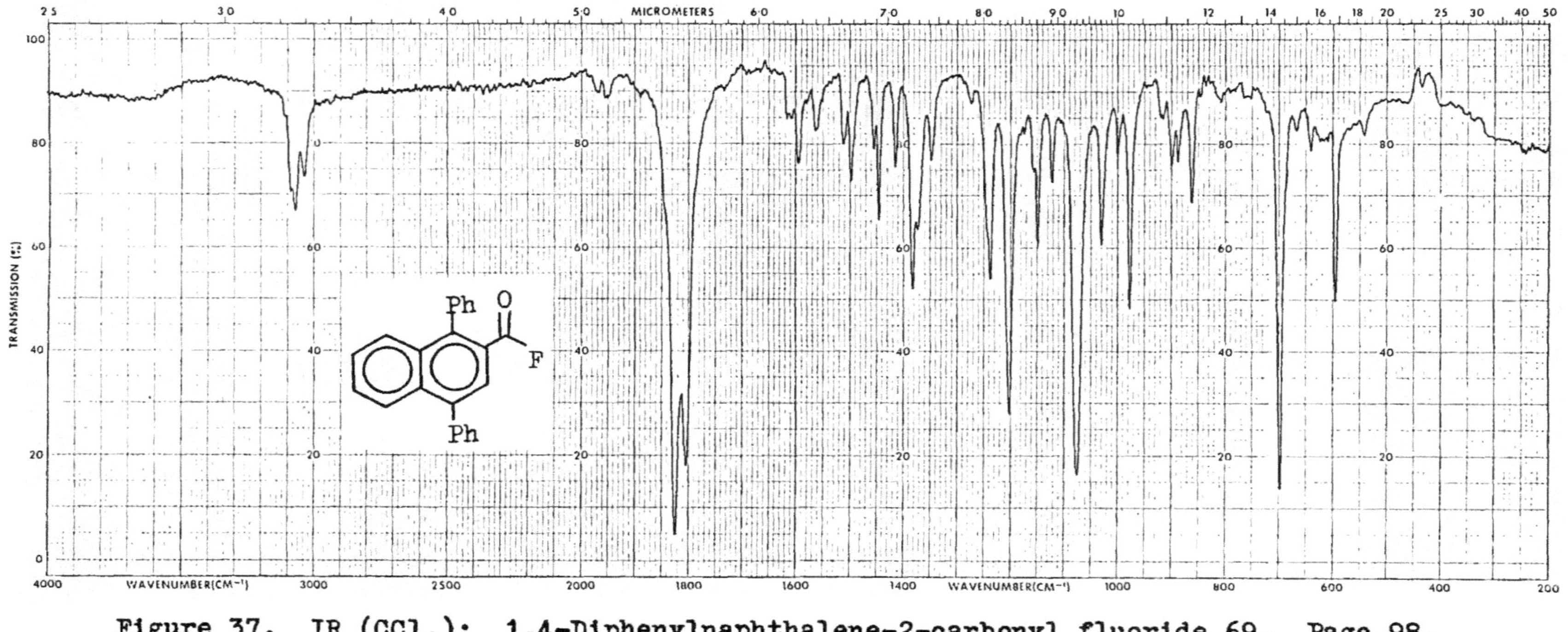

Figure 37. IR (CCl_4): 1,4-Diphenylnaphthalene-2-carbonyl fluoride 69. Page 98.

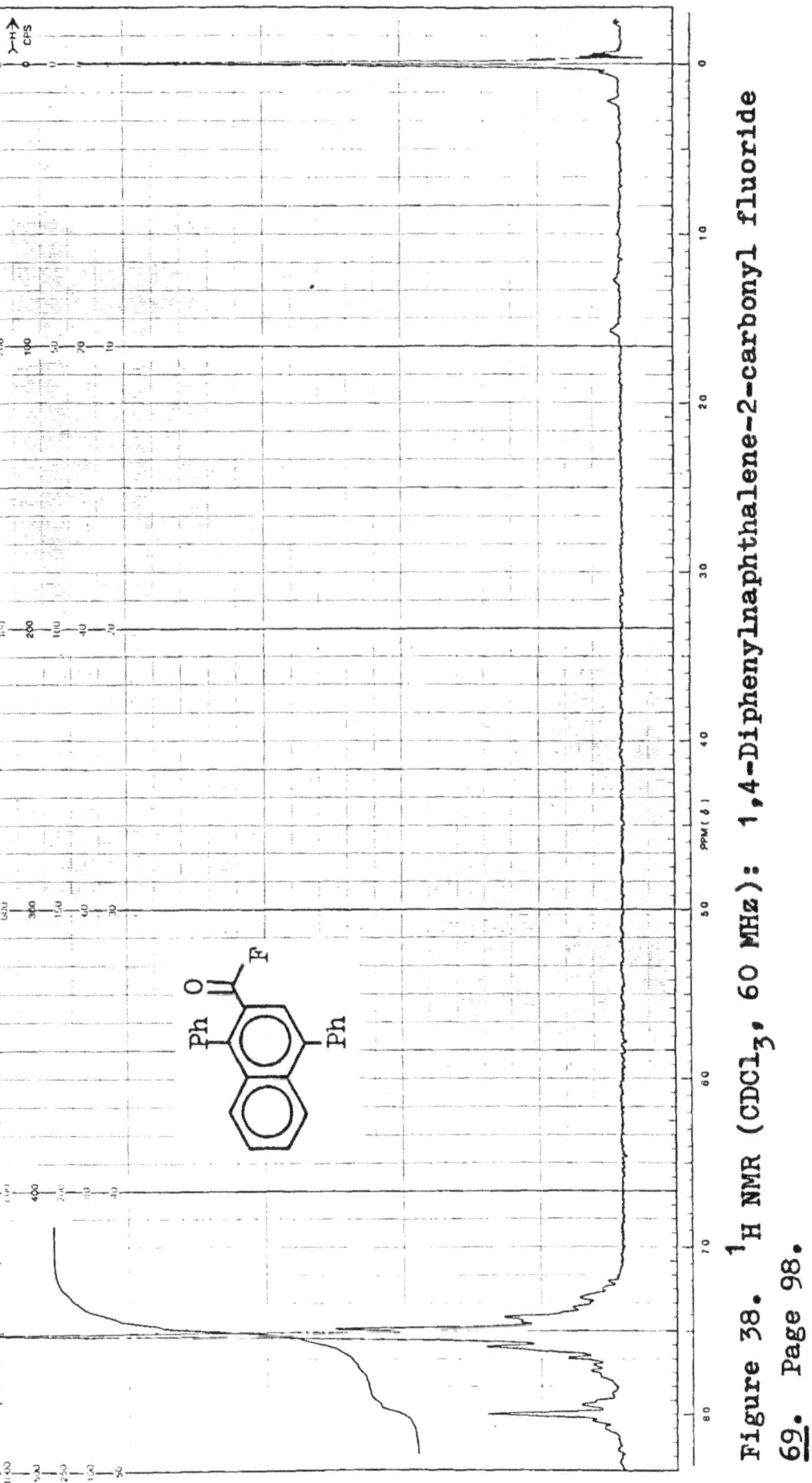

Figure 38. ^{1}H NMR ($CDCl_3$, 60 MHz): 1,4-Diphenylnaphthalene-2-carbonyl fluoride 69. Page 98.

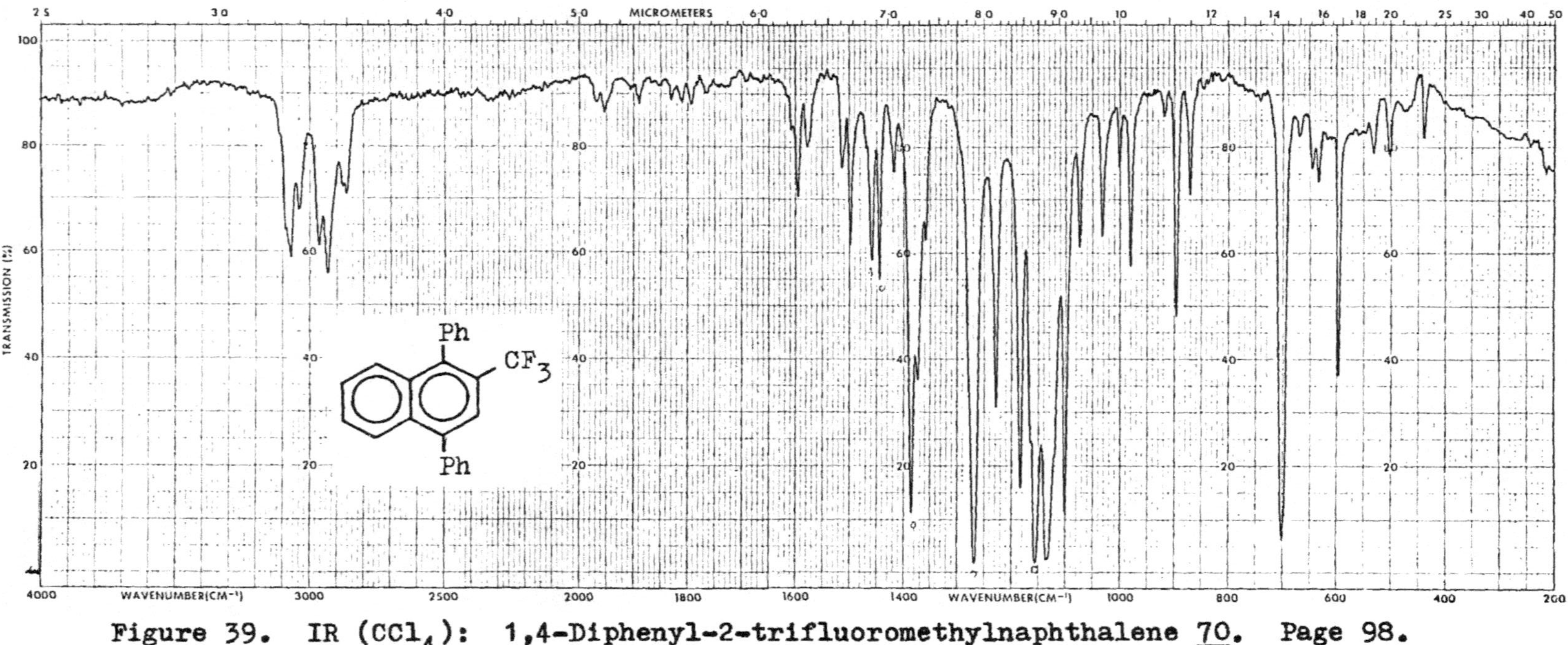

Figure 39. IR (CCl_4): 1,4-Diphenyl-2-trifluoromethylnaphthalene <u>70</u>. Page 98.

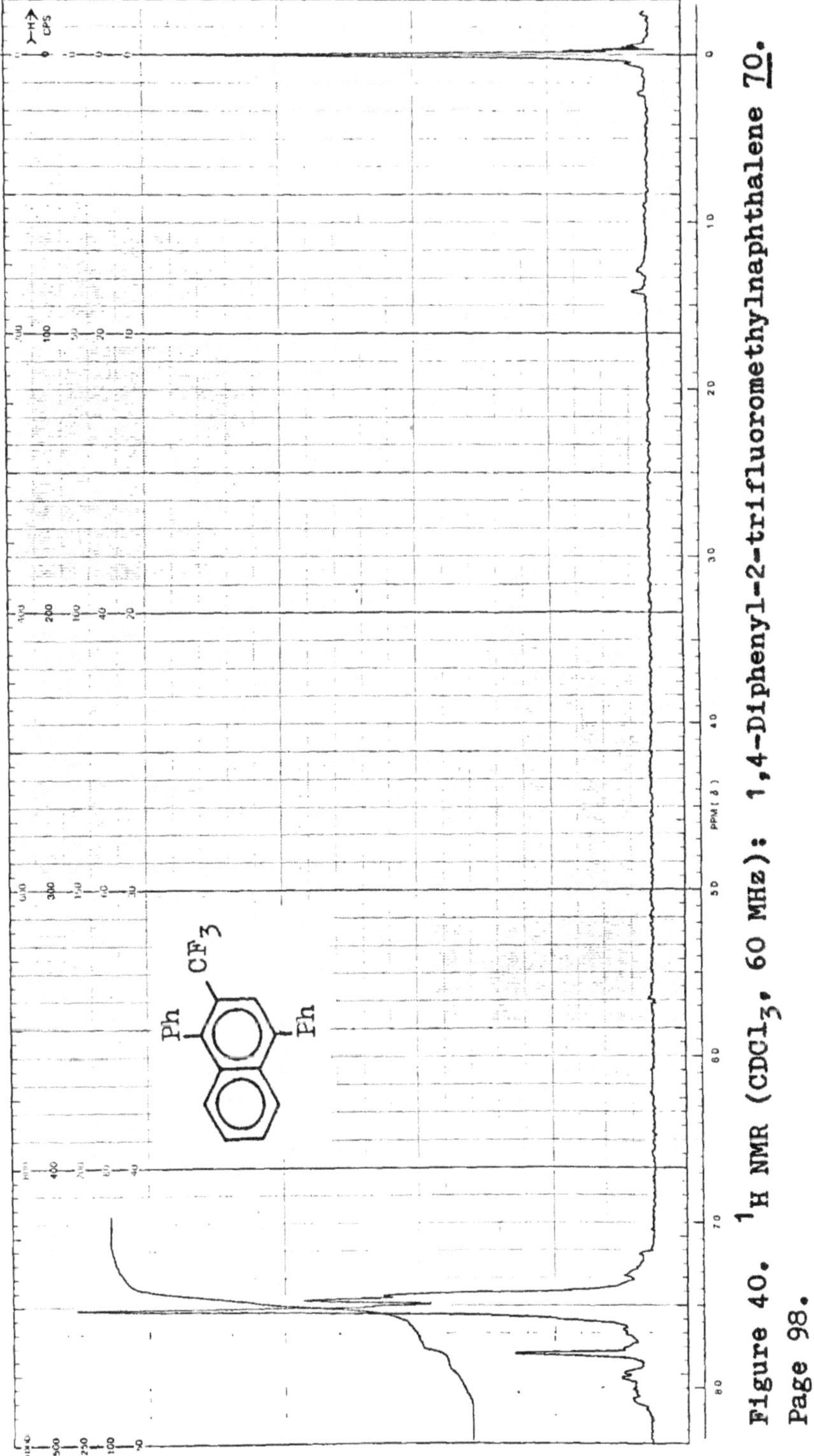

Figure 40. ^{1}H **NMR ($CDCl_3$, 60 MHz): 1,4-Diphenyl-2-trifluoromethylnaphthalene <u>70</u>. Page 98.**

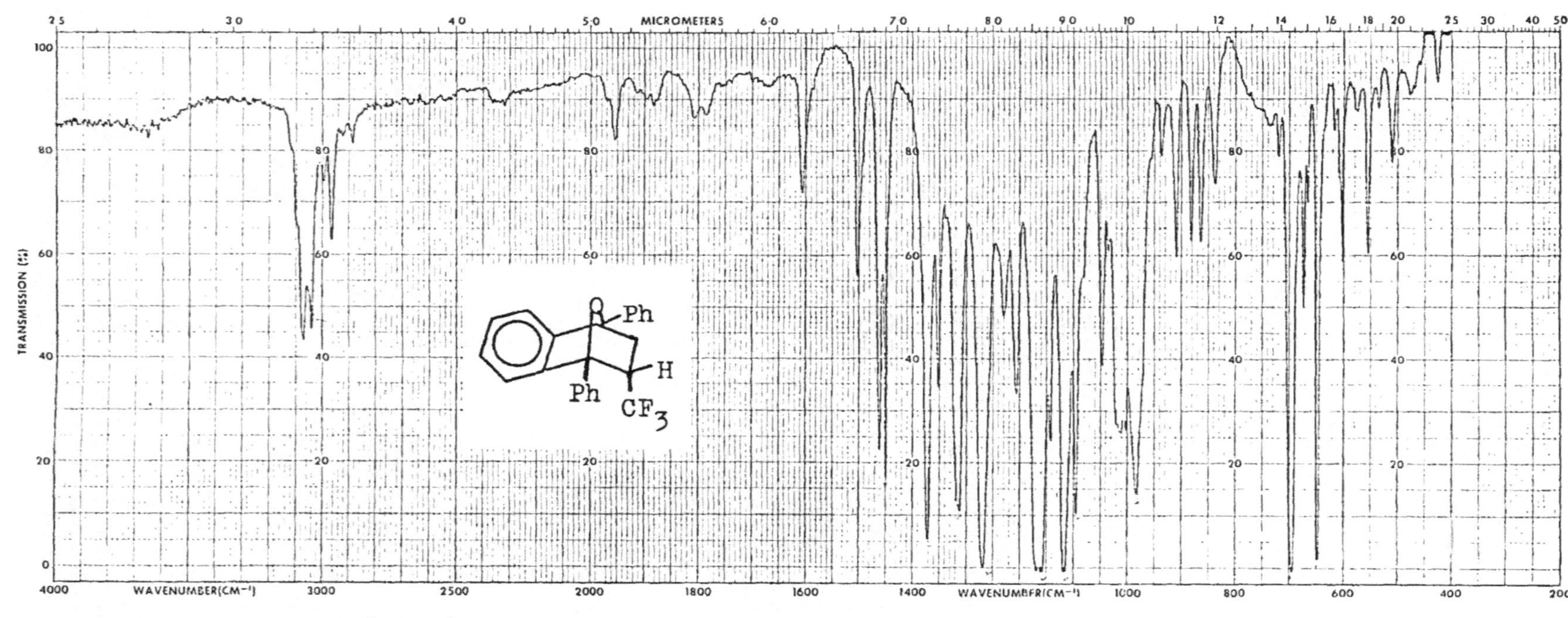

Figure 41. IR (CCl_4): Endo-1,2,3,4-Tetrahydro-1,4-diphenyl-2-trifluoromethyl-1,4-epoxynaphthalene <u>72</u>. Page 100.

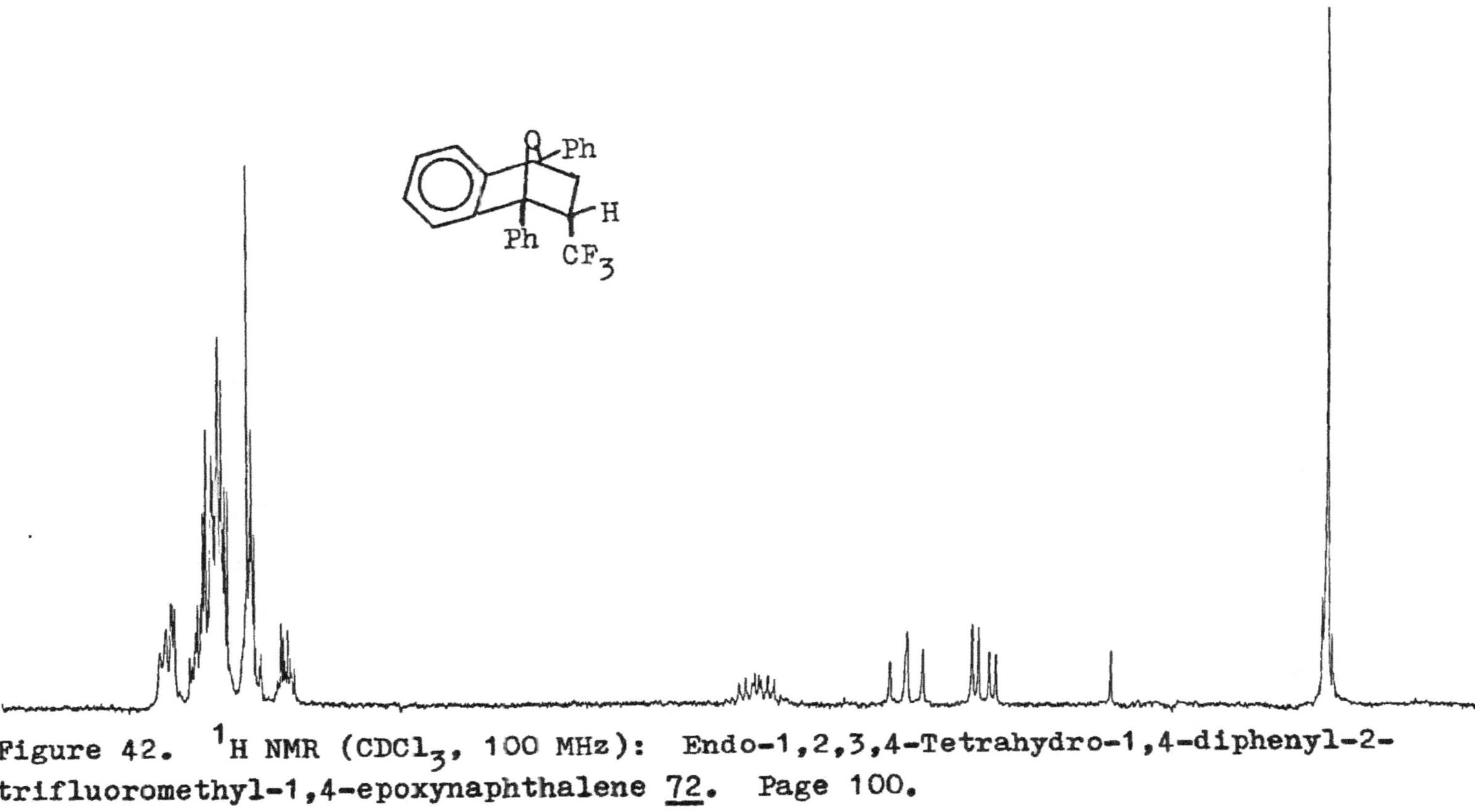

Figure 42. ^{1}H NMR ($CDCl_3$, 100 MHz): Endo-1,2,3,4-Tetrahydro-1,4-diphenyl-2-trifluoromethyl-1,4-epoxynaphthalene 72. Page 100.

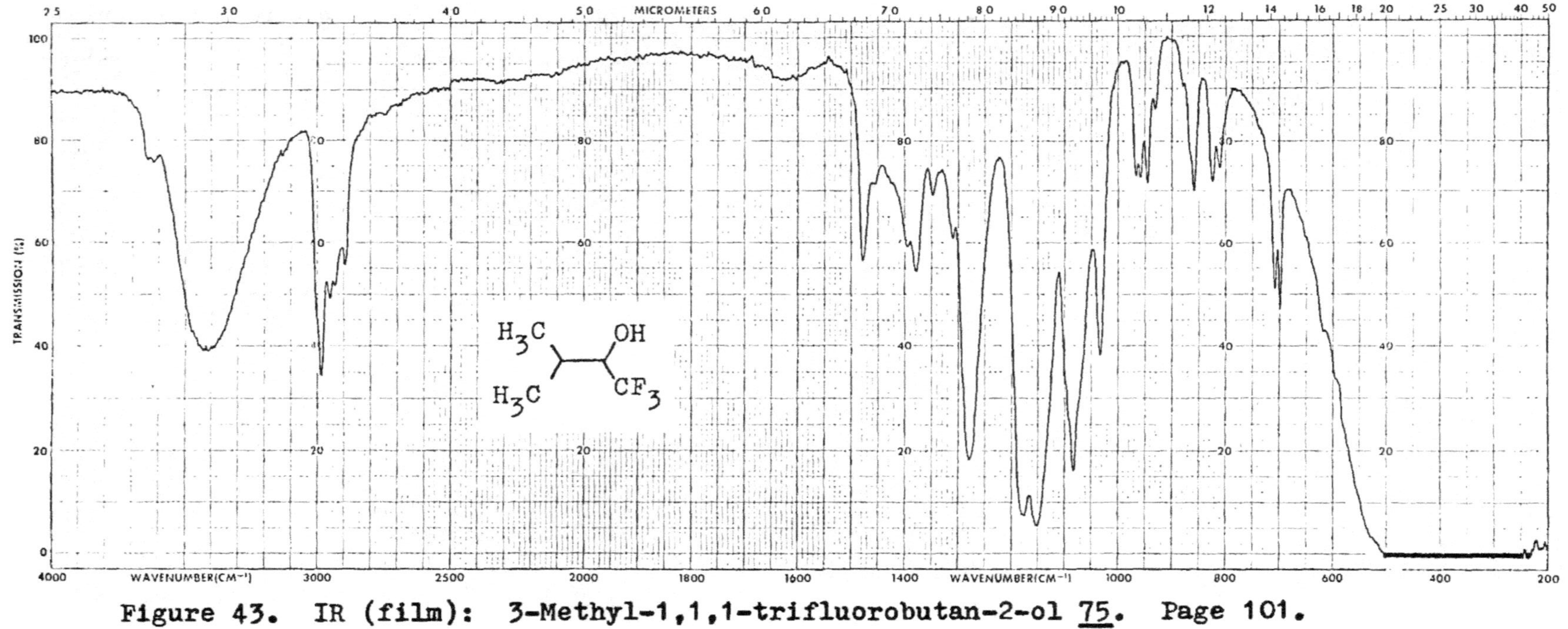

Figure 43. IR (film): 3-Methyl-1,1,1-trifluorobutan-2-ol <u>75</u>. Page 101.

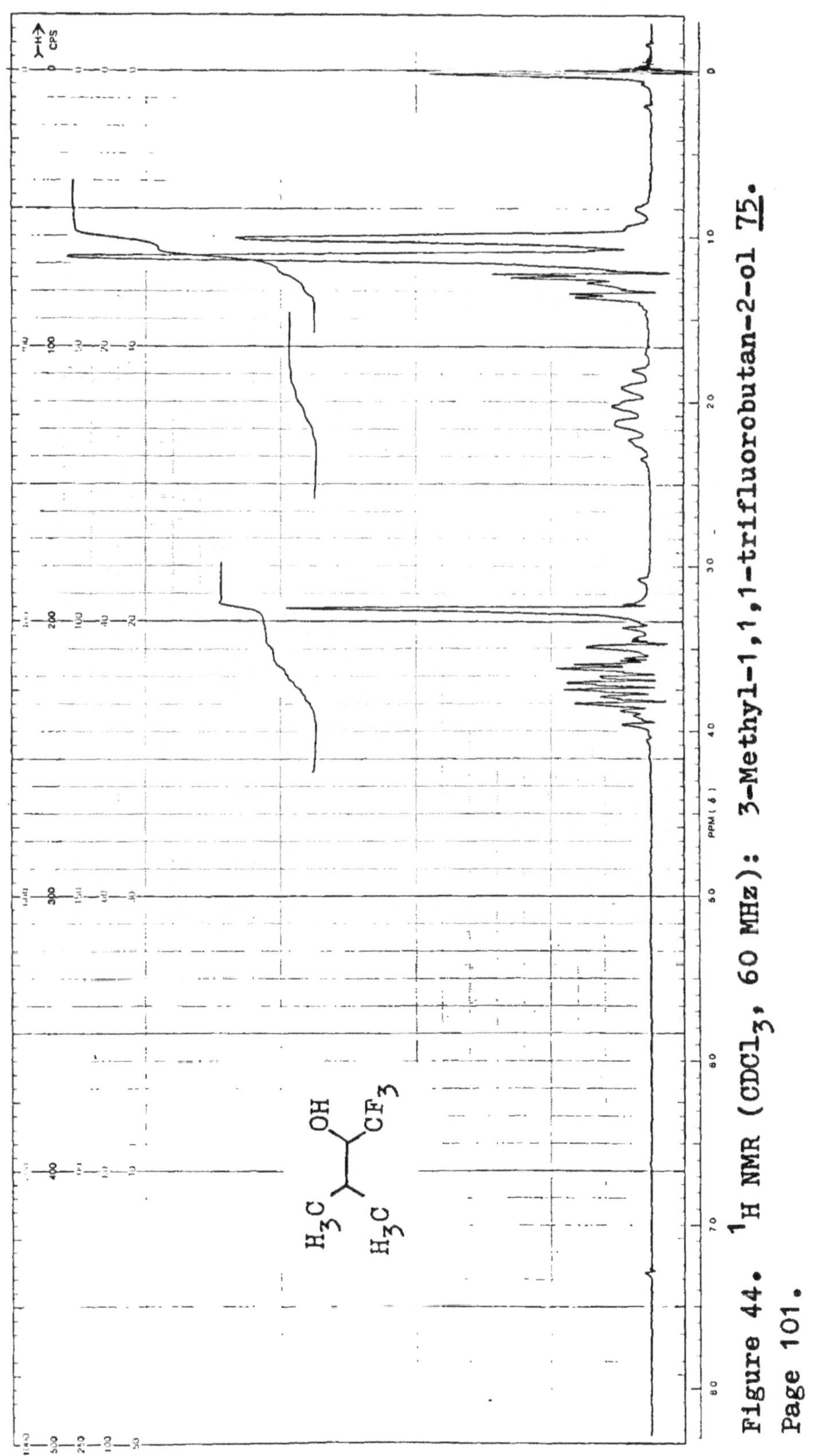

Figure 44. ^{1}H NMR ($CDCl_3$, 60 MHz): 3-Methyl-1,1,1-trifluorobutan-2-ol <u>75</u>. Page 101.

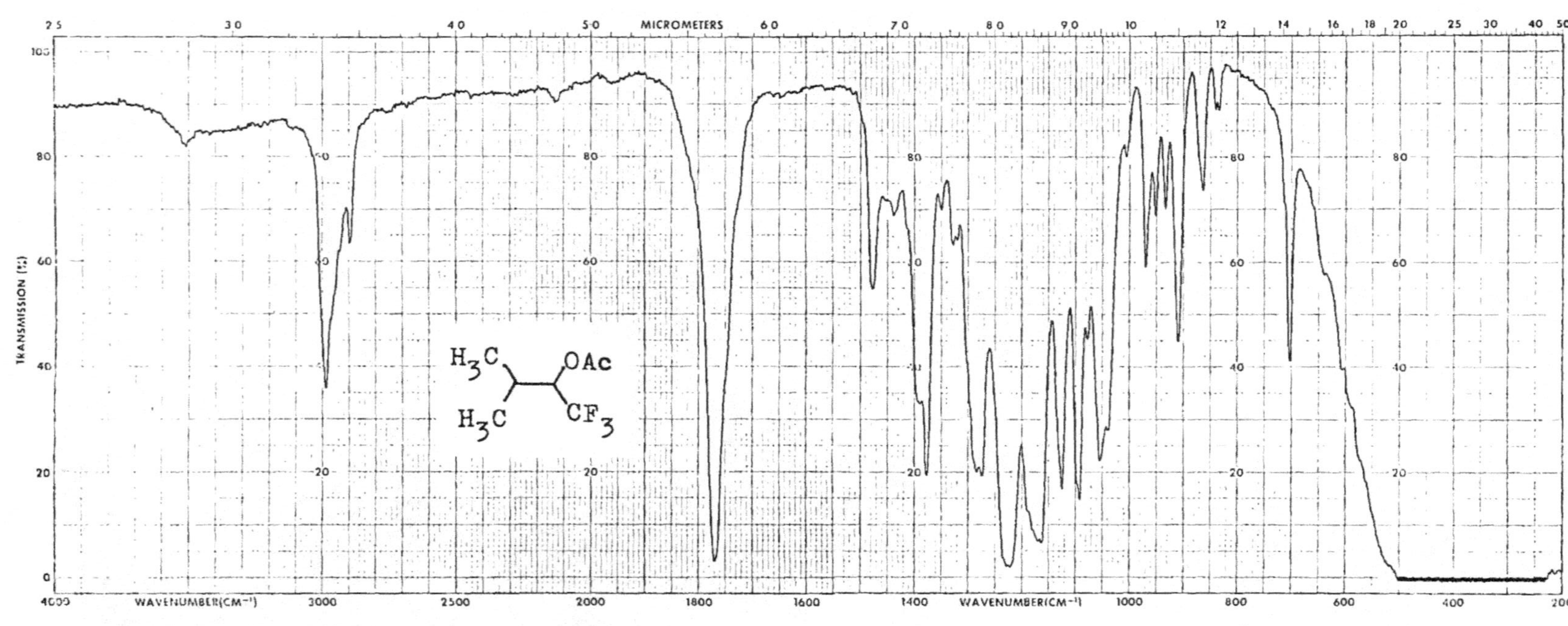

Figure 45. IR (film): 3-Methyl-1,1,1-trifluorobutan-2-ol acetate <u>76</u>. Page 102.

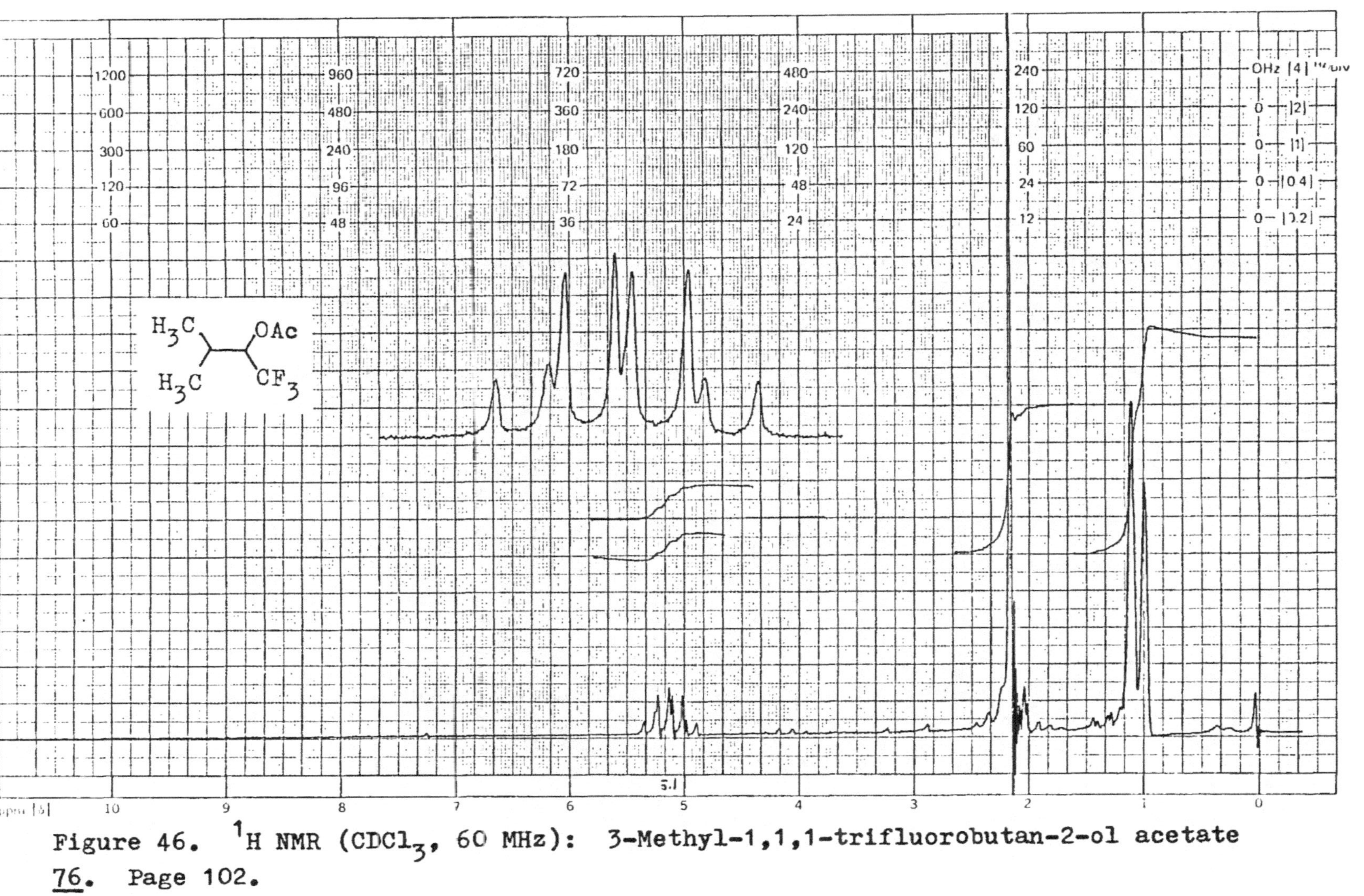

Figure 46. ^1H NMR ($CDCl_3$, 60 MHz): 3-Methyl-1,1,1-trifluorobutan-2-ol acetate <u>76</u>. Page 102.

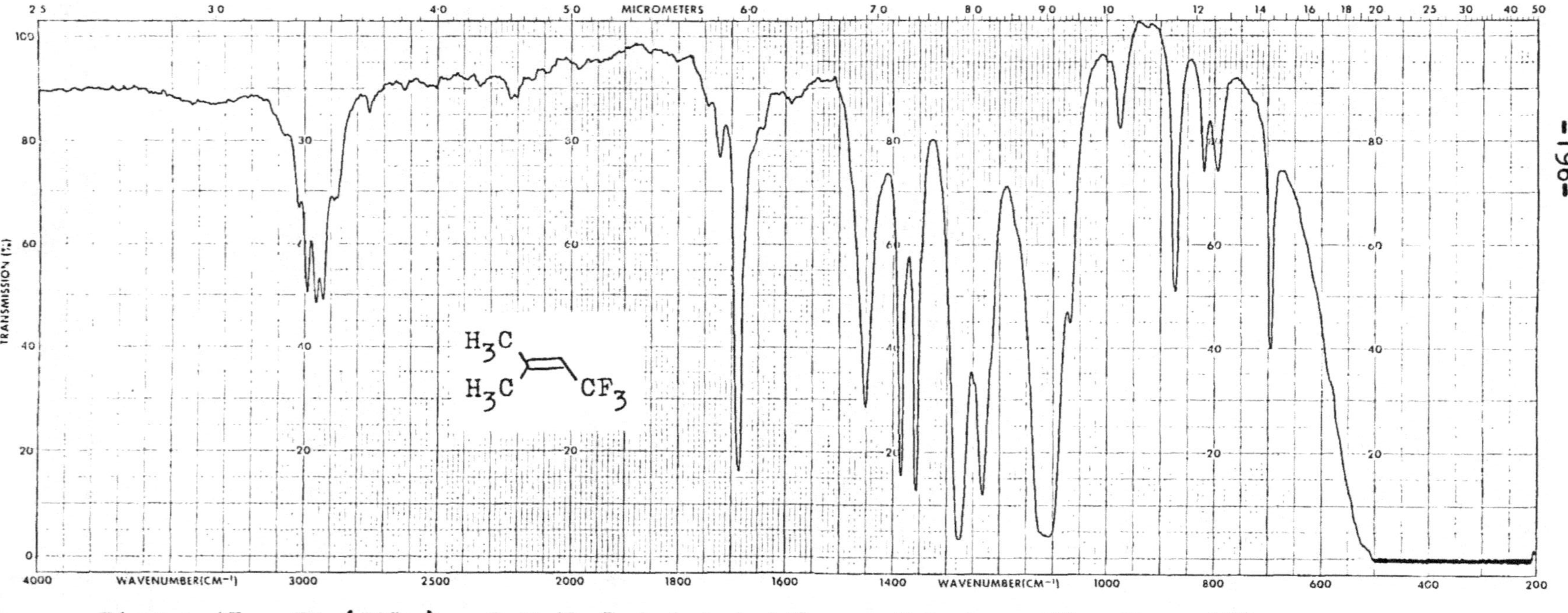

Figure 47. IR (film): 2-Methyl-4,4,4-trifluoro-2-butene <u>77</u>. Page 103.

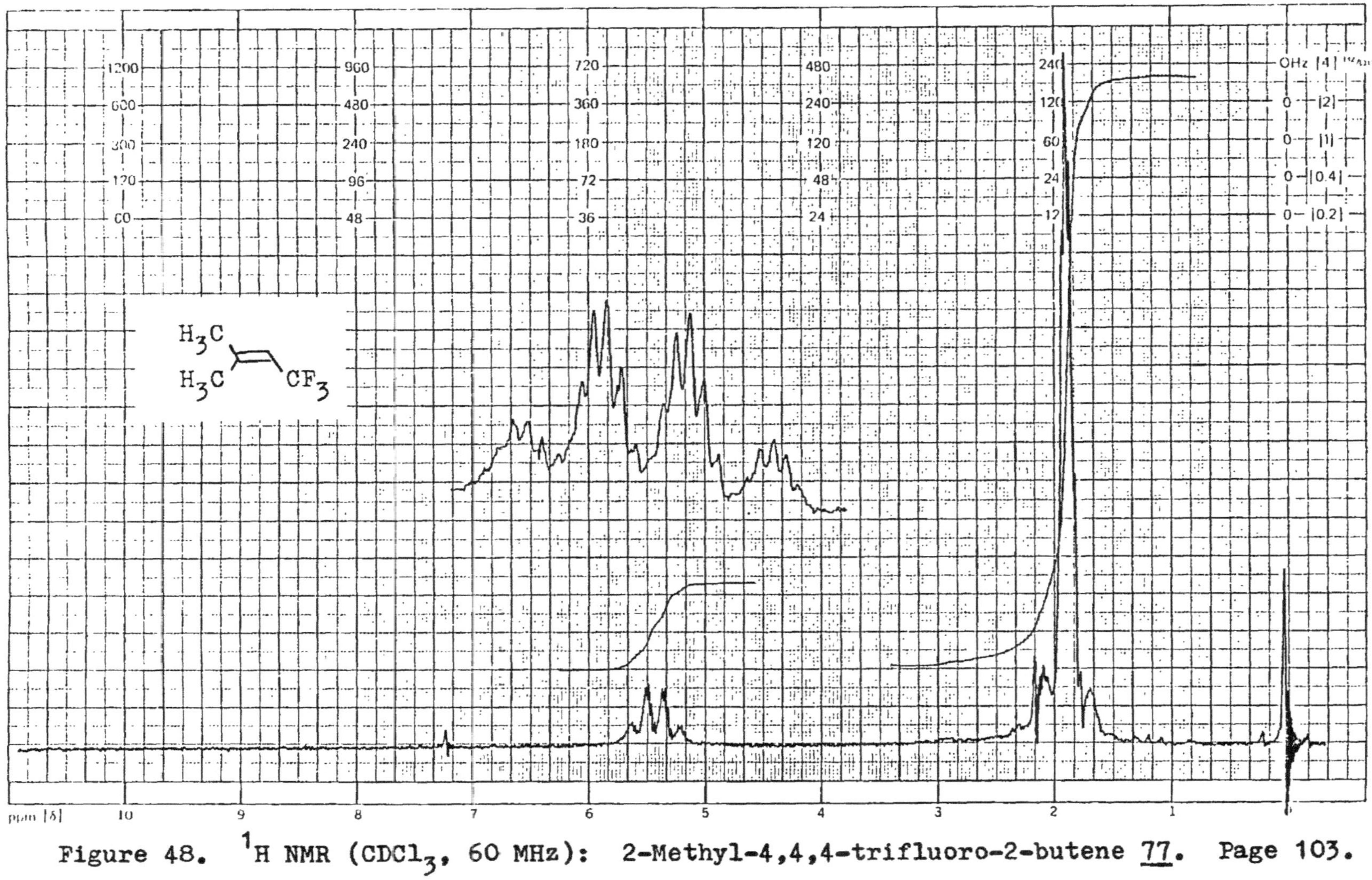

Figure 48. ^{1}H NMR ($CDCl_3$, 60 MHz): 2-Methyl-4,4,4-trifluoro-2-butene 77. Page 103.

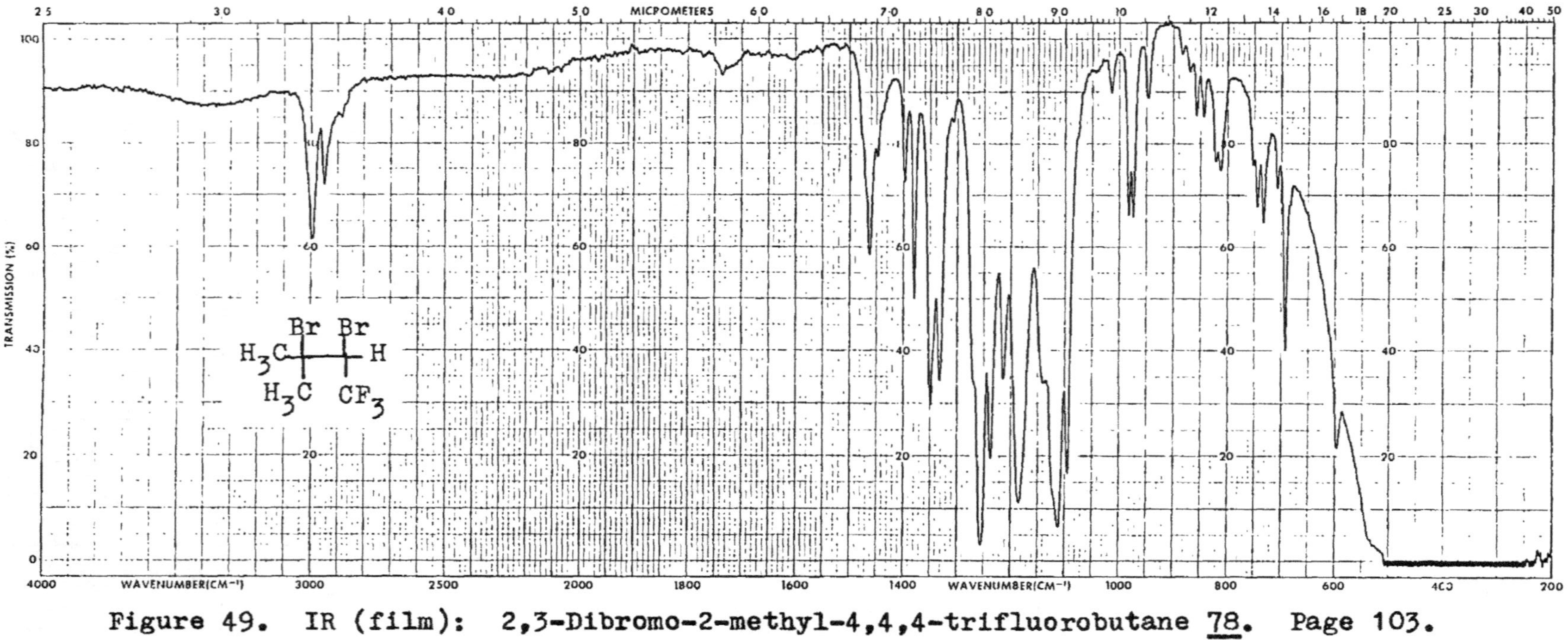

Figure 49. IR (film): 2,3-Dibromo-2-methyl-4,4,4-trifluorobutane <u>78</u>. Page 103.

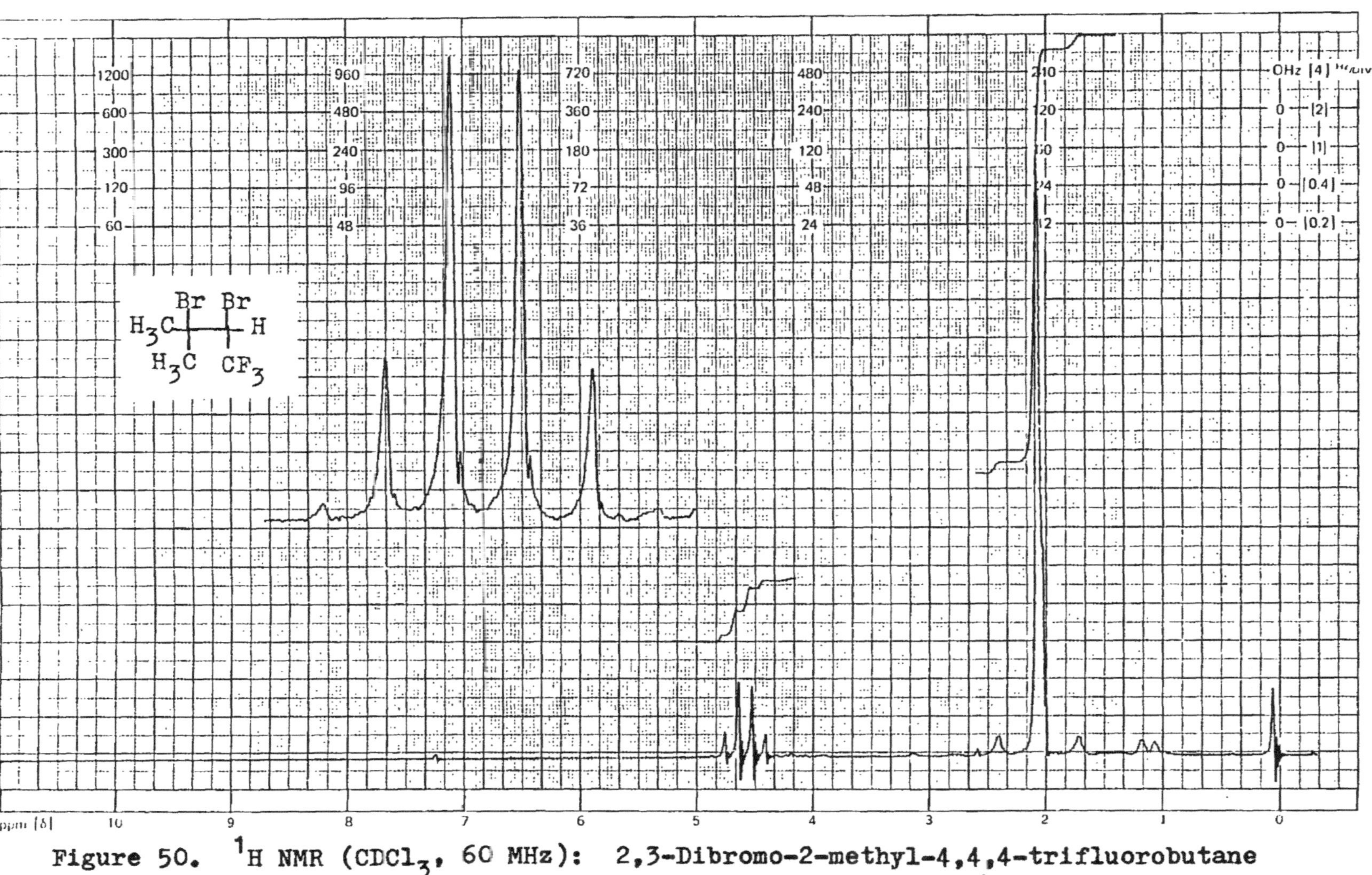

Figure 50. ^{1}H NMR ($CDCl_3$, 60 MHz): 2,3-Dibromo-2-methyl-4,4,4-trifluorobutane <u>78</u>. Page 103.

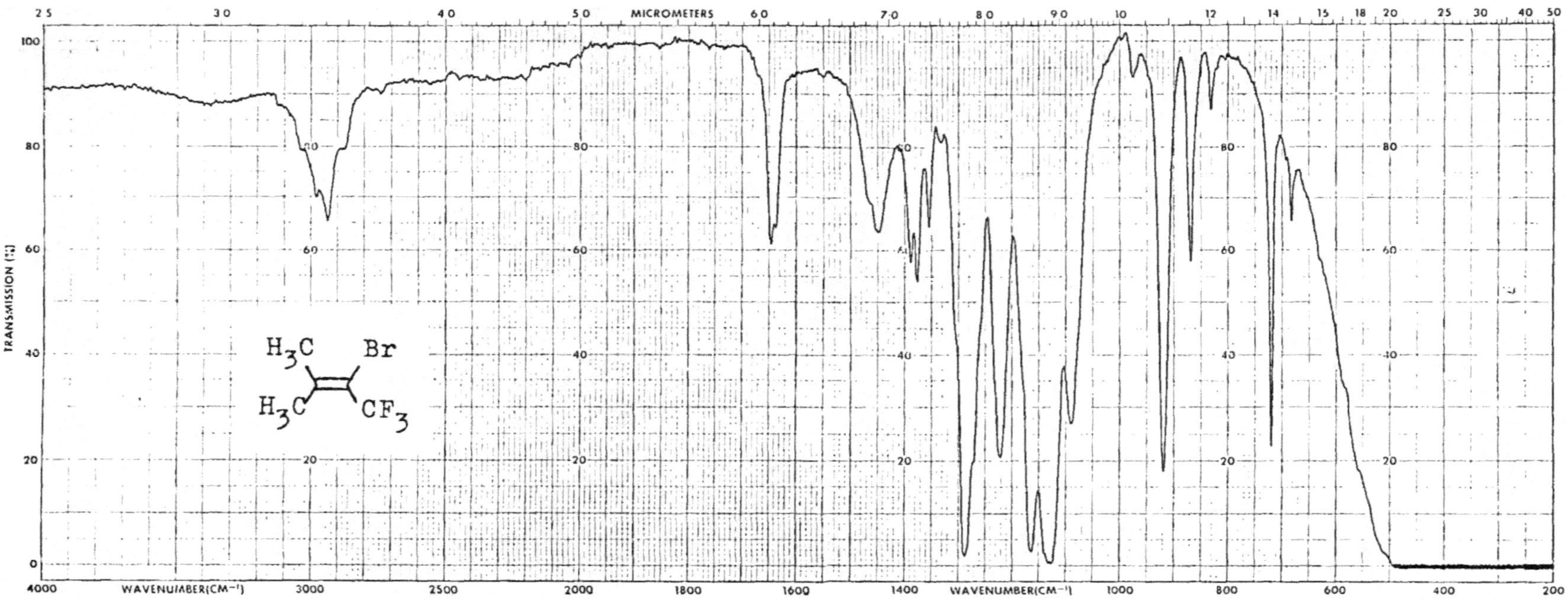

Figure 51. IR (film): 3-Bromo-2-methyl-4,4,4-trifluoro-2-butene 79. Page 104.

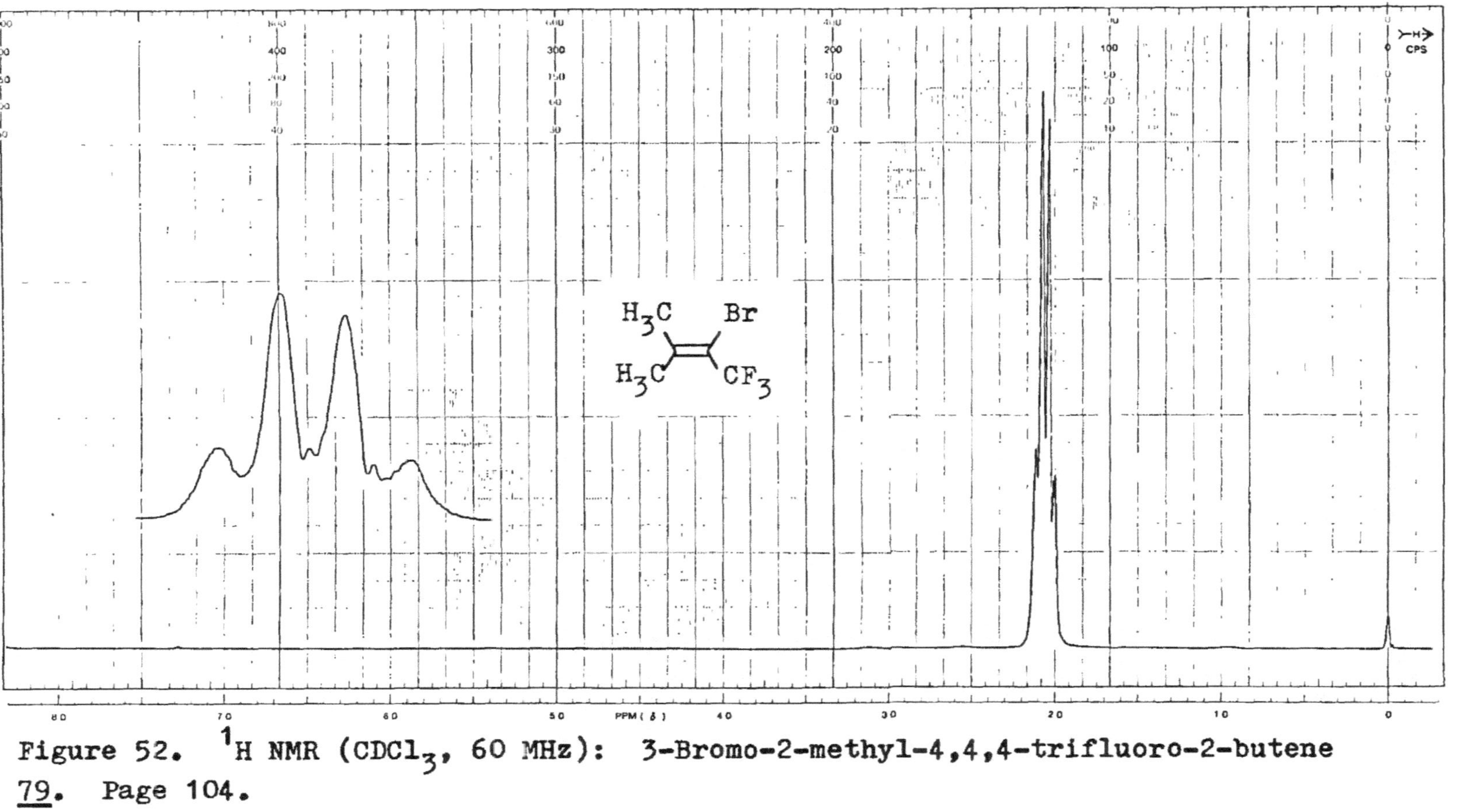

Figure 52. ^{1}H NMR ($CDCl_3$, 60 MHz): 3-Bromo-2-methyl-4,4,4-trifluoro-2-butene 79. Page 104.

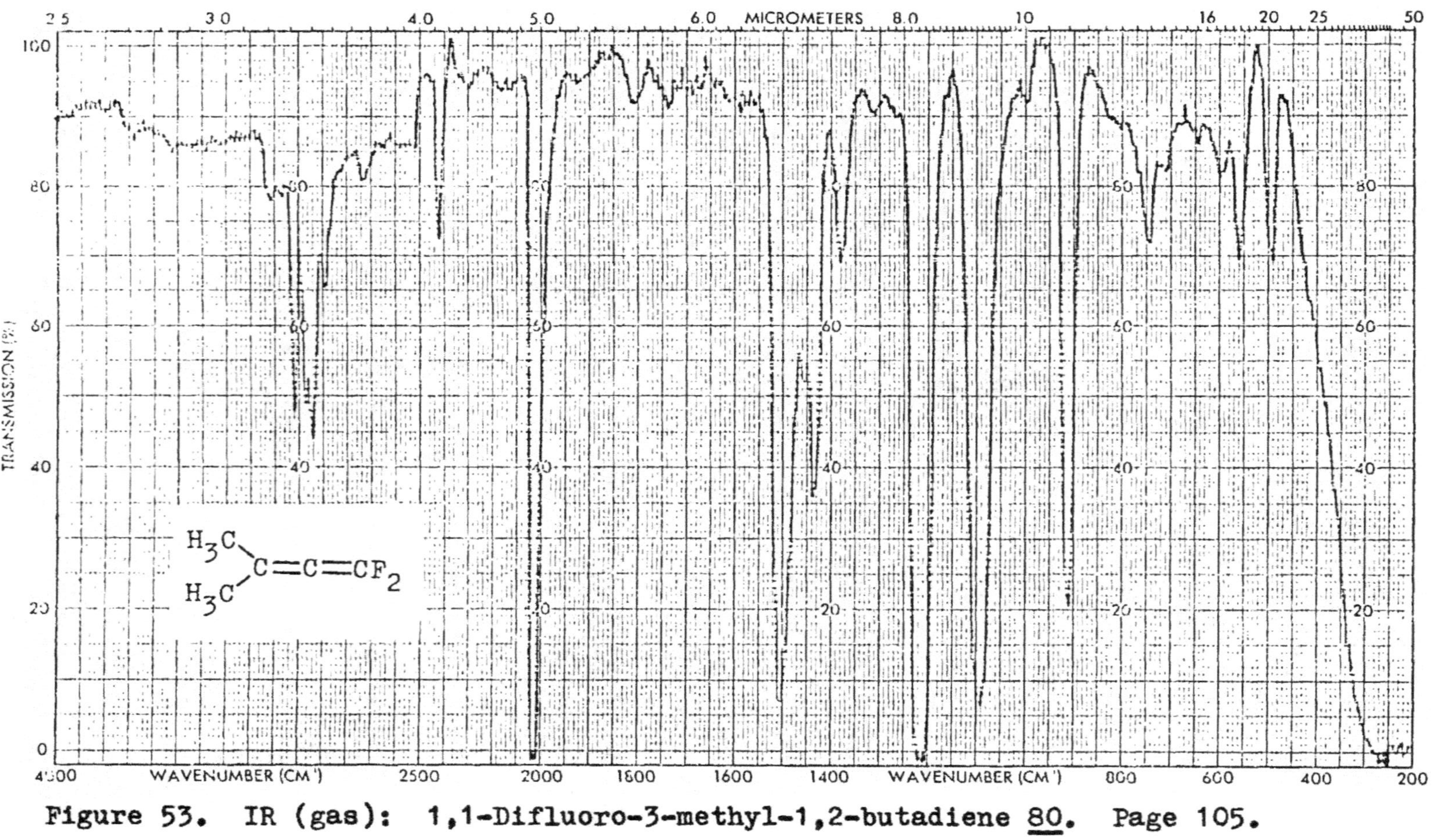

Figure 53. IR (gas): 1,1-Difluoro-3-methyl-1,2-butadiene <u>80</u>. Page 105.

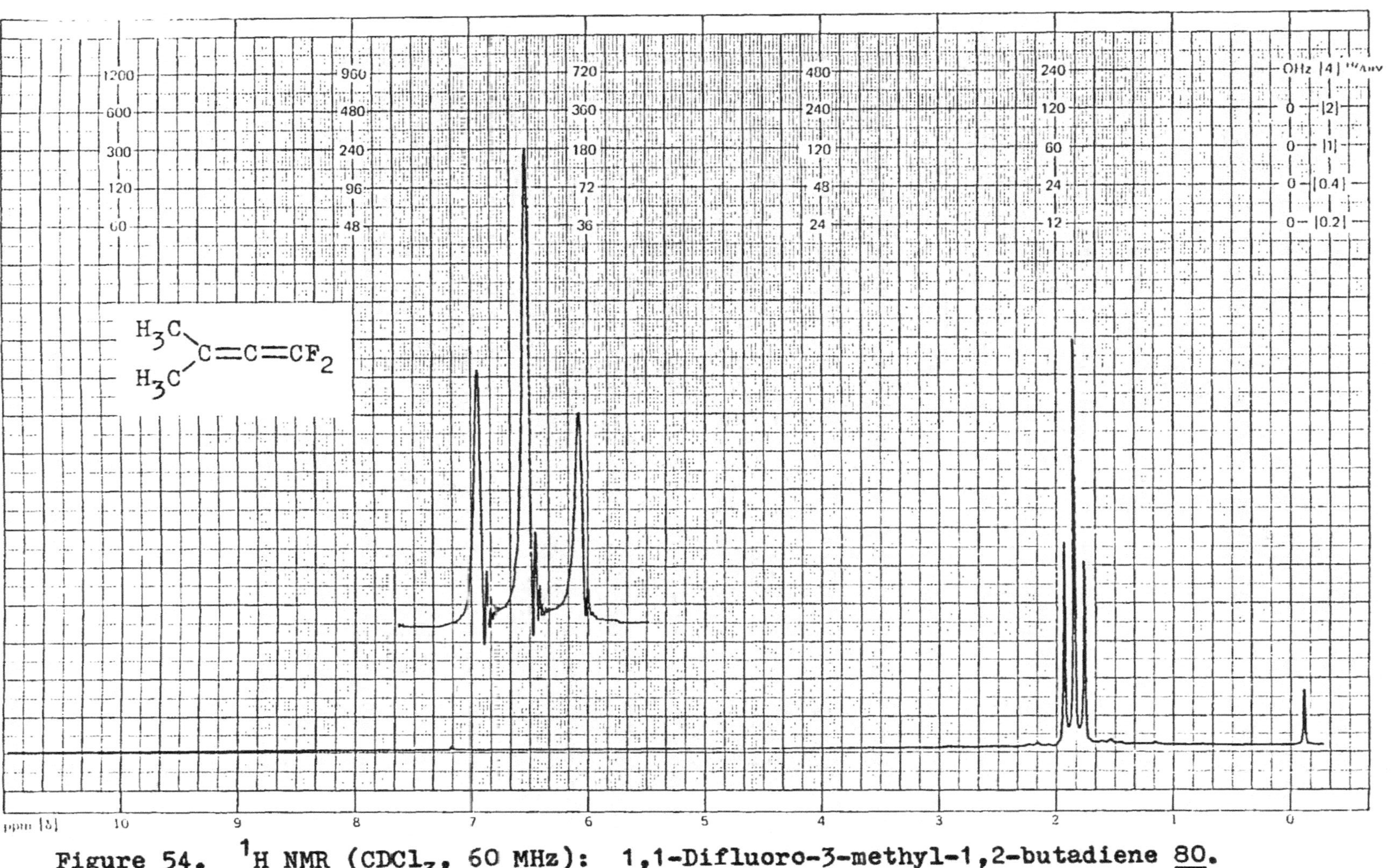

Figure 54. ^1H NMR ($CDCl_3$, 60 MHz): 1,1-Difluoro-3-methyl-1,2-butadiene 80. Page 105.

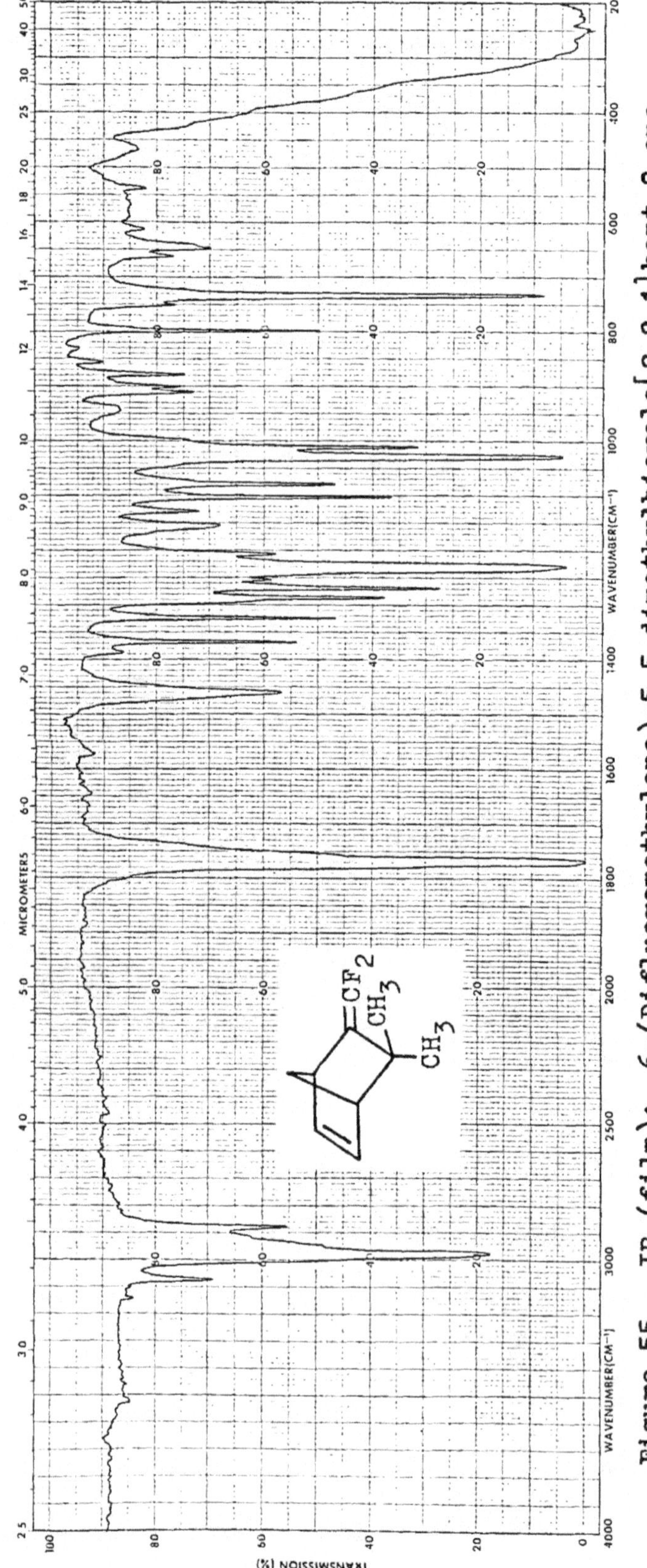

Figure 55. IR (film): 6-(Difluoromethylene)-5,5-dimethylbicyclo[2.2.1]hept-2-ene 81. Page 106.

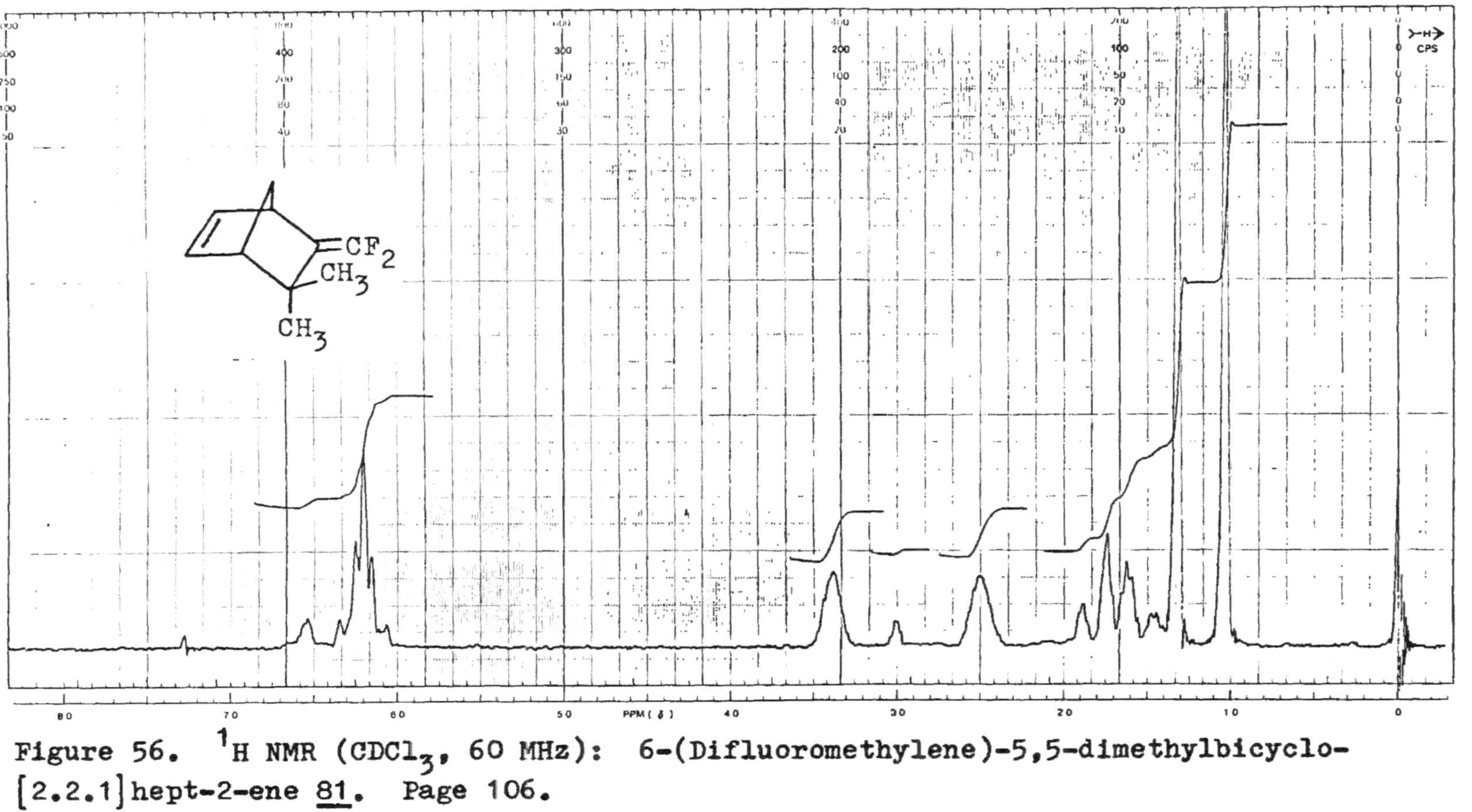

Figure 56. ^{1}H NMR ($CDCl_3$, 60 MHz): 6-(Difluoromethylene)-5,5-dimethylbicyclo-[2.2.1]hept-2-ene <u>81</u>. Page 106.

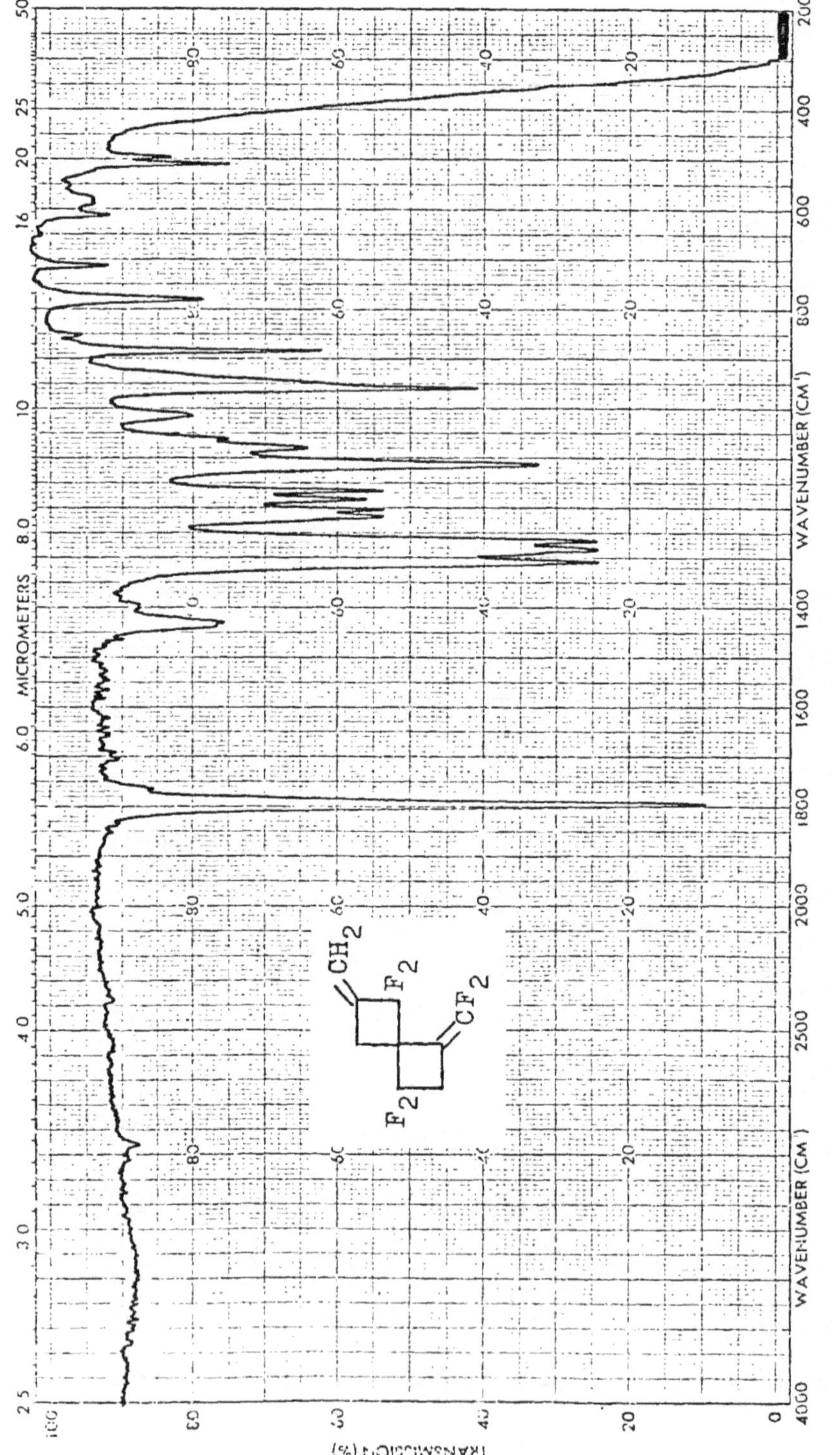

Figure 57. IR (film): 7-(Difluoromethylene)-2-methylene-1,1,5,5-tetrafluorospiro[3.3]heptane <u>82</u>. Page 107.

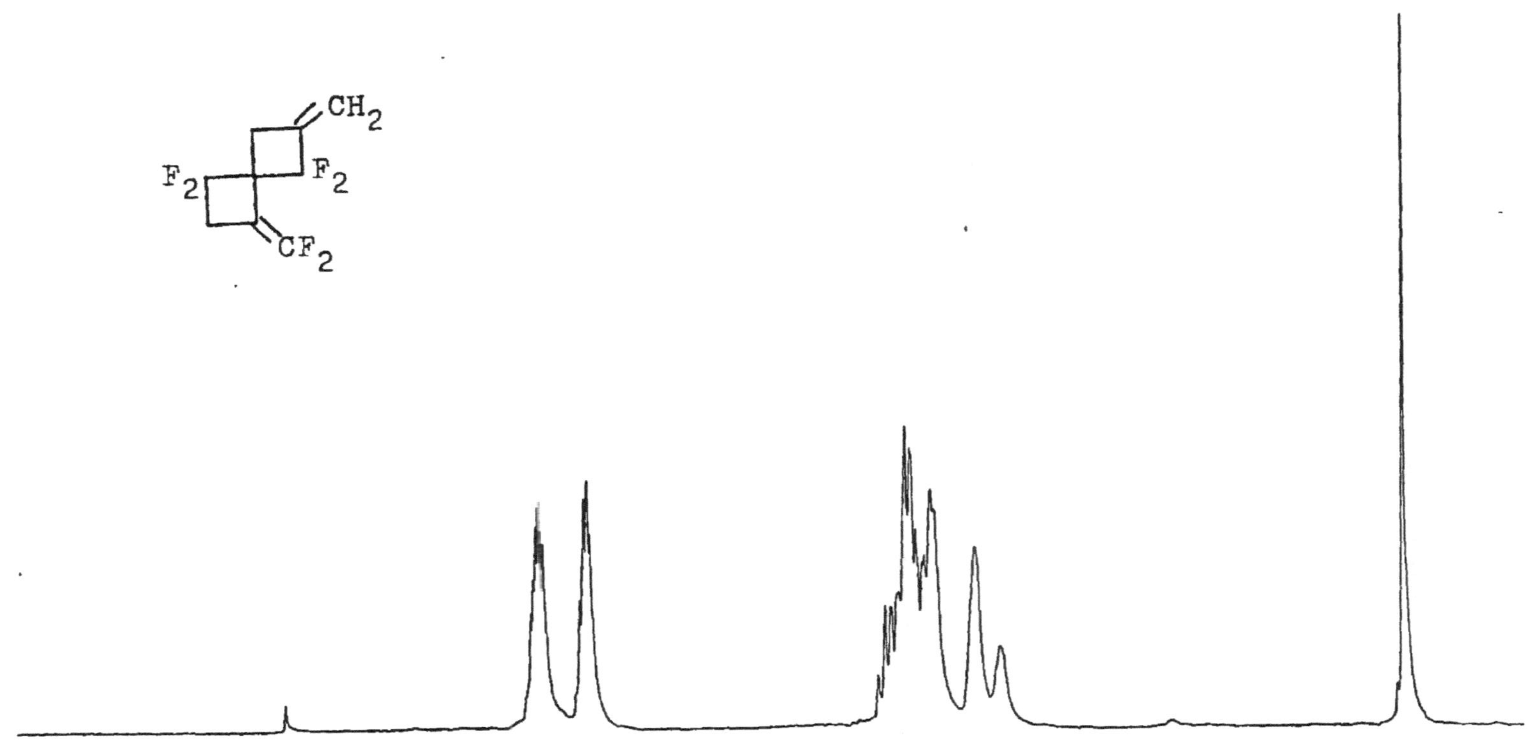

Figure 58. ^{1}H NMR ($CDCl_3$, 100 MHz): 7-(Difluoromethylene)-2-methylene-1,1,5,5-tetrafluorospiro[3.3]heptane 82. Page 107.

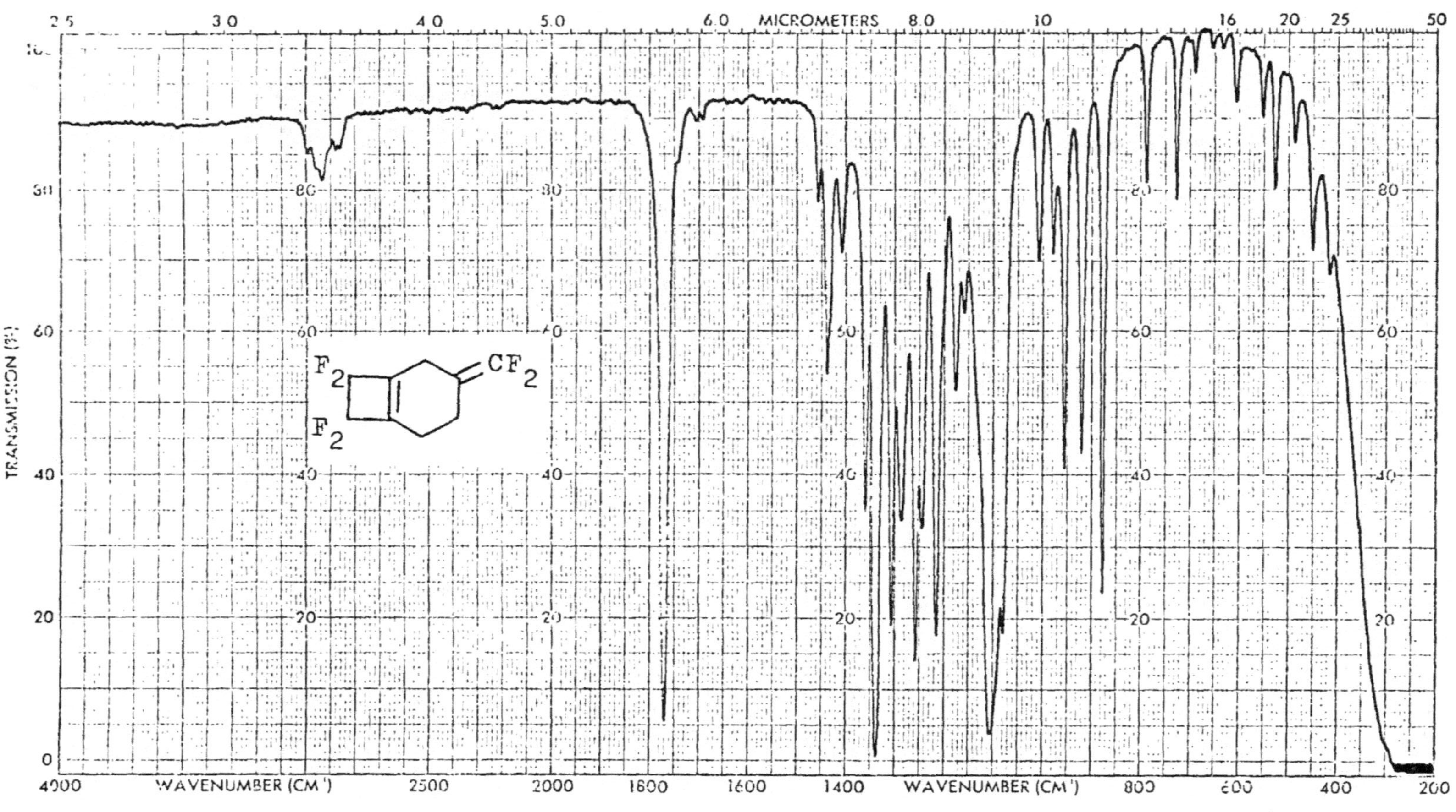

Figure 59. IR (film): 3-(Difluoromethylene)-7,7,8,8-tetrafluorobicyclo[4.2.0]-oct-1(6)-ene <u>83</u>. Page 107.

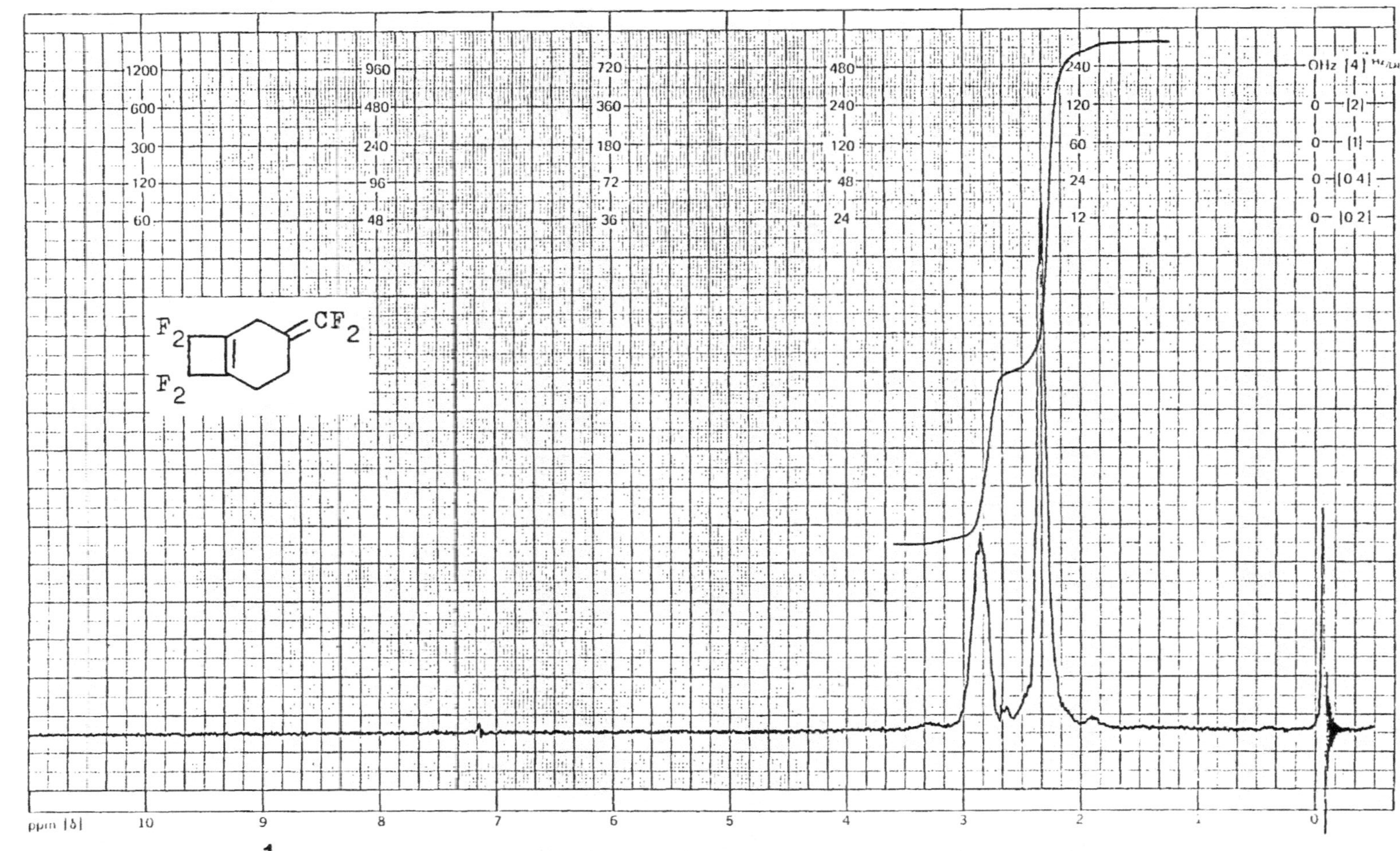

Figure 60. ^{1}H NMR ($CDCl_3$, 60 MHz): 3-(Difluoromethylene)-7,7,8,8-tetrafluorobicyclo[4.2.0]oct-1(6)-ene <u>83</u>. Page 107.

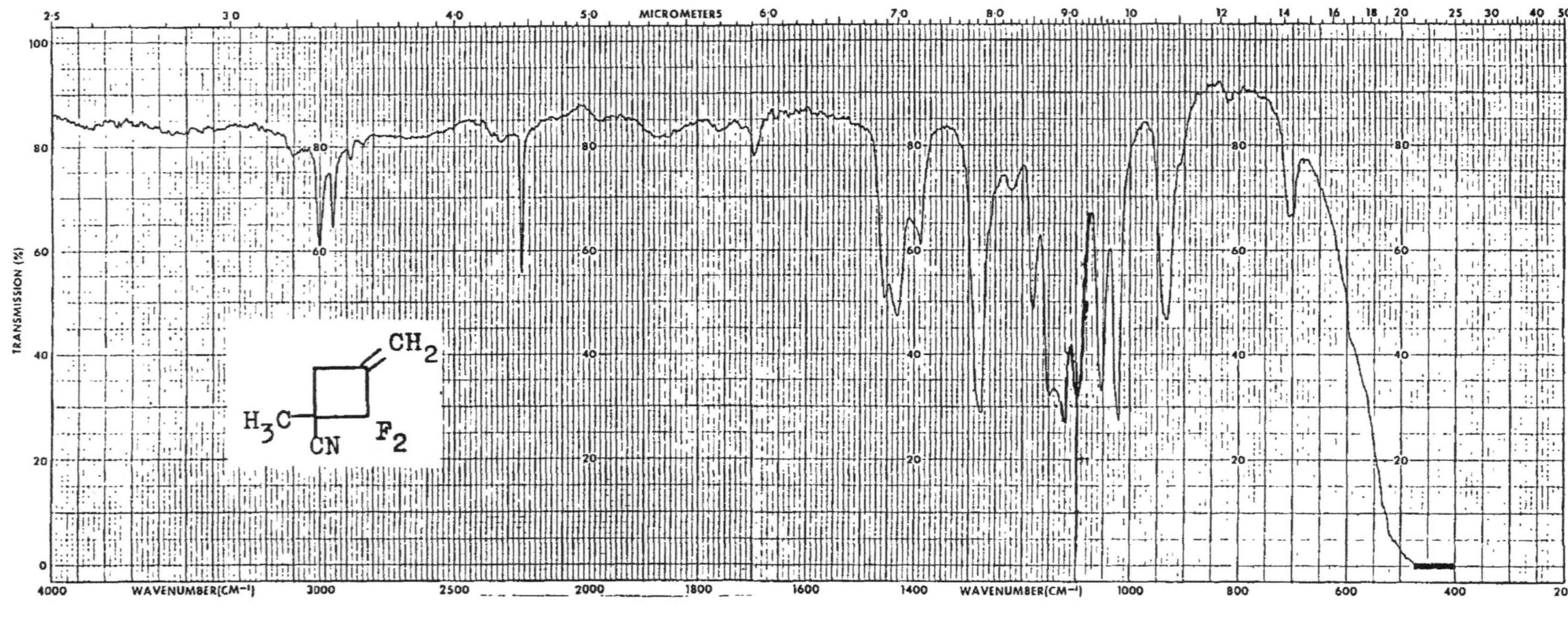

Figure 61. IR (film): 2,2-Difluoro-1-methyl-3-methylenecyclobutanecarbonitrile <u>86</u>. Page 110.

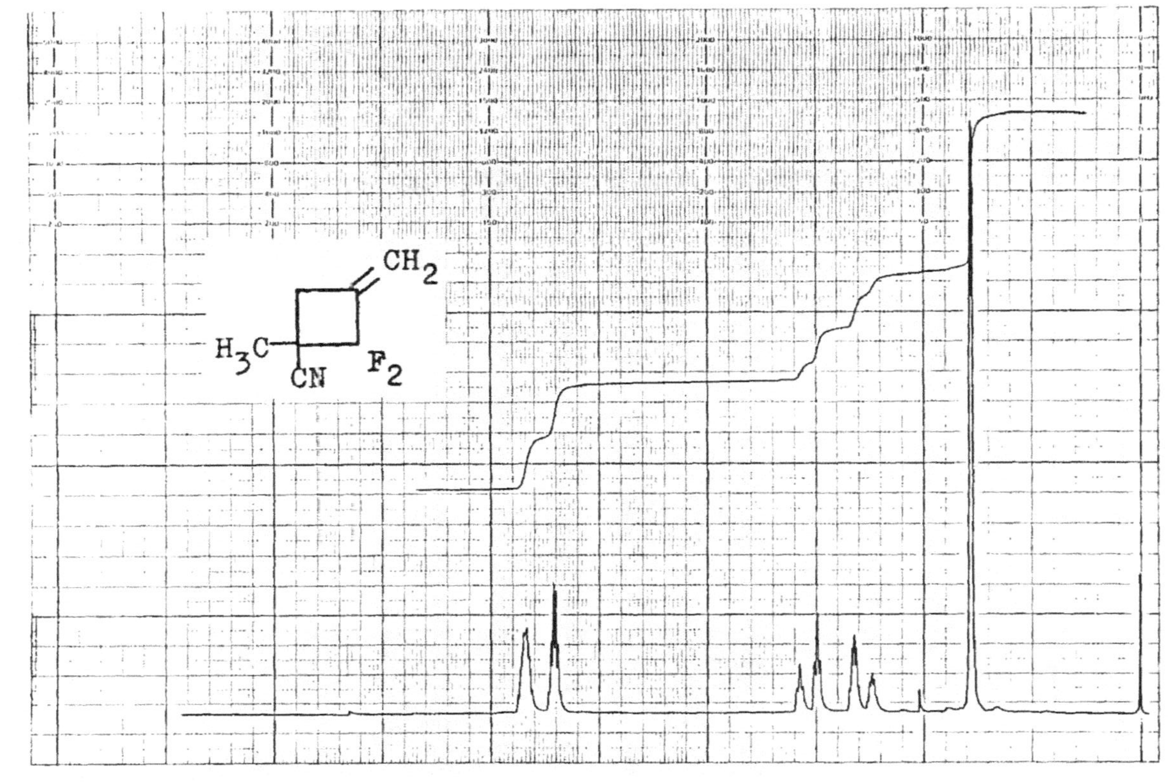

Figure 62. ^{1}H NMR ($CDCl_3$, 100 MHz): 2,2-Difluoro-1-methyl-3-methylene-cyclobutanecarbonitrile <u>86</u>. Page 110.

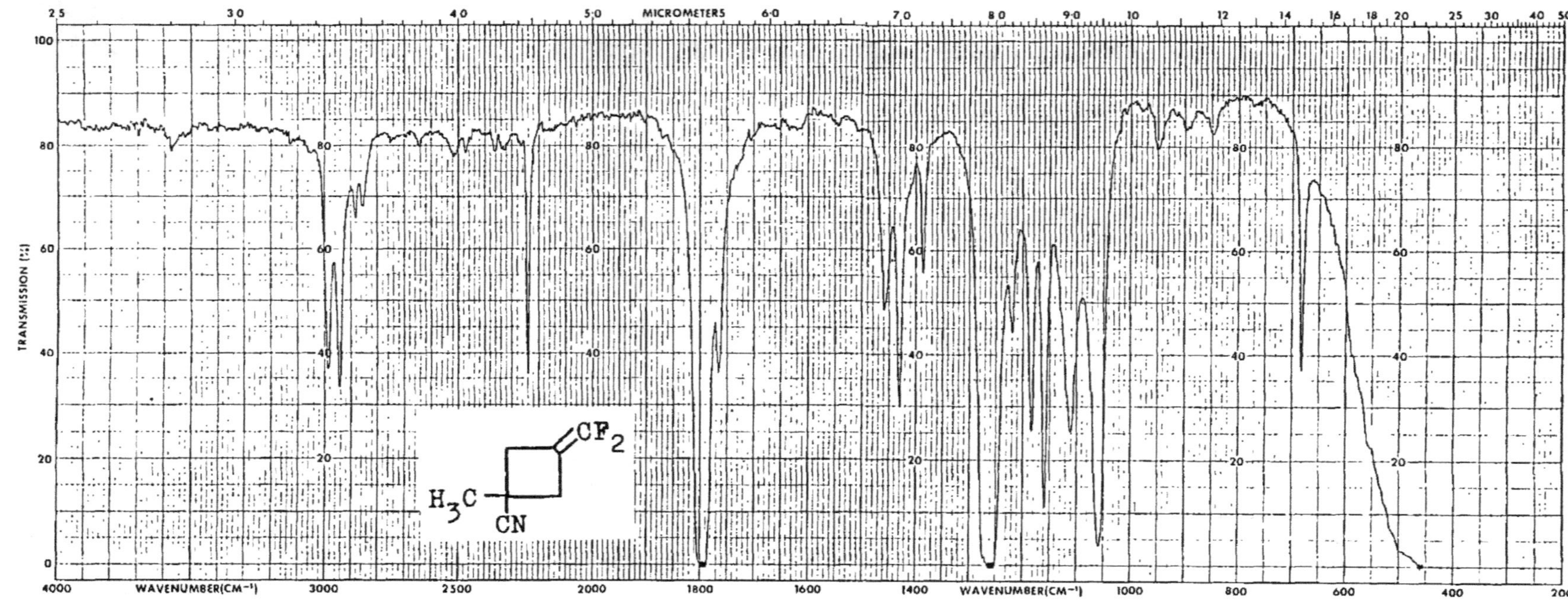

Figure 63. IR (film): 1-Methyl-3-(difluoromethylene)cyclobutanecarbonitrile 87. Page 110.

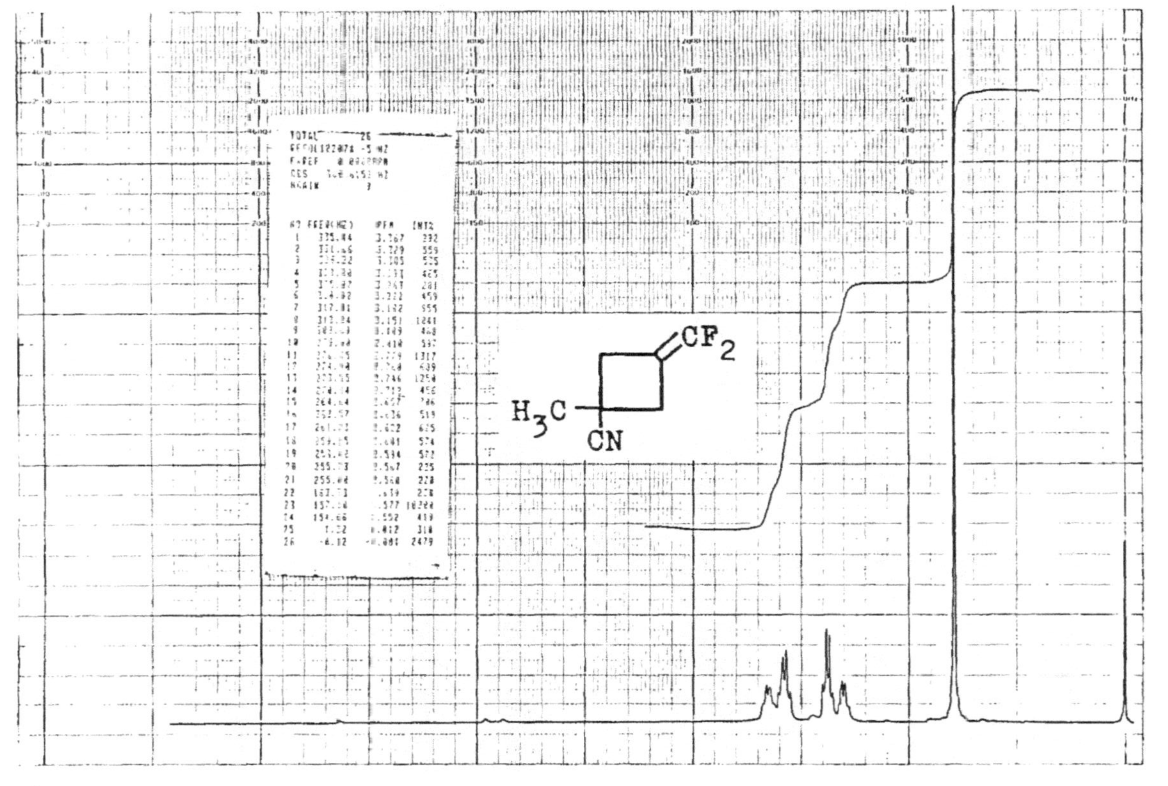

Figure 64. ^{1}H NMR ($CDCl_3$, 100 MHz): 1-Methyl-3-(difluoromethylene)cyclobutanecarbonitrile 87. Page 110.

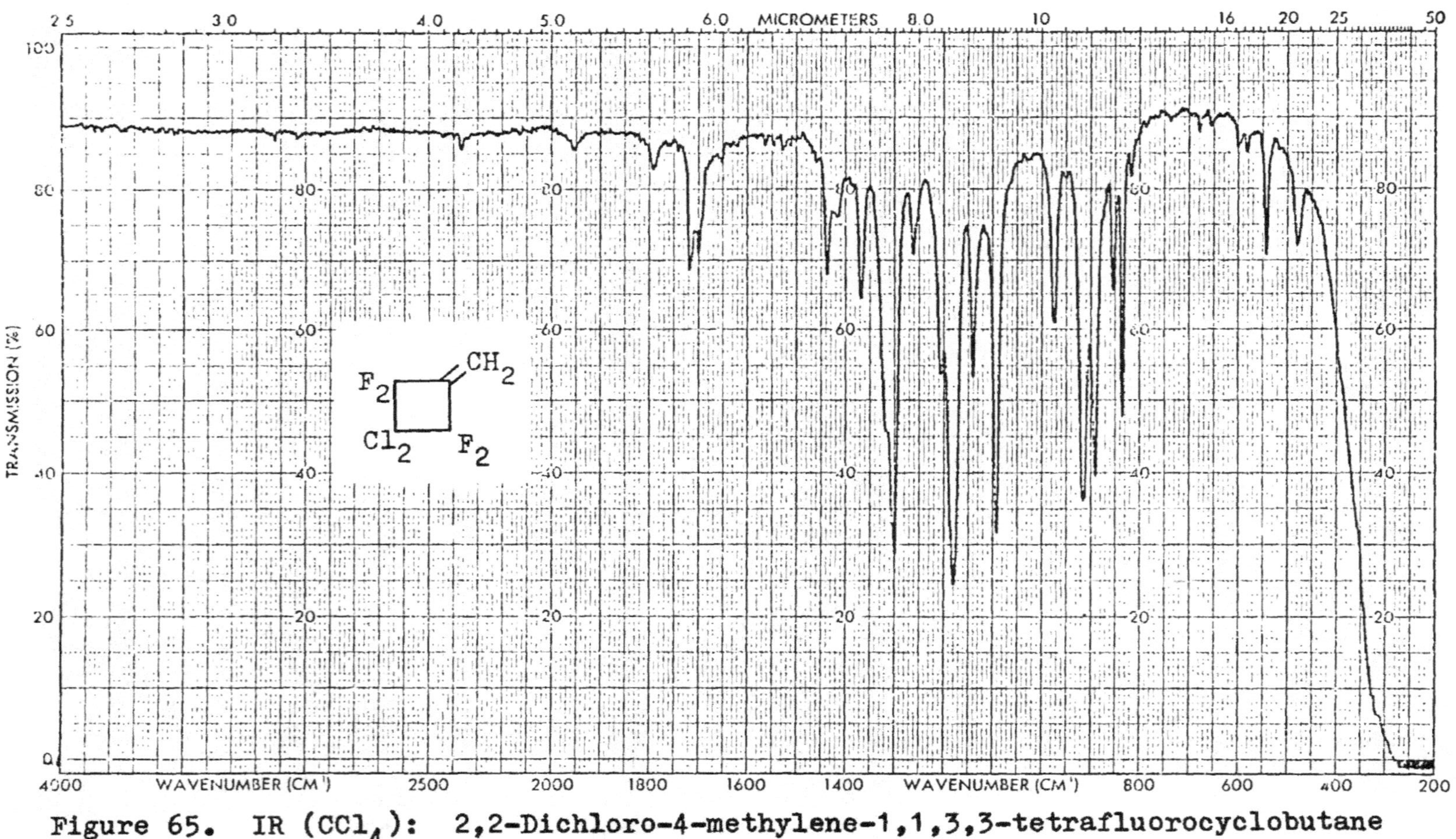

Figure 65. IR (CCl_4): 2,2-Dichloro-4-methylene-1,1,3,3-tetrafluorocyclobutane <u>88</u>. Page 112.

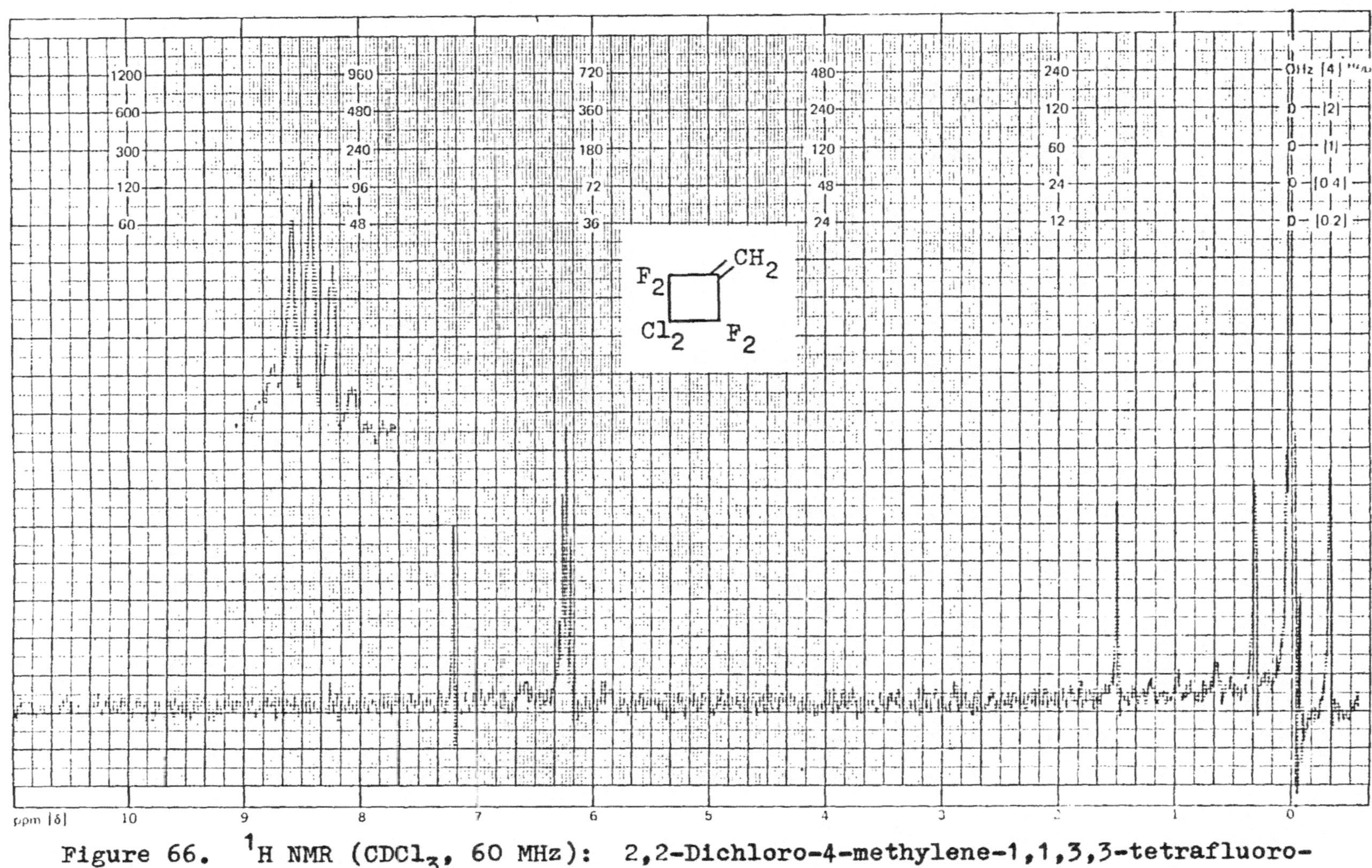

Figure 66. ^{1}H NMR ($CDCl_3$, 60 MHz): 2,2-Dichloro-4-methylene-1,1,3,3-tetrafluorocyclobutane <u>88</u>. Page 112.

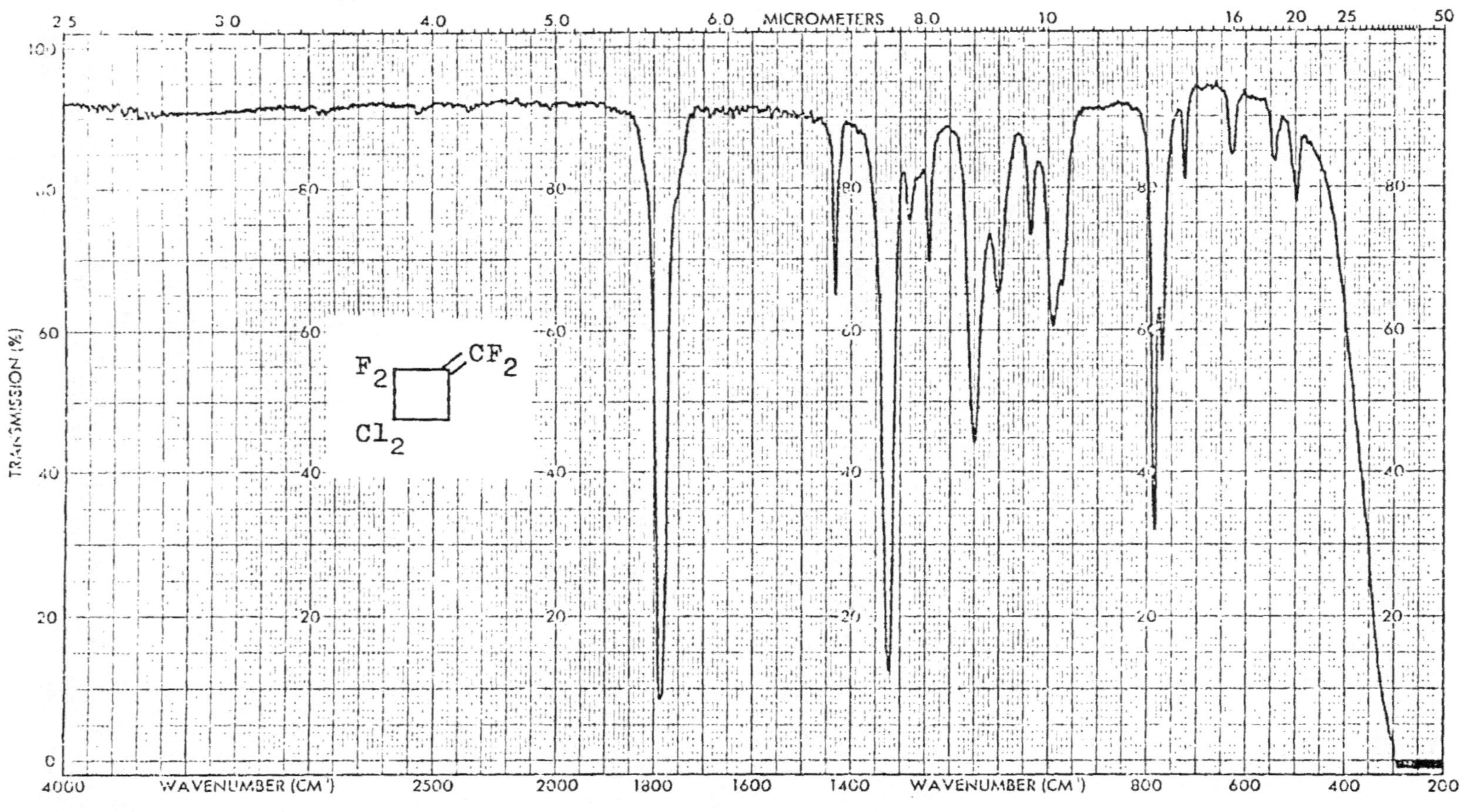

Figure 67. IR (CCl_4): 1,1-Dichloro-2,2-difluoro-3-(difluoromethylene)cyclobutane <u>89</u>. Page 112.

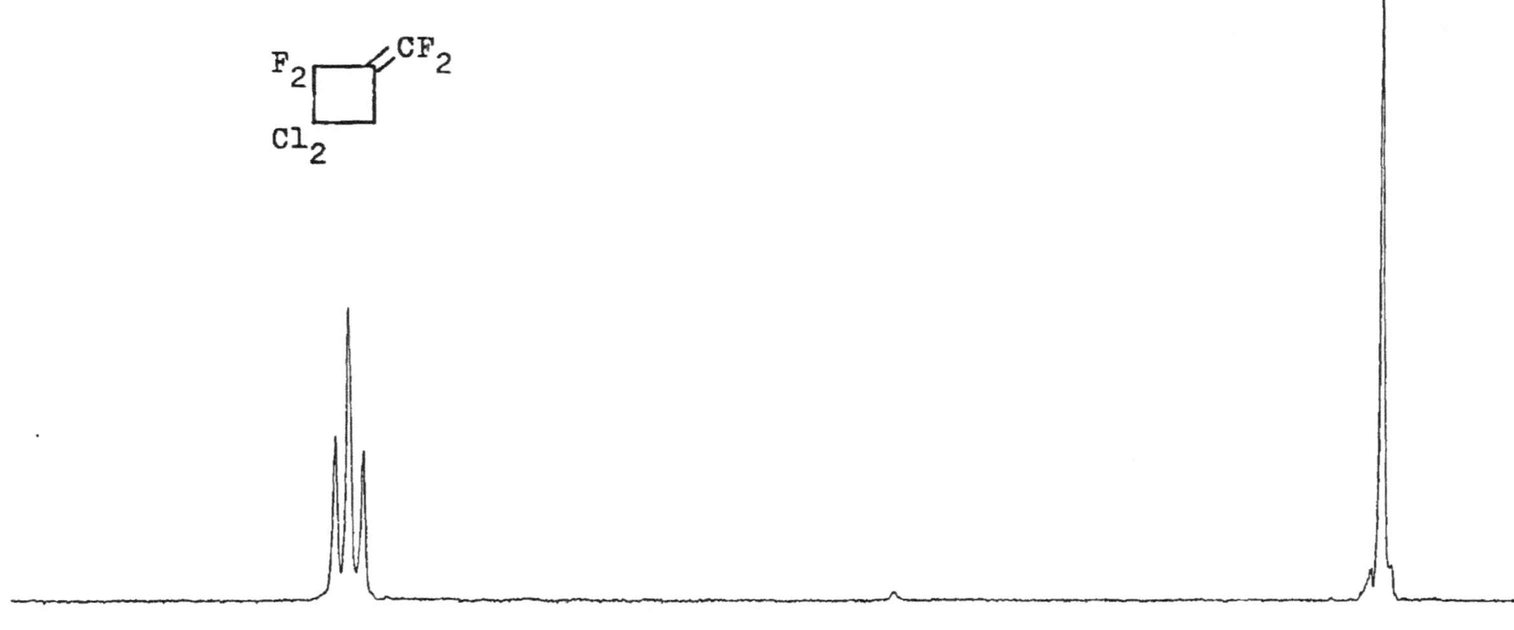

Figure 68. ^{1}H NMR ($CDCl_3$, 100 MHz): 1,1-Dichloro-2,2-difluoro-3-(difluoromethylene)cyclobutane 89; Expansion. Page 112.

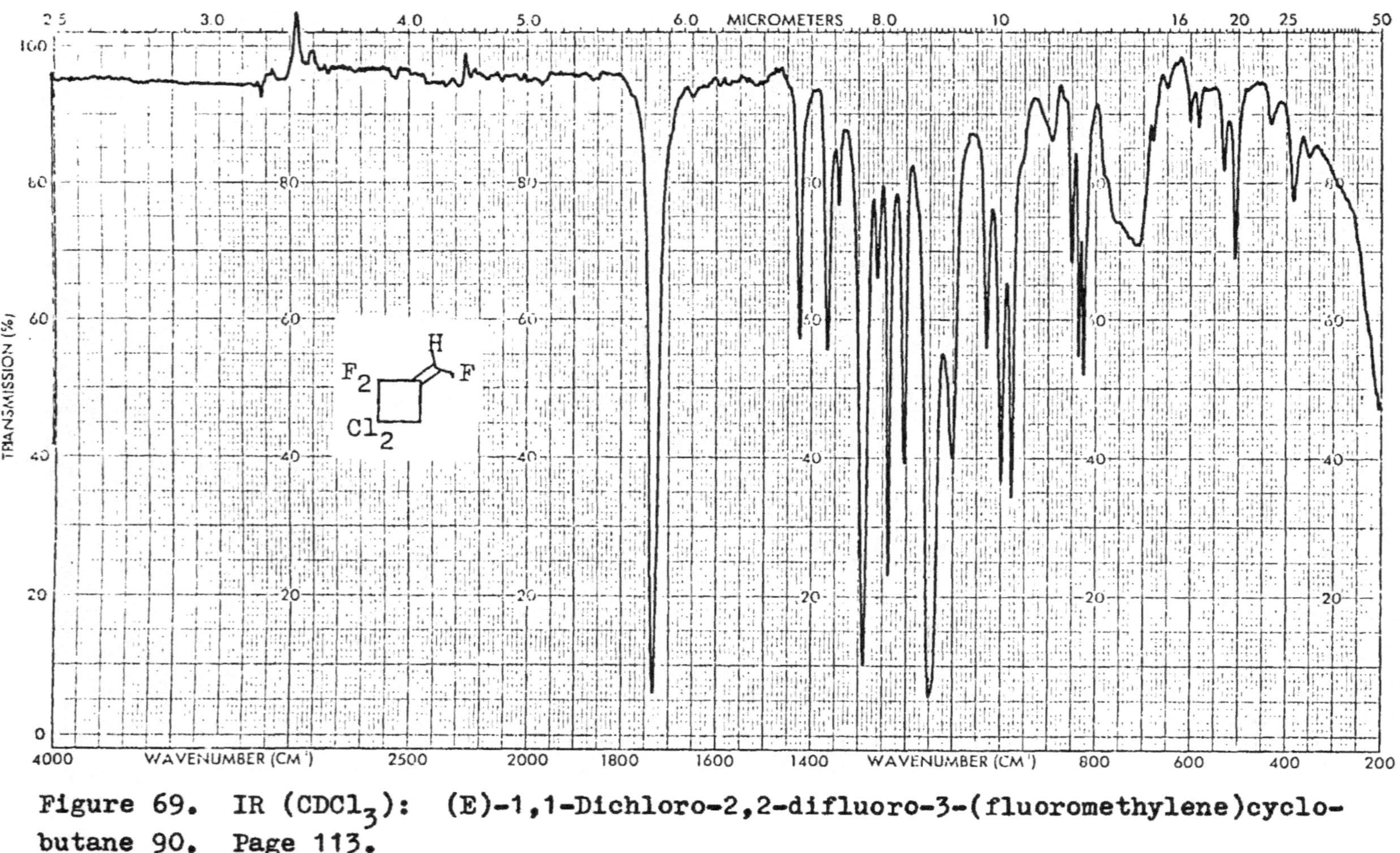

Figure 69. IR ($CDCl_3$): (E)-1,1-Dichloro-2,2-difluoro-3-(fluoromethylene)cyclobutane 90. Page 113.

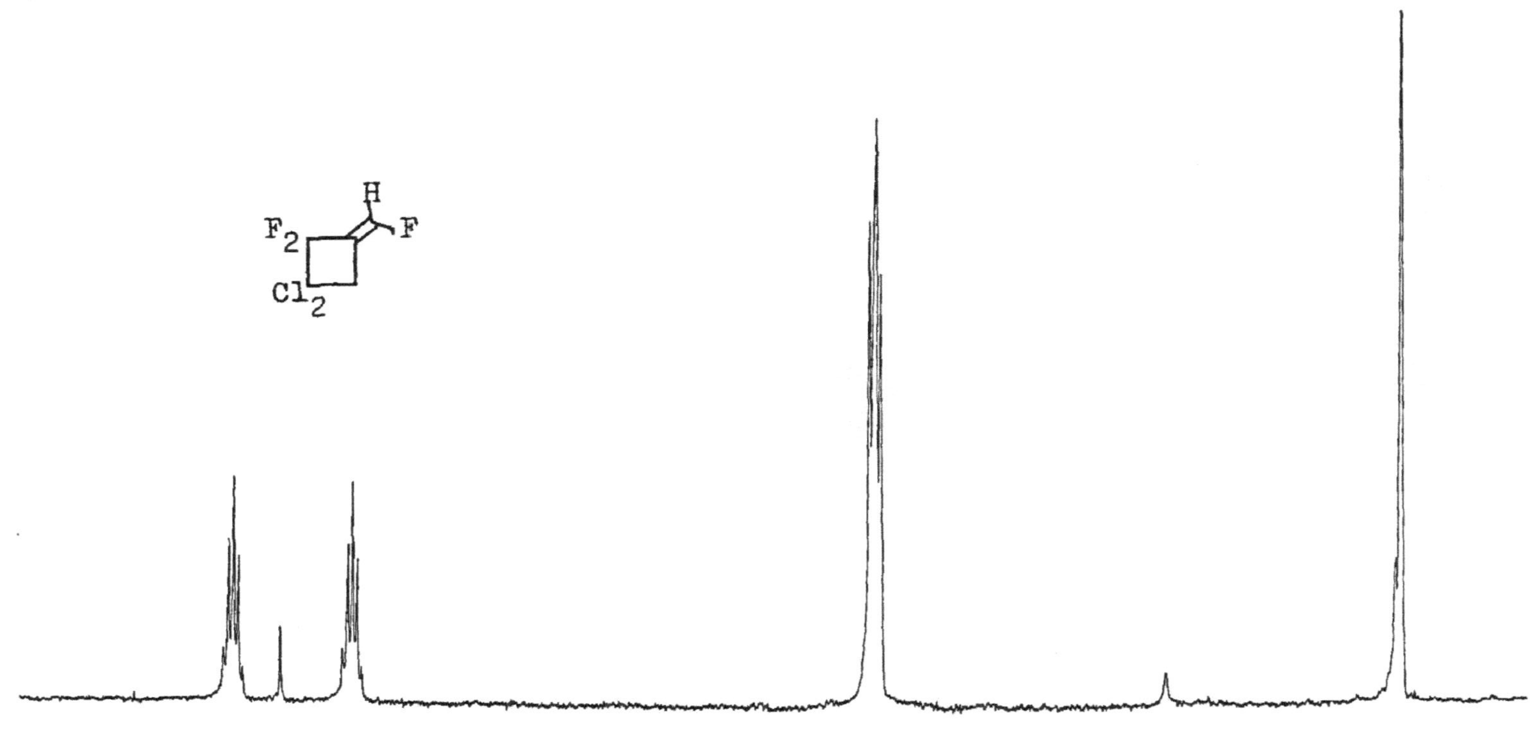

Figure 70. ^{1}H NMR ($CDCl_3$, 100 MHz): (E)-1,1-Dichloro-2,2-difluoro-3-(fluoromethylene)cyclobutane 90. Page 113.

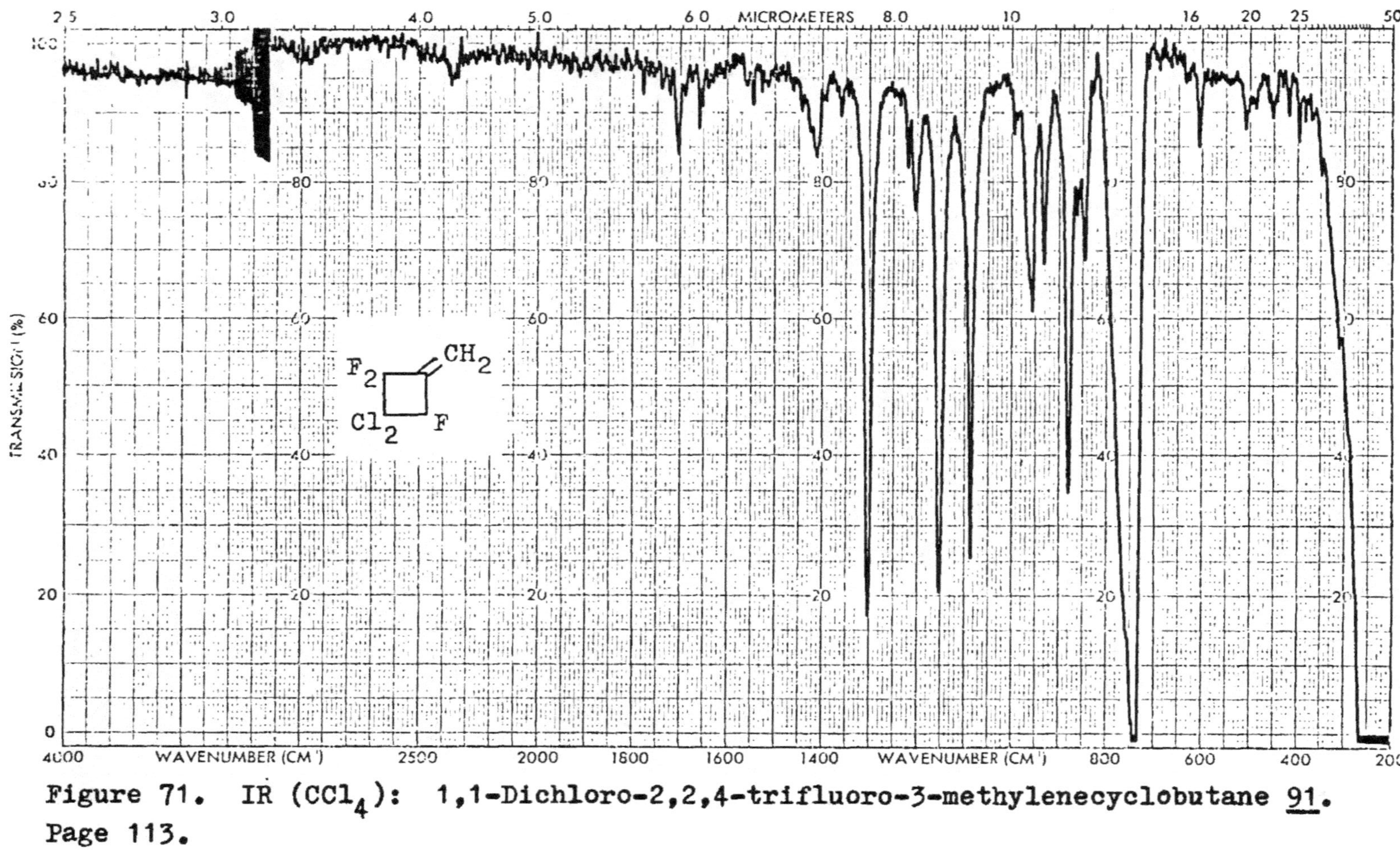

Figure 71. IR (CCl_4): 1,1-Dichloro-2,2,4-trifluoro-3-methylenecyclobutane <u>91</u>. Page 113.

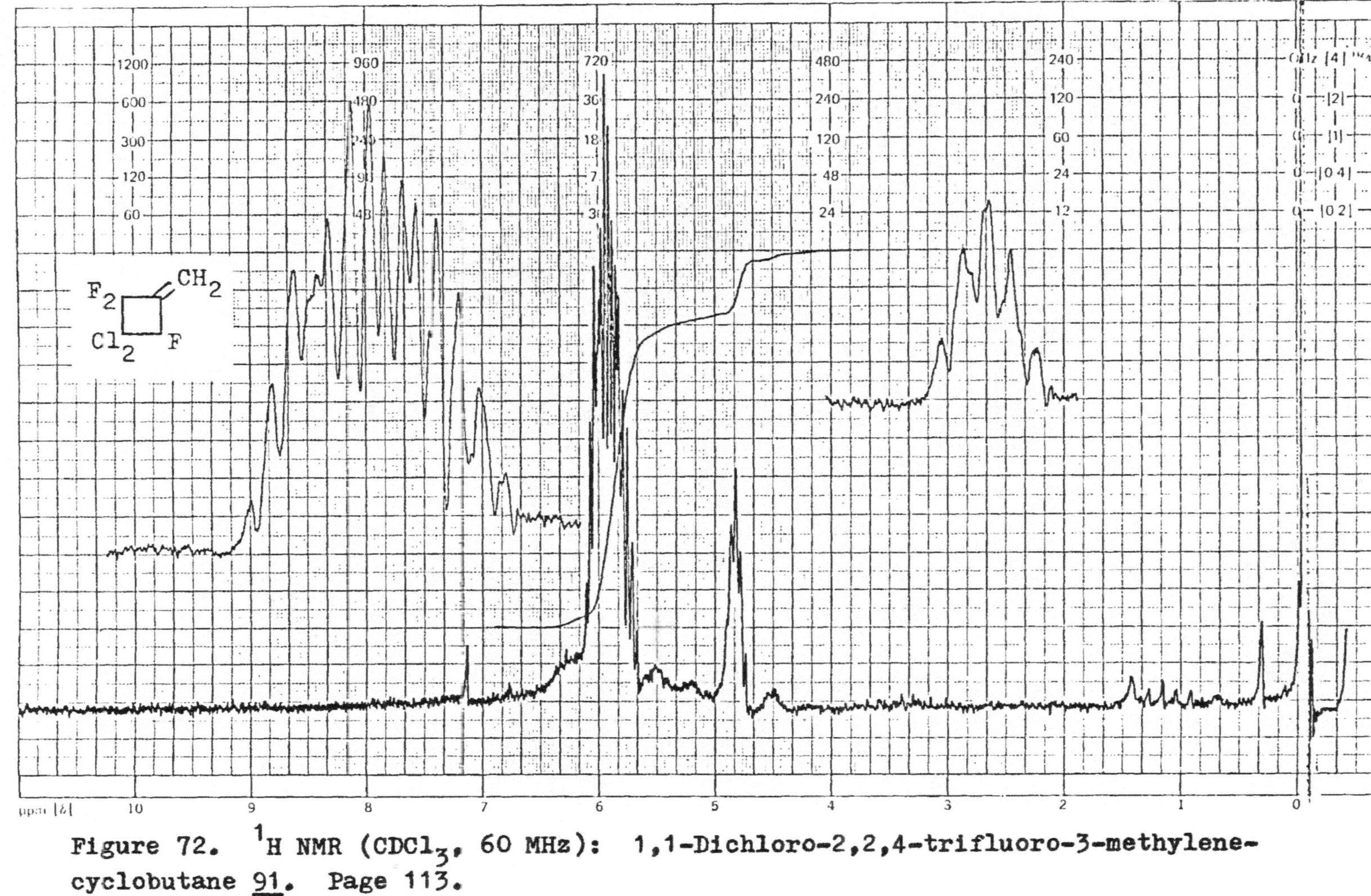

Figure 72. ^{1}H NMR ($CDCl_3$, 60 MHz): 1,1-Dichloro-2,2,4-trifluoro-3-methylene-cyclobutane <u>91</u>. Page 113.

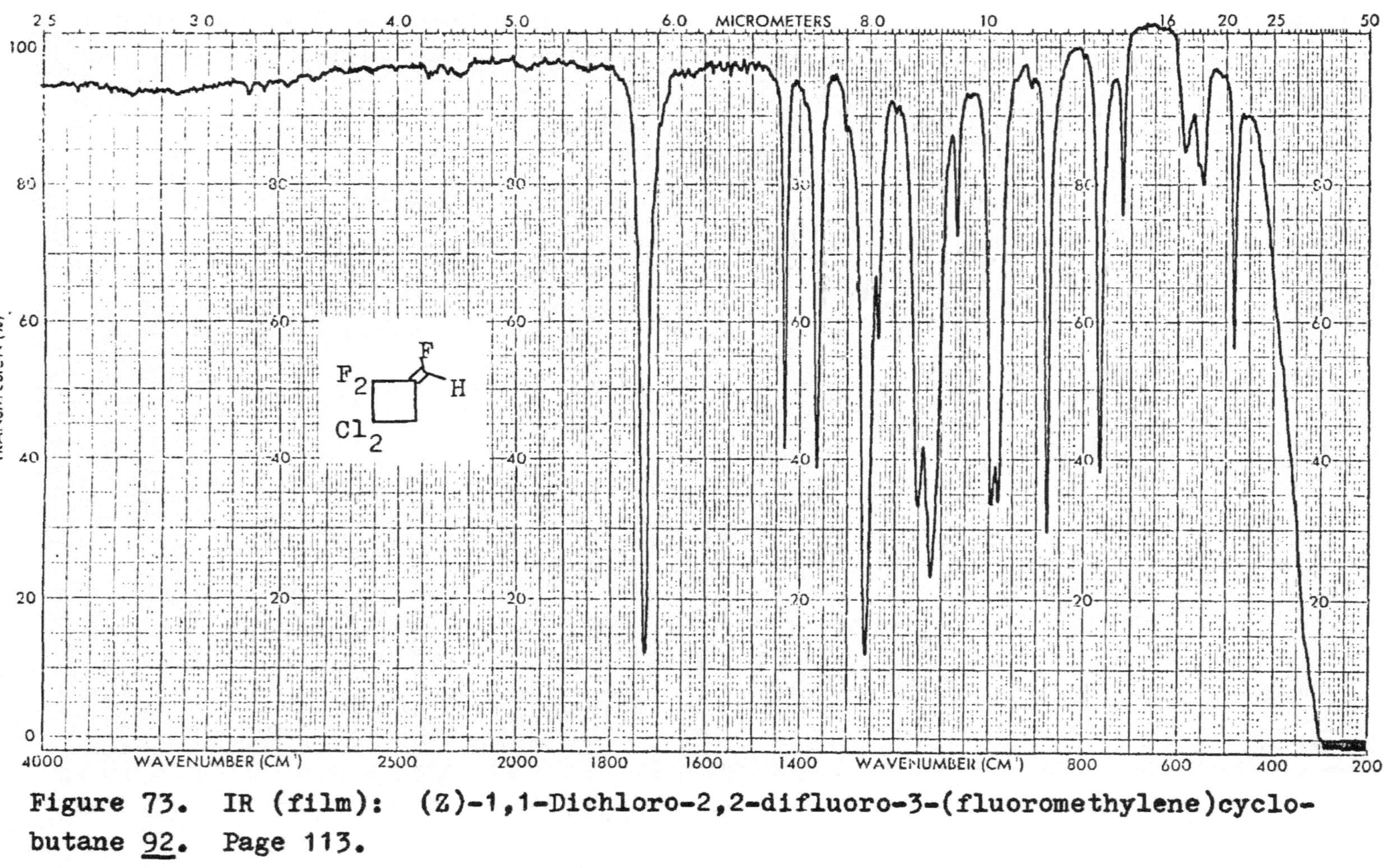

Figure 73. IR (film): (Z)-1,1-Dichloro-2,2-difluoro-3-(fluoromethylene)cyclobutane 92. Page 113.

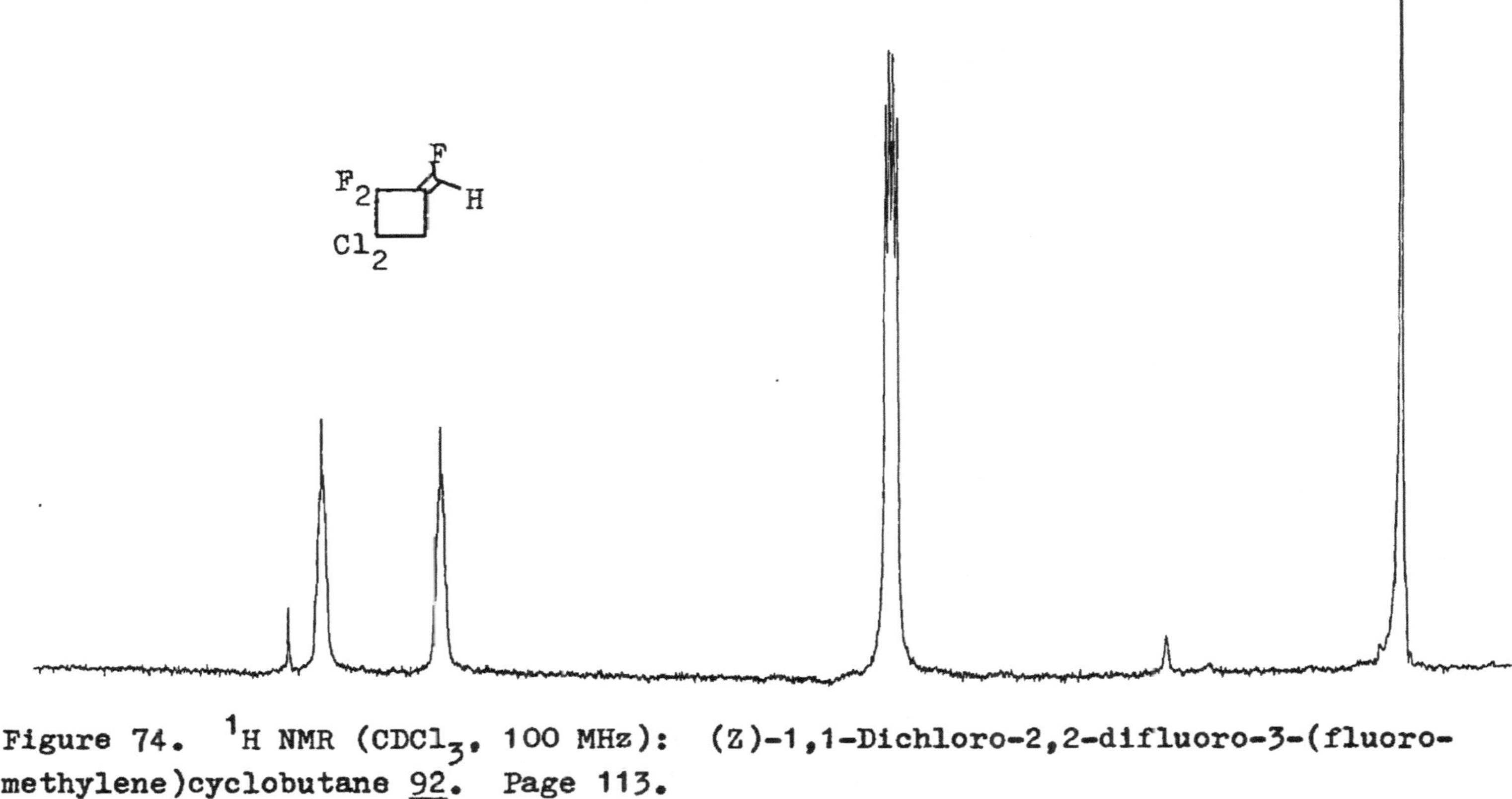

Figure 74. ^{1}H NMR ($CDCl_3$, 100 MHz): (Z)-1,1-Dichloro-2,2-difluoro-3-(fluoromethylene)cyclobutane 92. Page 113.

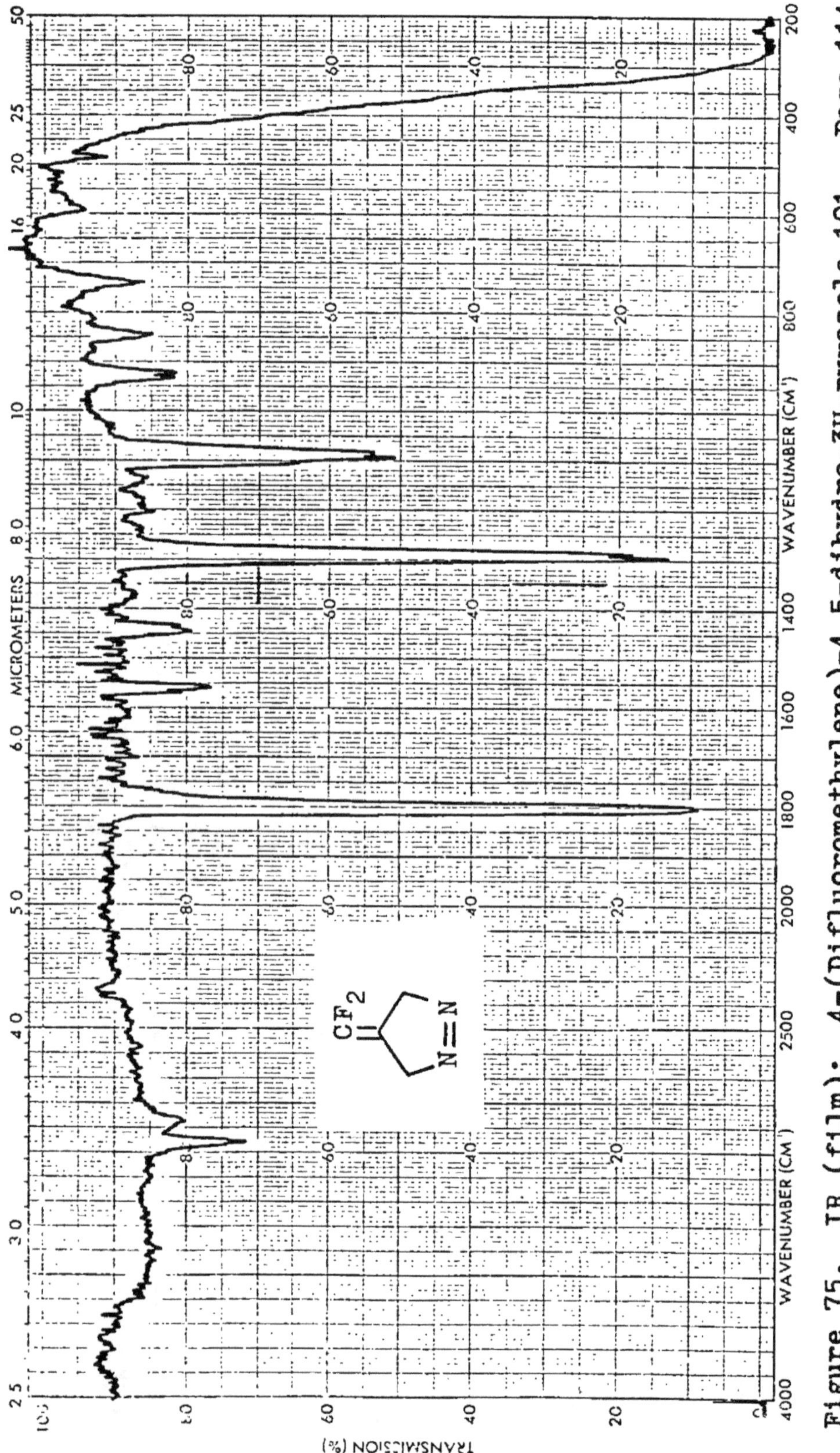

Figure 75. IR (film): 4-(Difluoromethylene)-4,5-dihydro-3H-pyrazole <u>101</u>. Page 114.

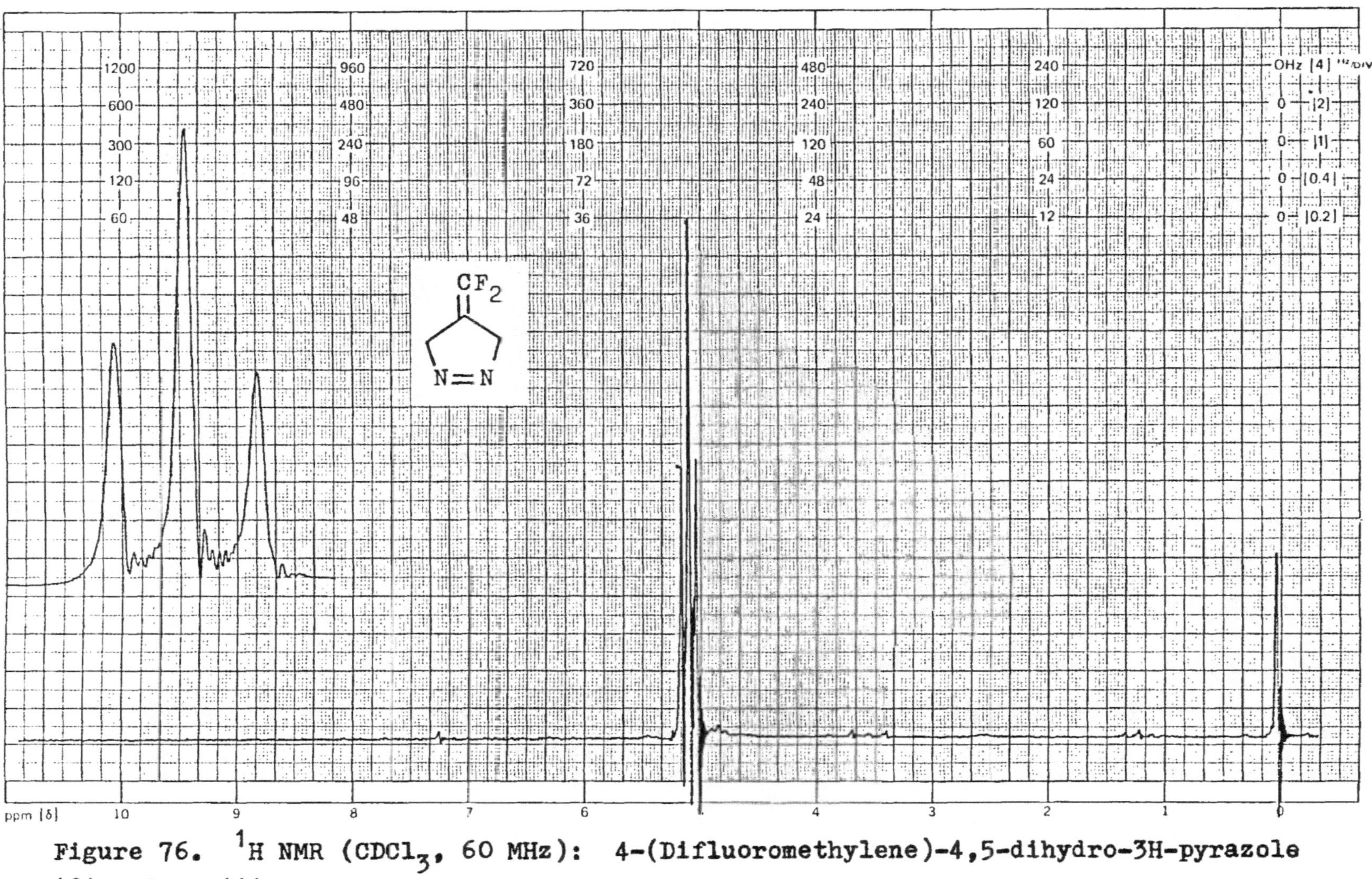

Figure 76. ${}^{1}H$ NMR ($CDCl_3$, 60 MHz): 4-(Difluoromethylene)-4,5-dihydro-3H-pyrazole <u>101</u>. Page 114.

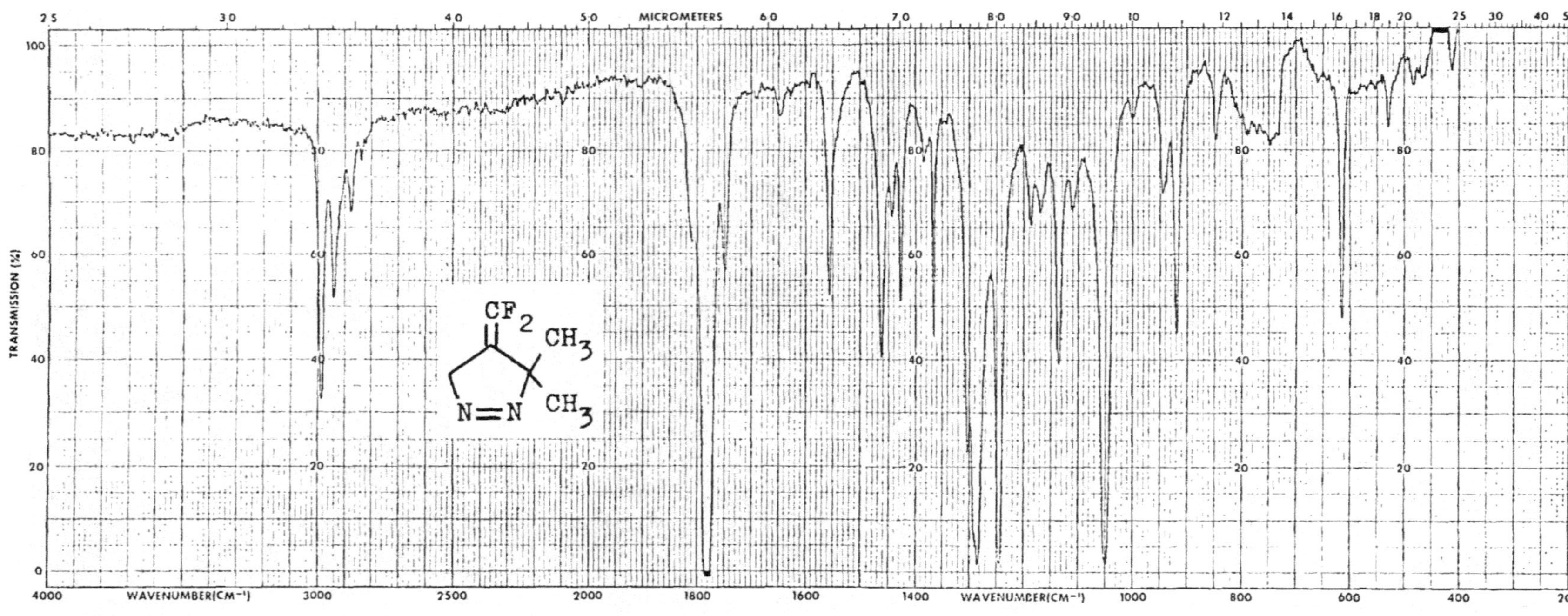

Figure 77. IR (CCl_4): 4-(Difluoromethylene)-4,5-dihydro-3,3-dimethyl-3H-pyrazole **104**. Page 117.

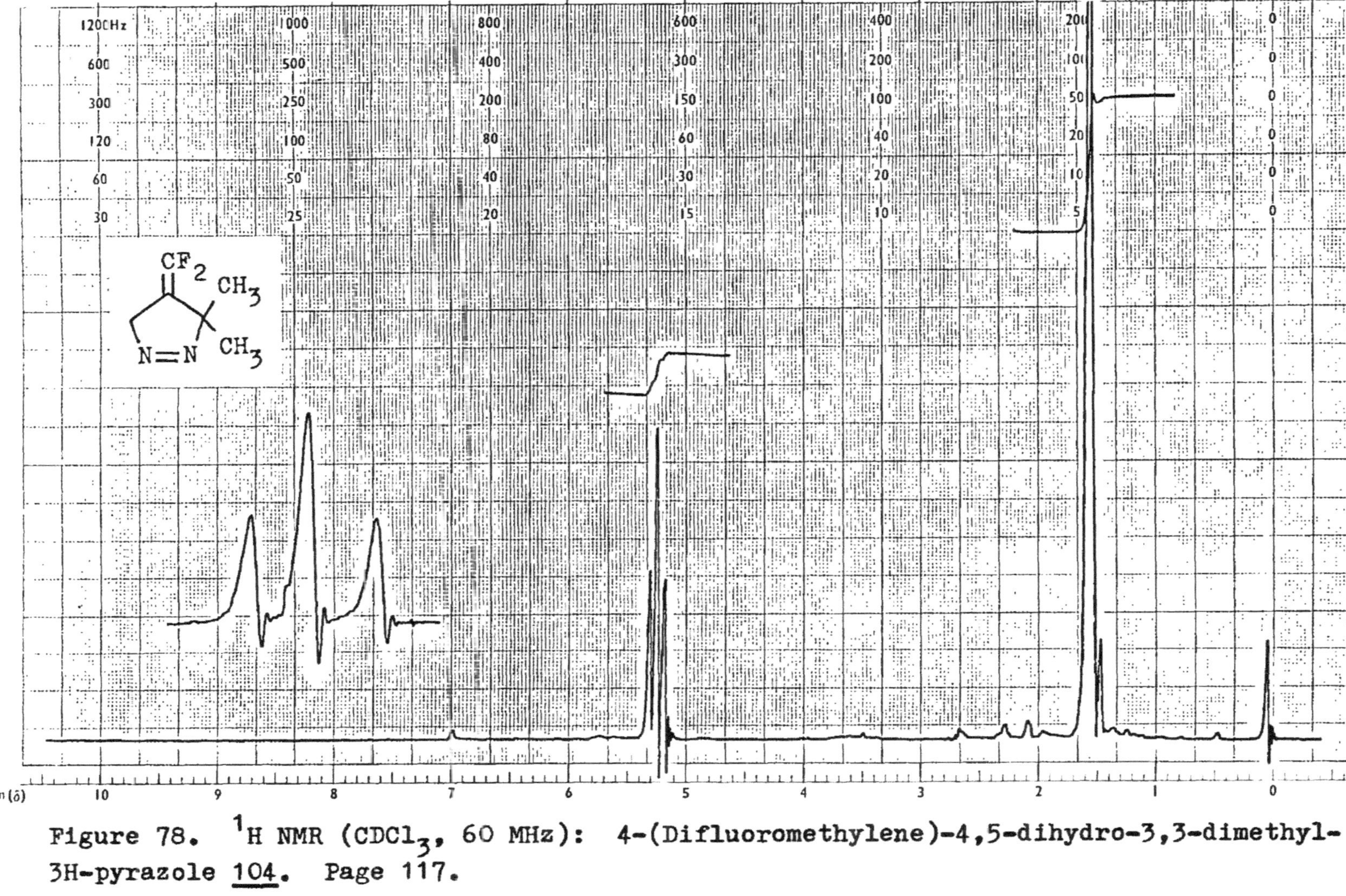

Figure 78. ^{1}H NMR ($CDCl_3$, 60 MHz): 4-(Difluoromethylene)-4,5-dihydro-3,3-dimethyl-3H-pyrazole 104. Page 117.

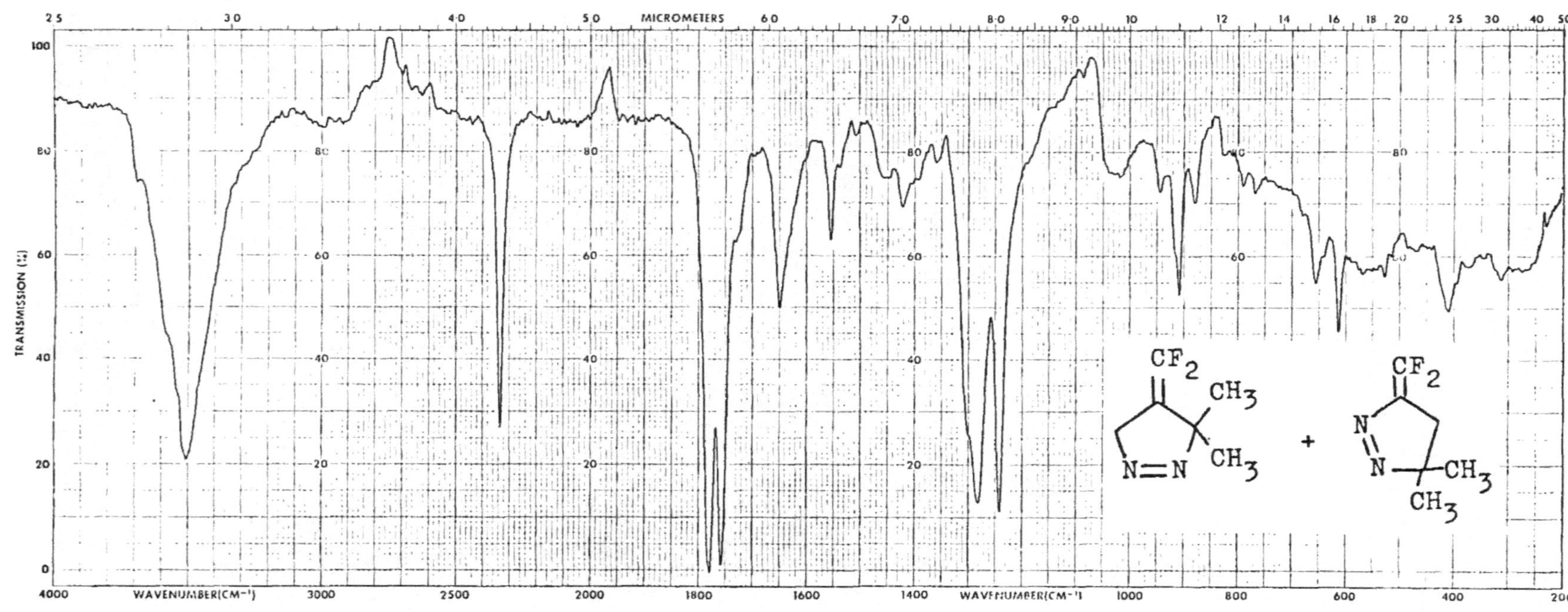

Figure 79. IR (CCl_4): 4-(Difluoromethylene)-4,5-dihydro-3,3-dimethyl-3H-pyrazole 104, 5-(Difluoromethylene)-4,5-dihydro-3,3-dimethyl-3H-pyrazole 105, and acetone azine. Page 116.

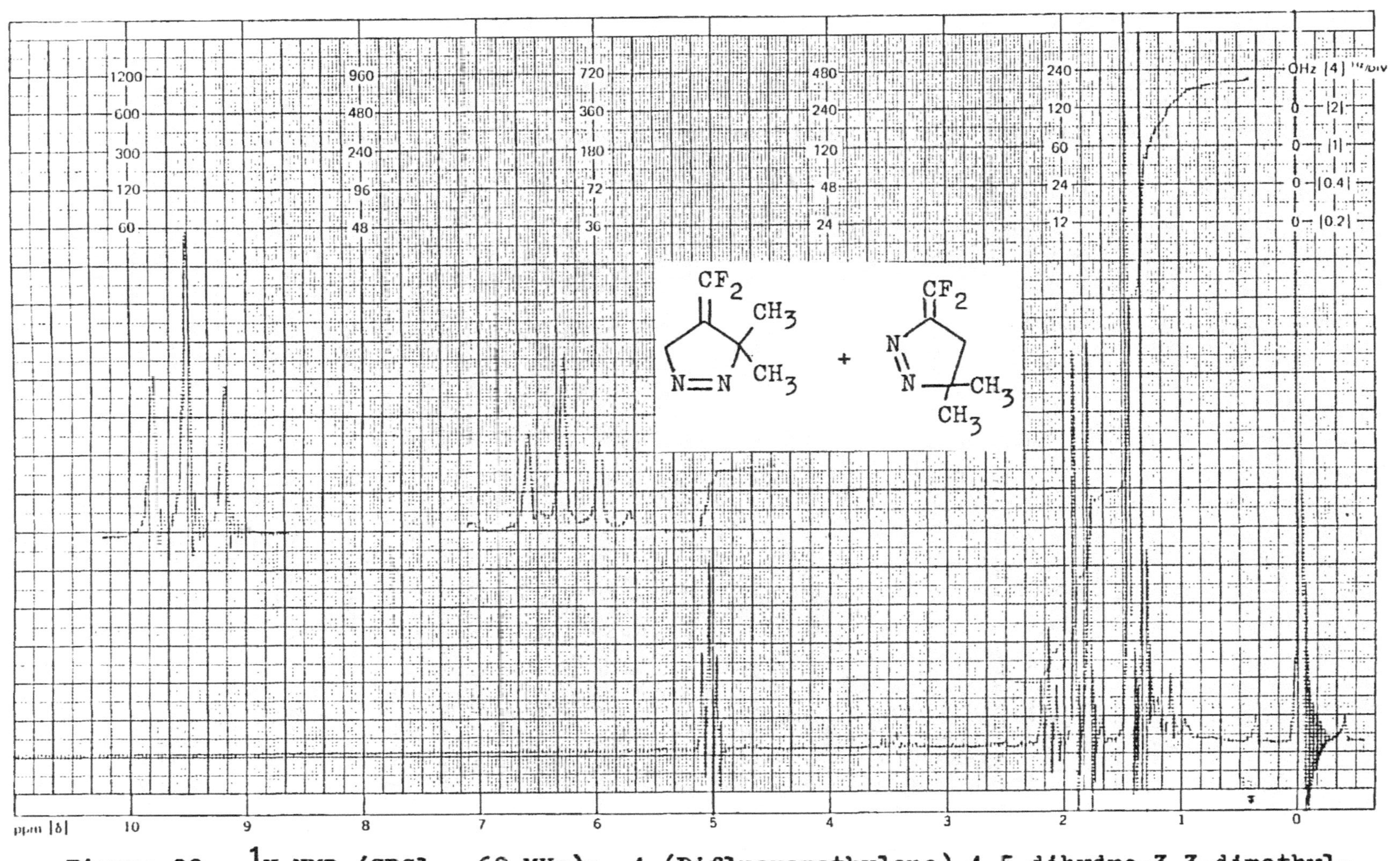

Figure 80. ^{1}H NMR ($CDCl_3$, 60 MHz): 4-(Difluoromethylene)-4,5-dihydro-3,3-dimethyl-3H-pyrazole <u>104</u>, 5-(Difluoromethylene)-4,5-dihydro-3,3-dimethyl-3H-pyrazole <u>105</u>, and acetone azine. Page 116.

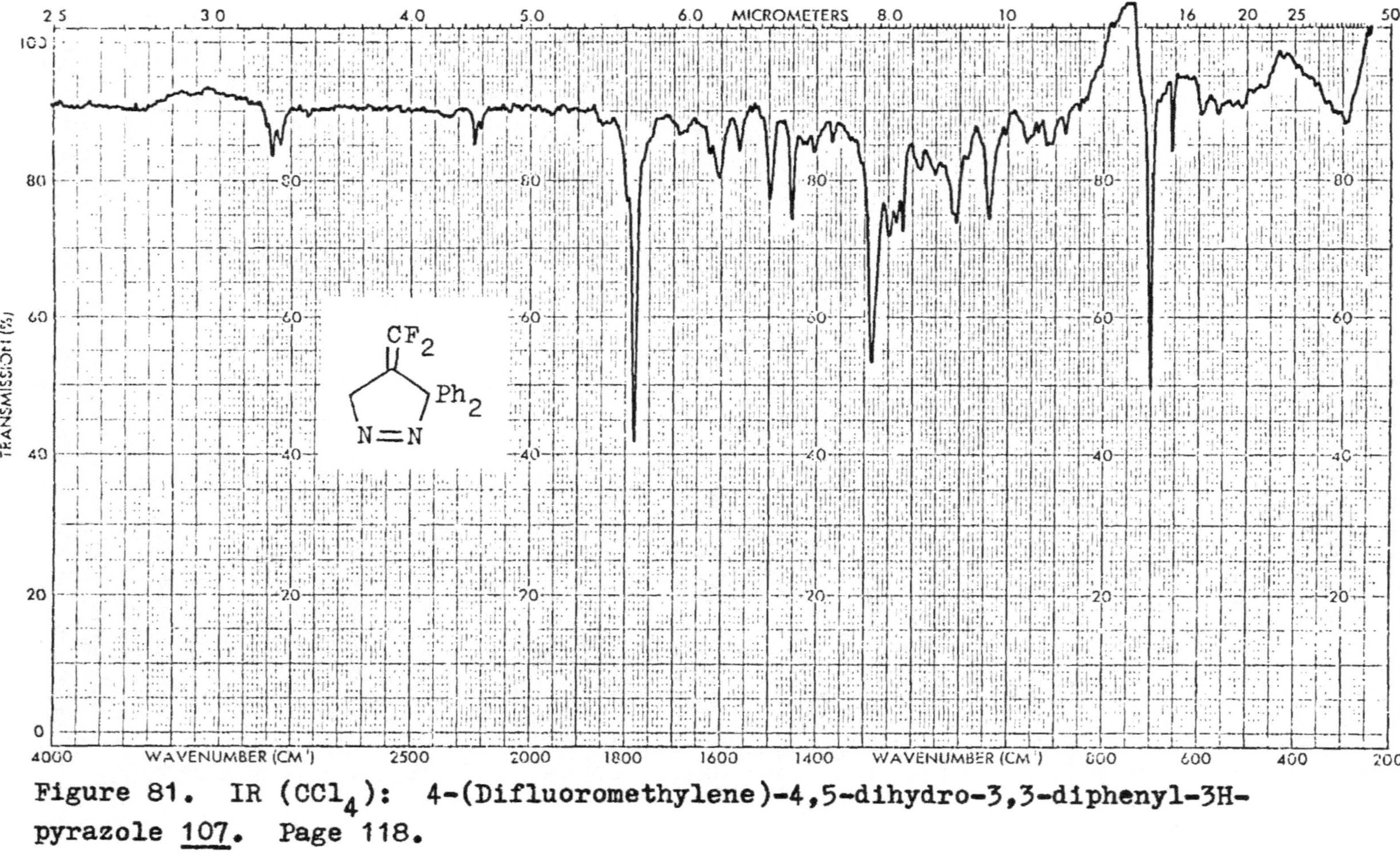

Figure 81. IR (CCl_4): 4-(Difluoromethylene)-4,5-dihydro-3,3-diphenyl-3H-pyrazole <u>107</u>. Page 118.

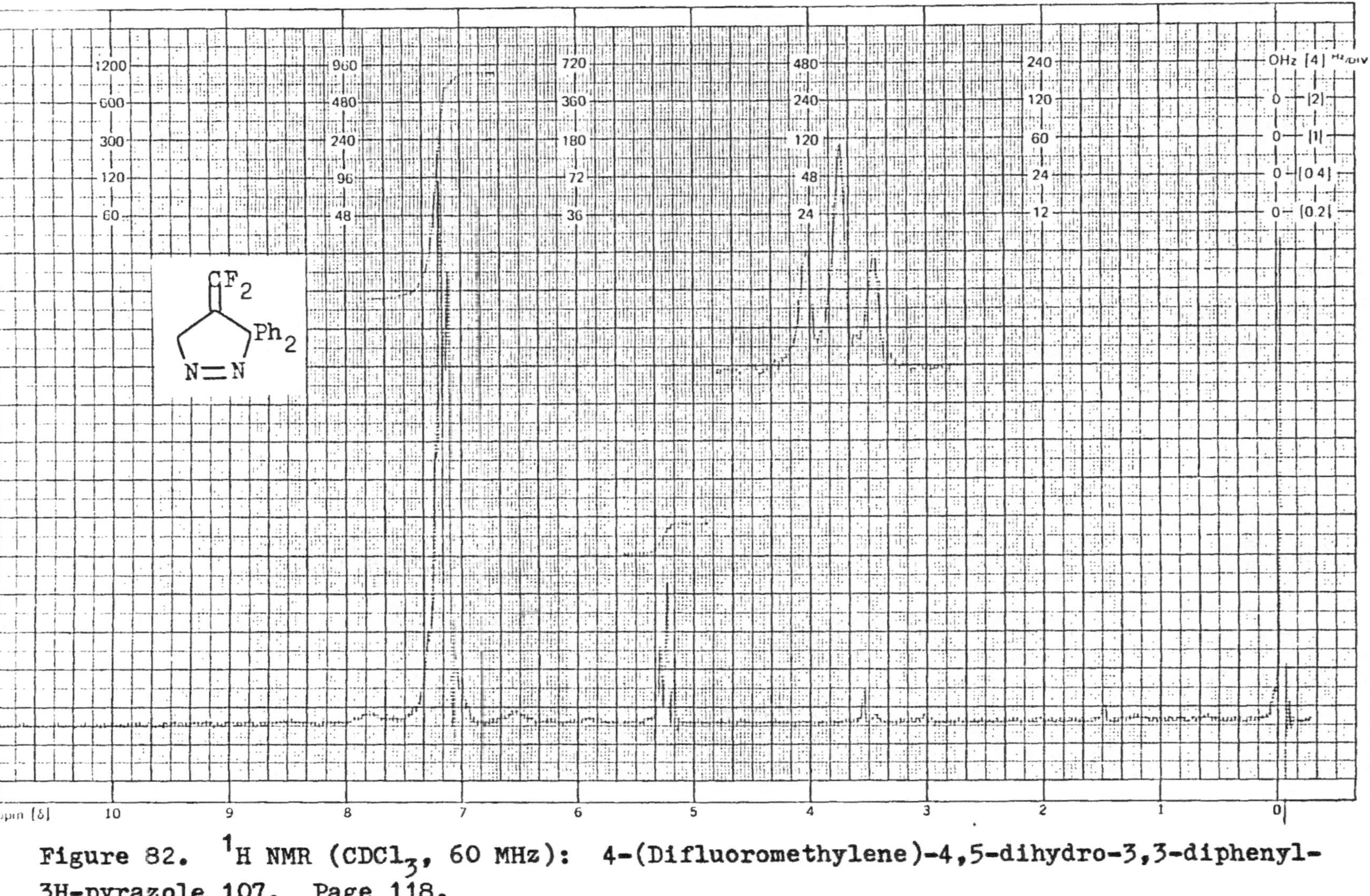

Figure 82. ^{1}H NMR ($CDCl_3$, 60 MHz): 4-(Difluoromethylene)-4,5-dihydro-3,3-diphenyl-3H-pyrazole <u>107</u>. Page 118.

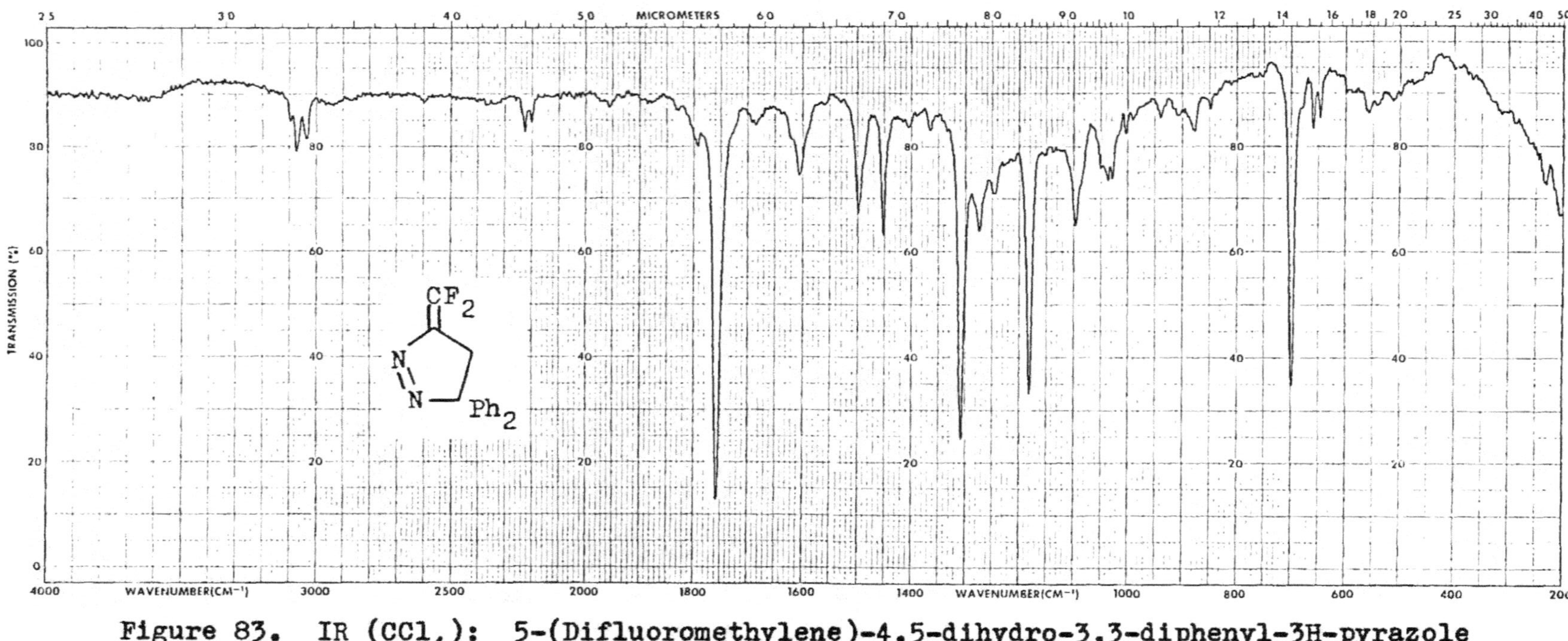

Figure 83. IR (CCl_4): 5-(Difluoromethylene)-4,5-dihydro-3,3-diphenyl-3H-pyrazole <u>108</u>. Page 118.

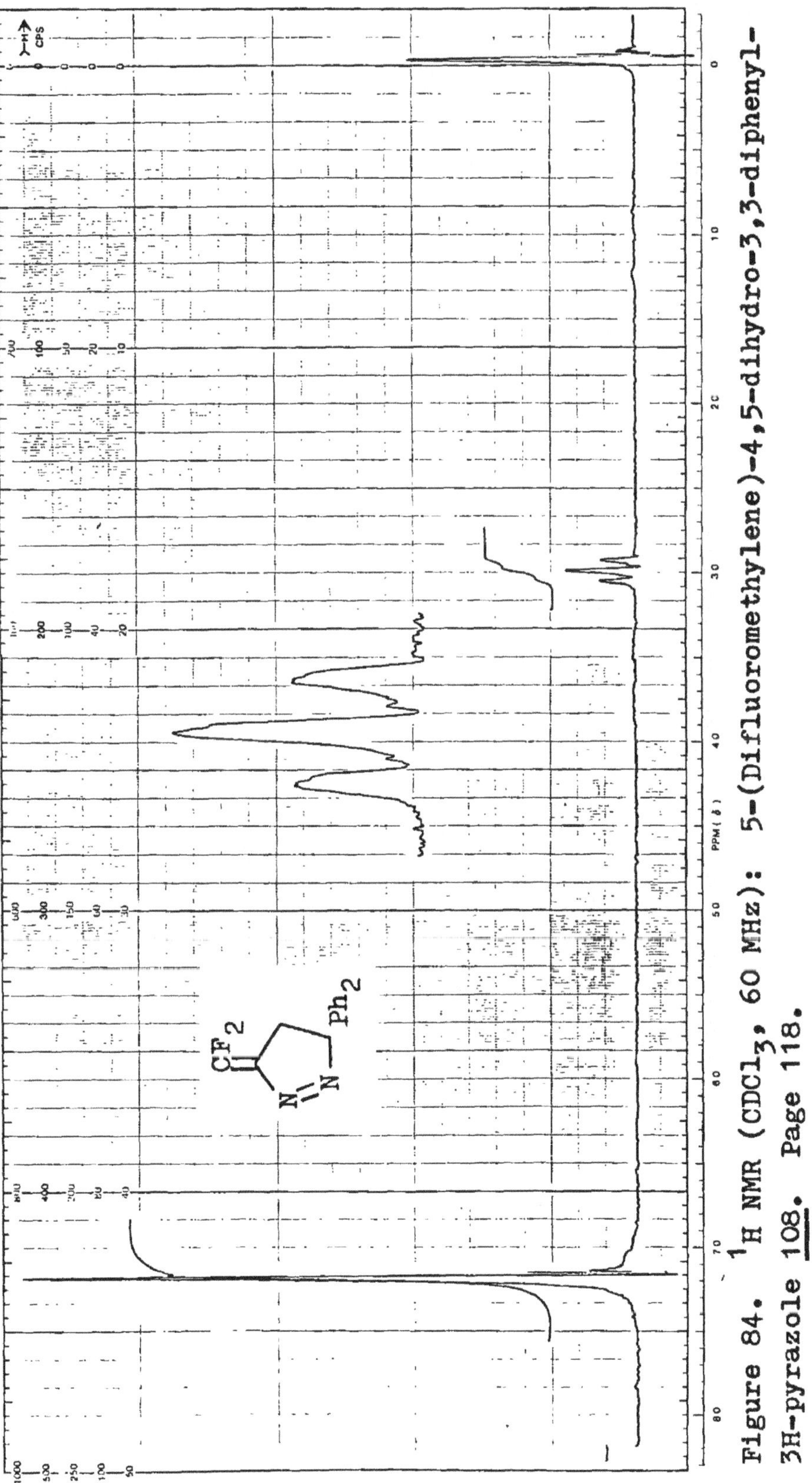

Figure 84. ^{1}H NMR ($CDCl_3$, 60 MHz): 5-(Difluoromethylene)-4,5-dihydro-3,3-diphenyl-3H-pyrazole <u>108</u>. Page 118.

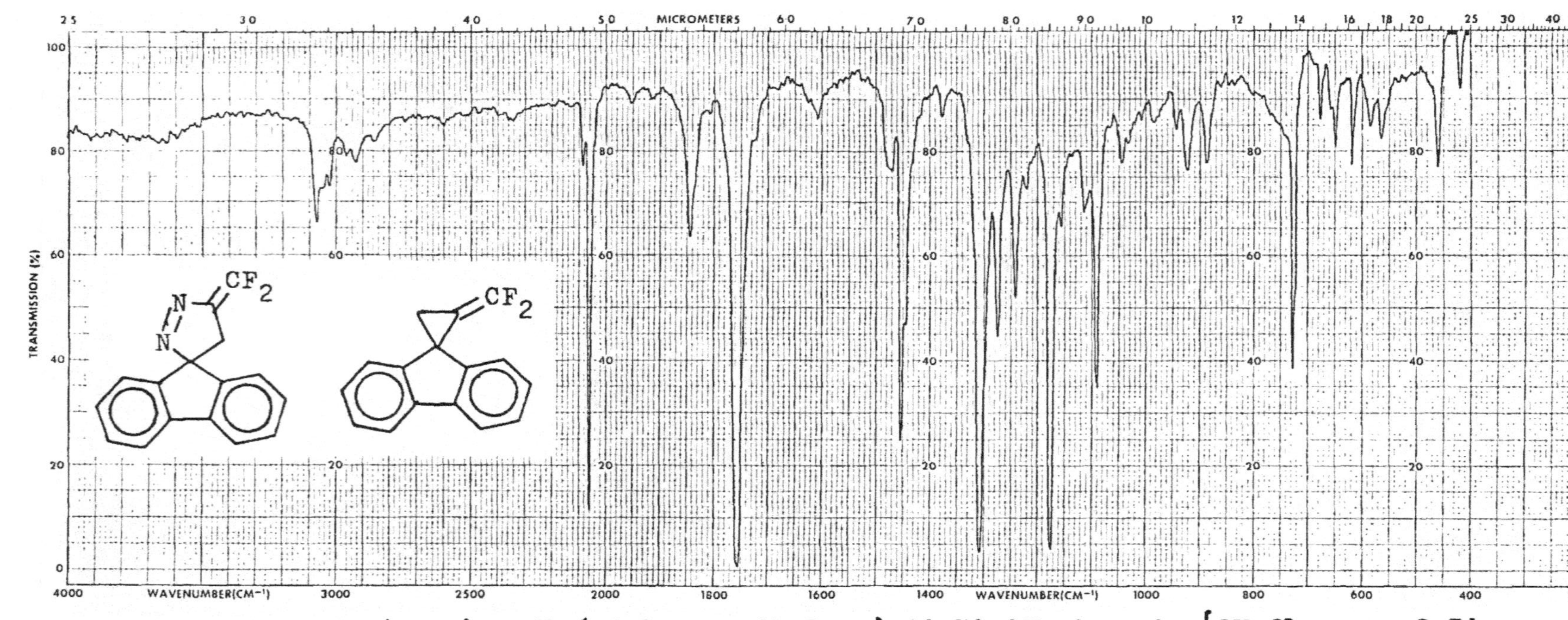

Figure 85. IR (CCl_4): 5'-(Difluoromethylene)-4',5'-dihydrospiro[9H-fluorene-9,3'-[3H]pyrazole] 111 and 2-(Difluoromethylene)spiro[cyclopropane-1,9'-[9H]fluorene] 112. Page 120.

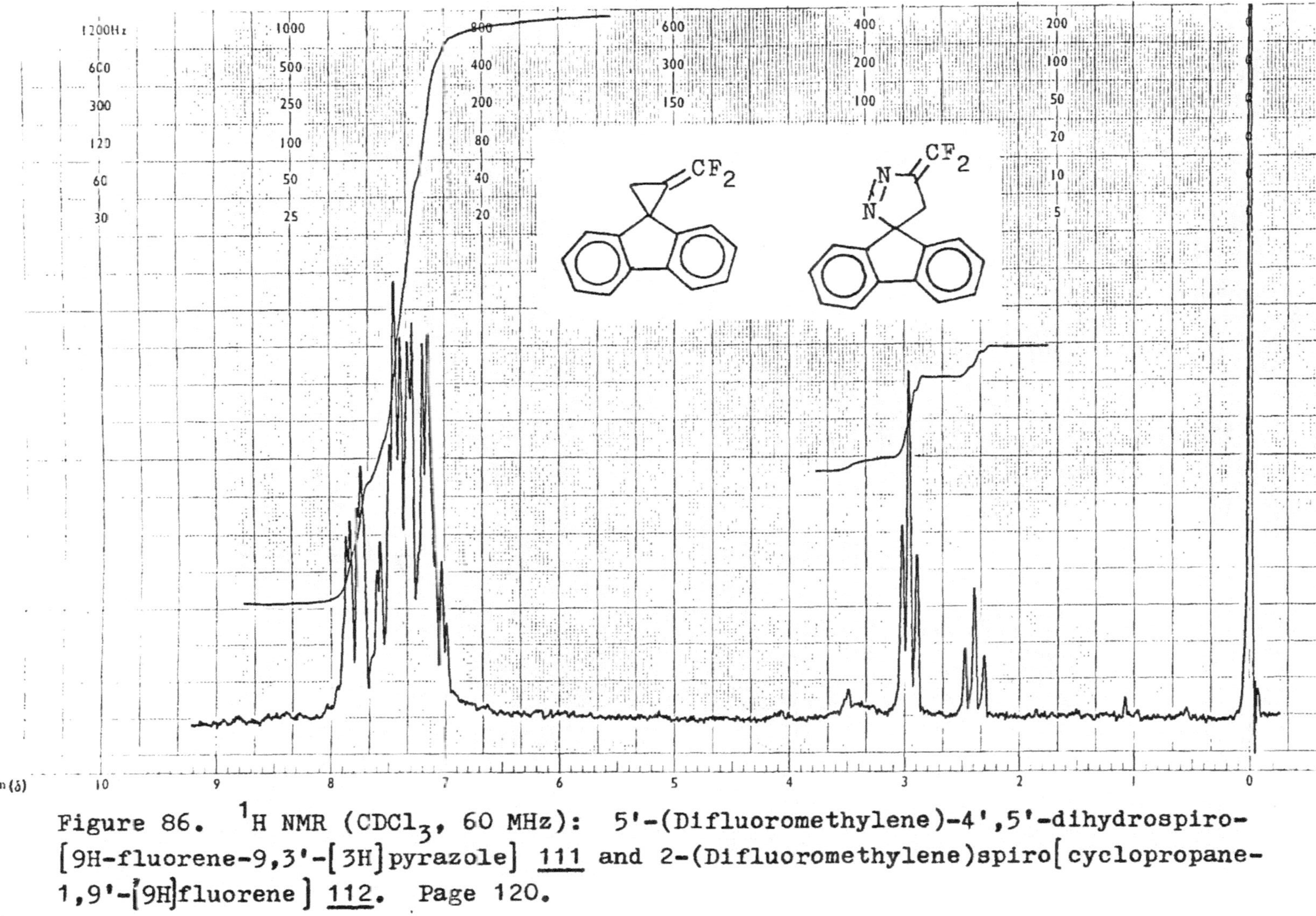

Figure 86. ^{1}H NMR ($CDCl_3$, 60 MHz): 5'-(Difluoromethylene)-4',5'-dihydrospiro-[9H-fluorene-9,3'-[3H]pyrazole] 111 and 2-(Difluoromethylene)spiro[cyclopropane-1,9'-[9H]fluorene] 112. Page 120.

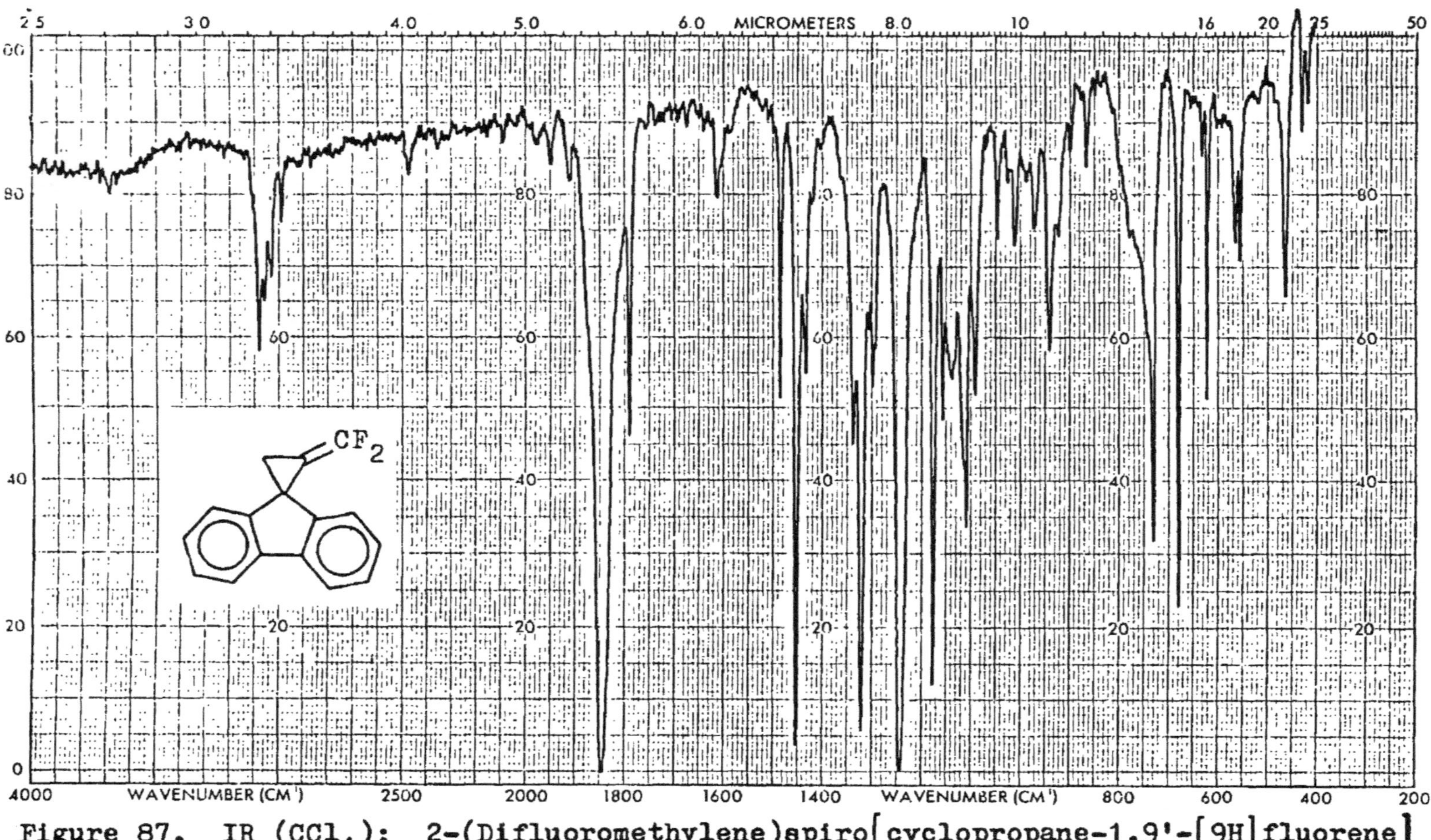

Figure 87. IR (CCl_4): 2-(Difluoromethylene)spiro[cyclopropane-1,9'-[9H]fluorene] 112. Pages 120 and 133.

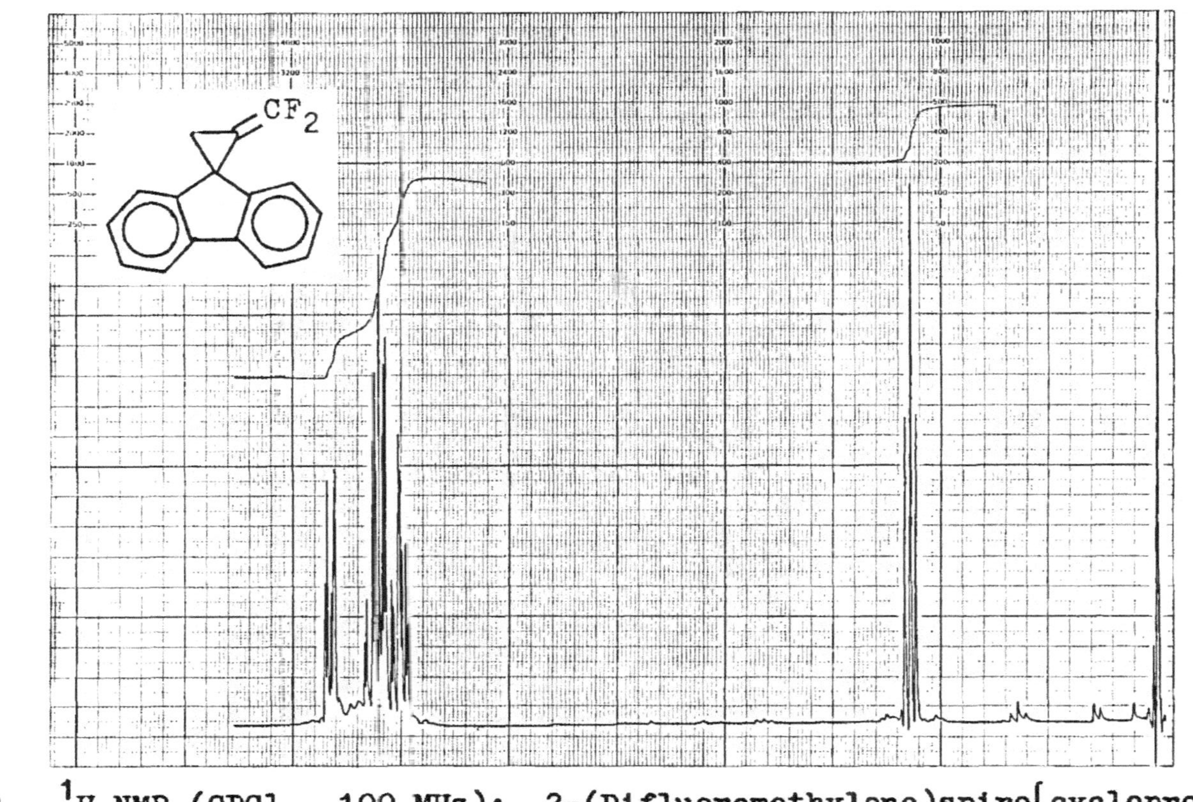

Figure 88. ^{1}H NMR ($CDCl_3$, 100 MHz): 2-(Difluoromethylene)spiro[cyclopropane-1,9'-[9H]fluorene] <u>112</u>. Pages 120 and 133.

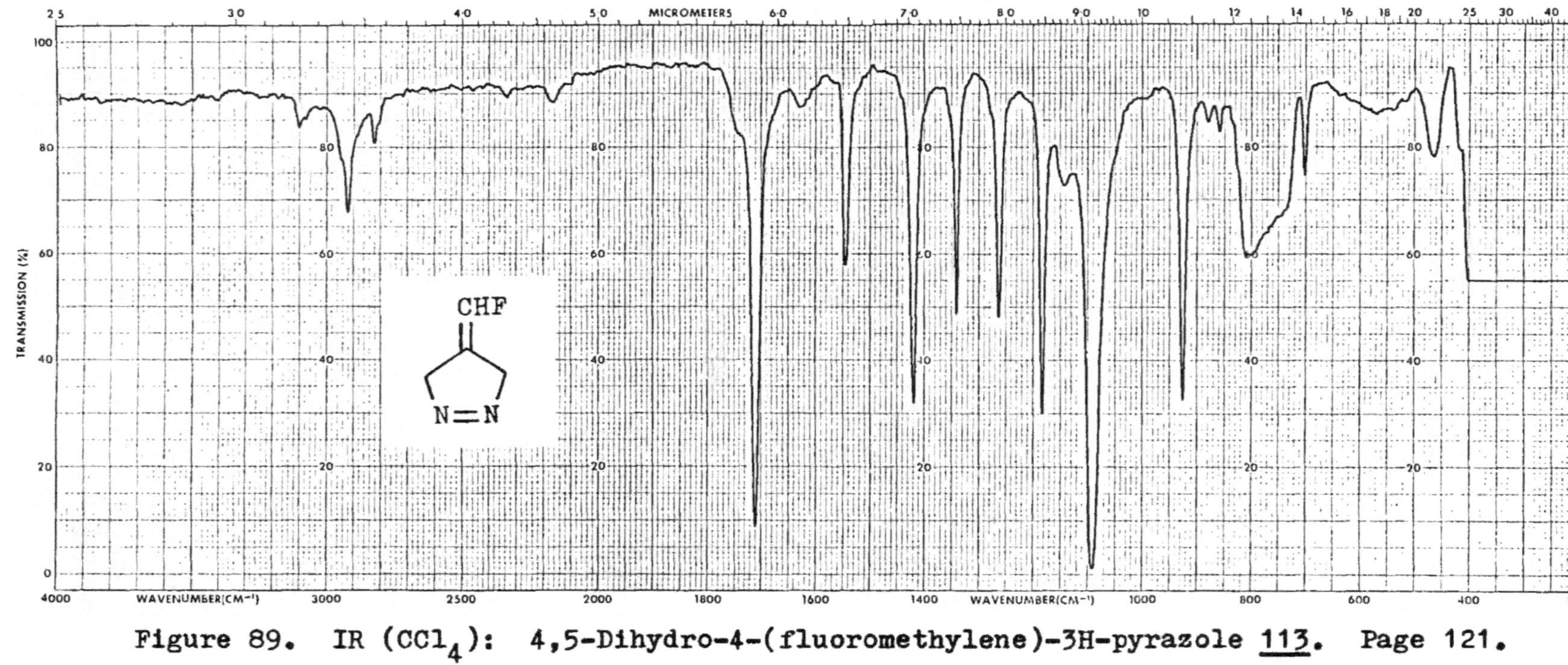

Figure 89. IR (CCl_4): 4,5-Dihydro-4-(fluoromethylene)-3H-pyrazole <u>113</u>. Page 121.

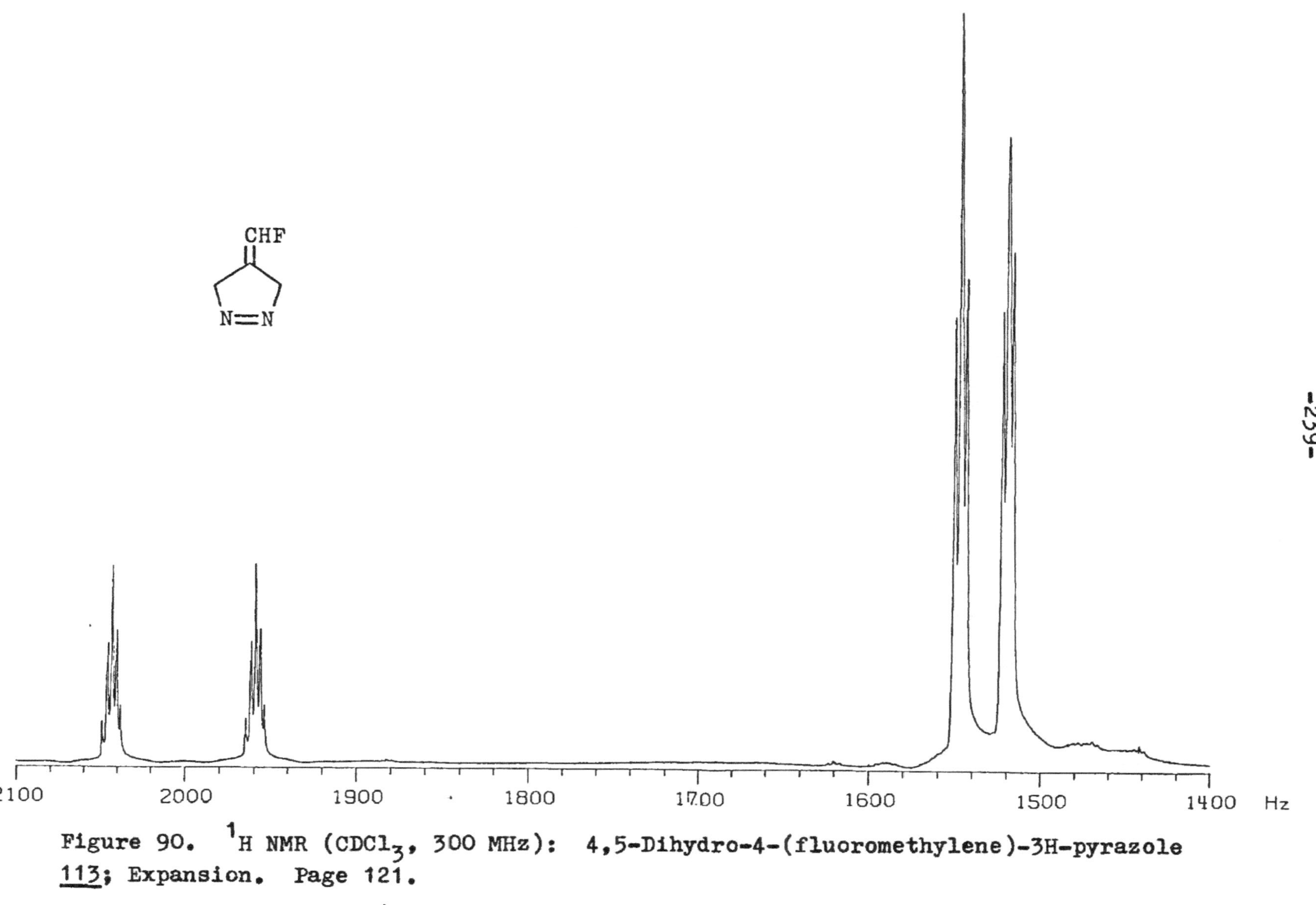

Figure 90. ^{1}H NMR ($CDCl_3$, 300 MHz): 4,5-Dihydro-4-(fluoromethylene)-3H-pyrazole 113; Expansion. Page 121.

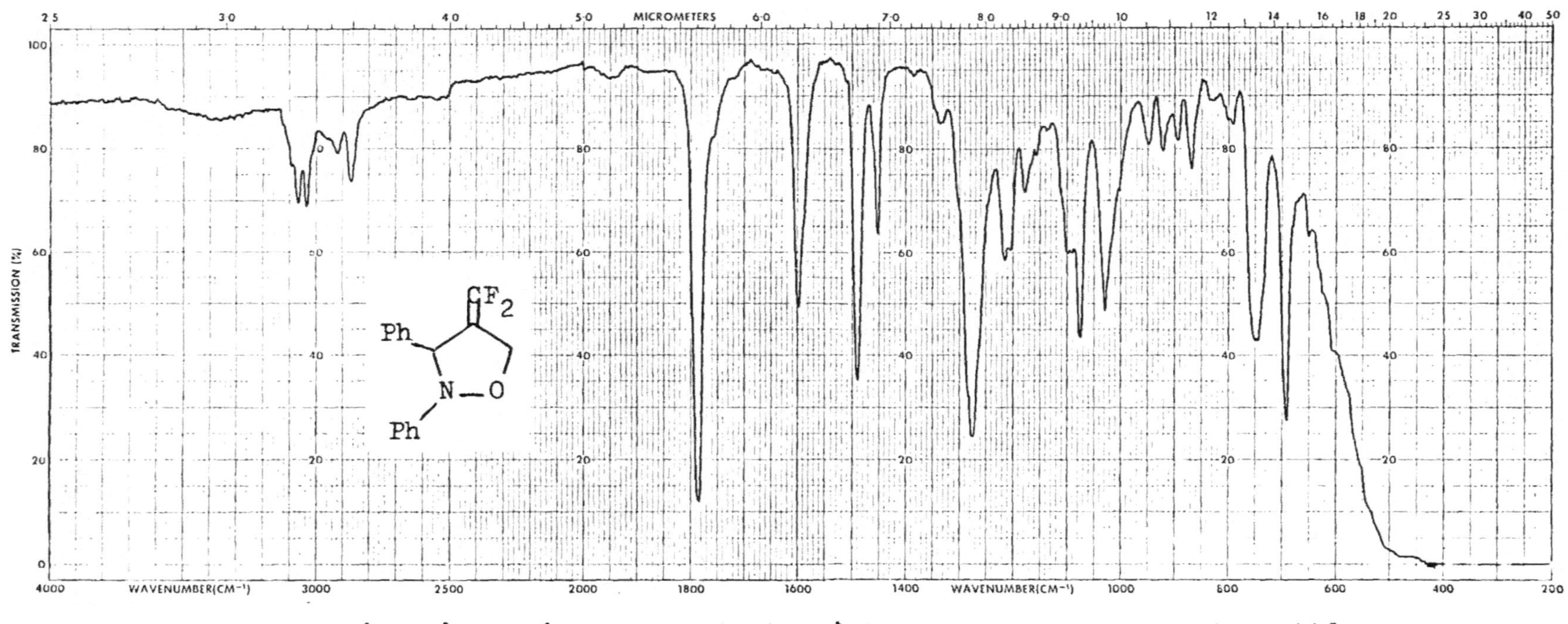

Figure 91. IR (CCl_4): 4-(Difluoromethylene)-2,3-diphenylisoxazolidine <u>116</u>. Page 122.

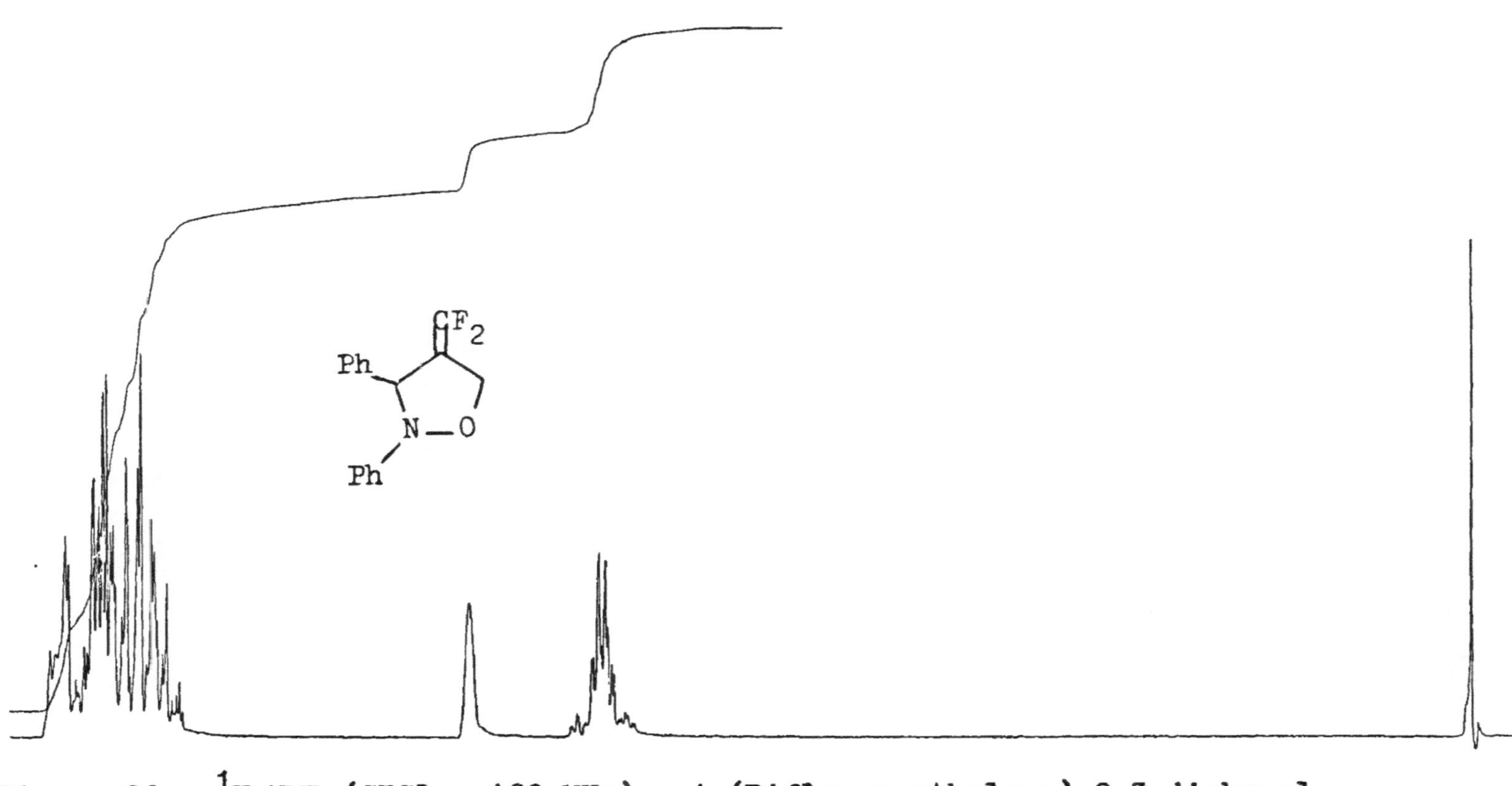

Figure 92. ^{1}H NMR ($CDCl_3$, 100 MHz): 4-(Difluoromethylene)-2,3-diphenylisoxazolidine 116. Page 122.

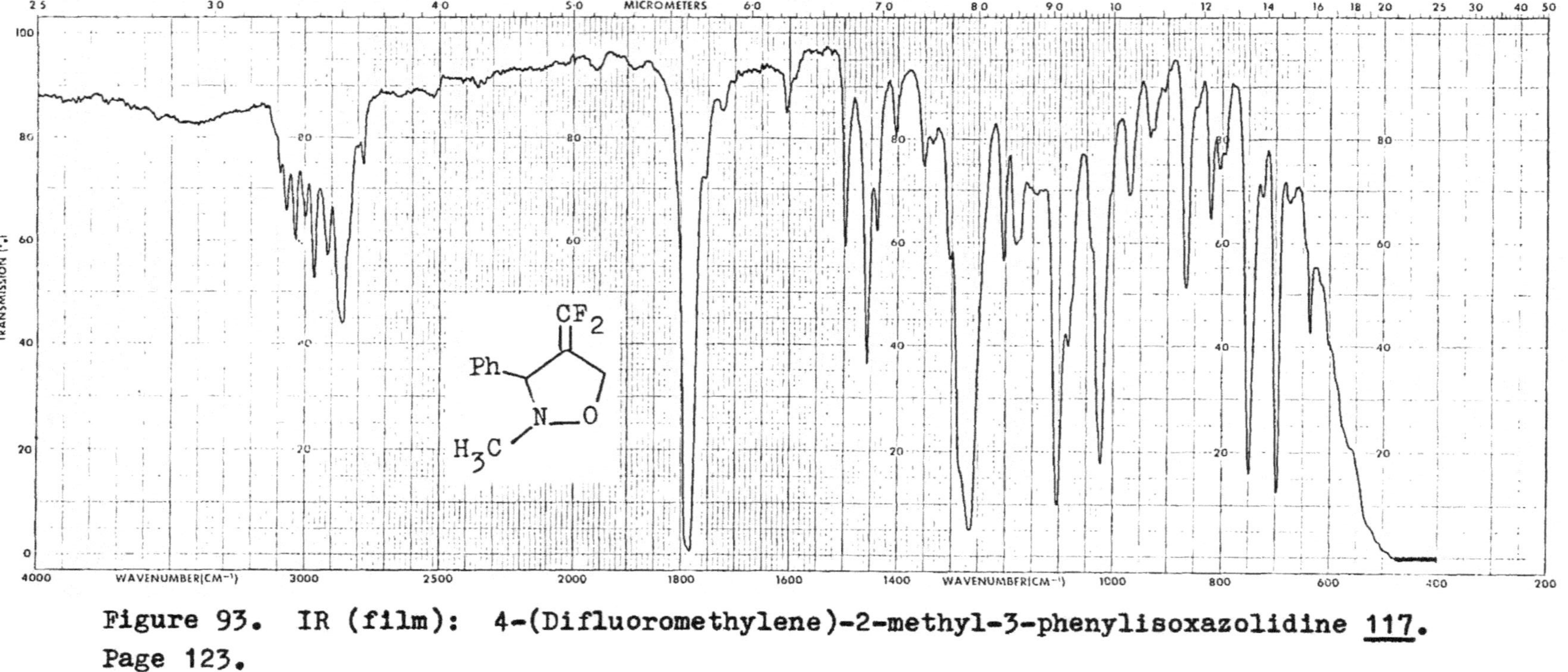

Figure 93. IR (film): 4-(Difluoromethylene)-2-methyl-3-phenylisoxazolidine **117**. Page 123.

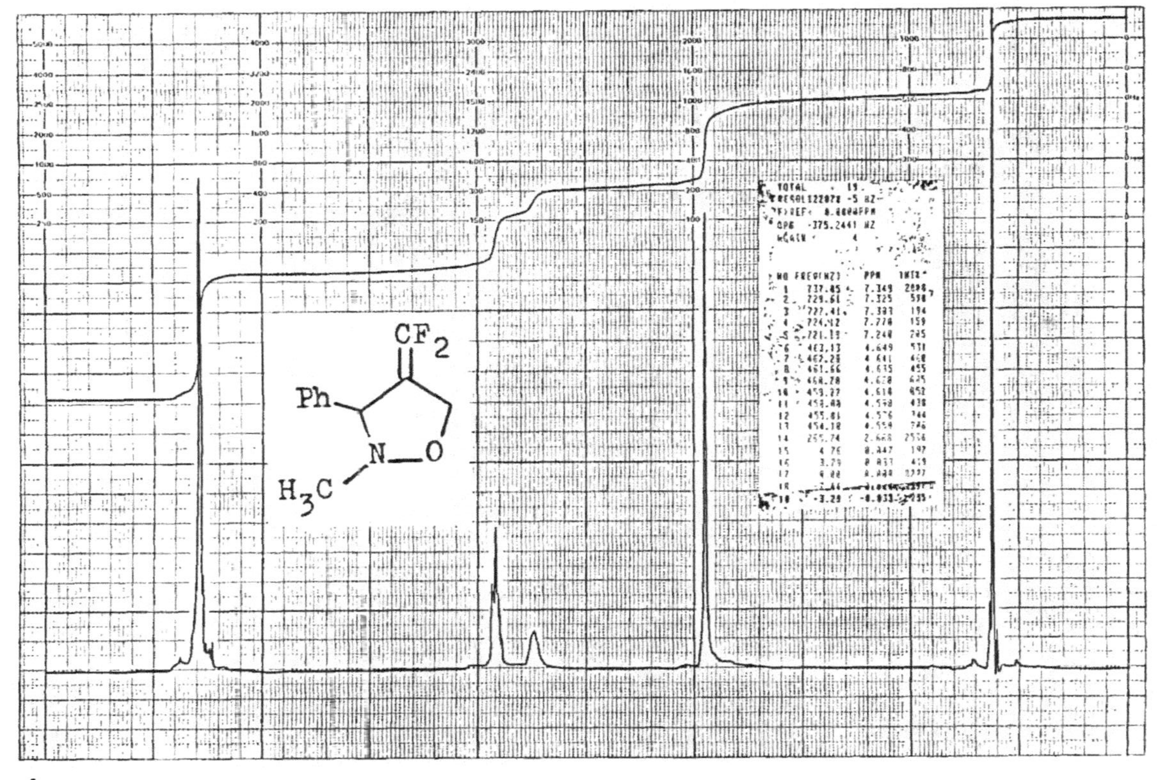

Figure 94. ^{1}H NMR ($CDCl_3$, 100 MHz): 4-(Difluoromethylene)-2-methyl-3-phenylisoxazolidine <u>117</u>. Page 123.

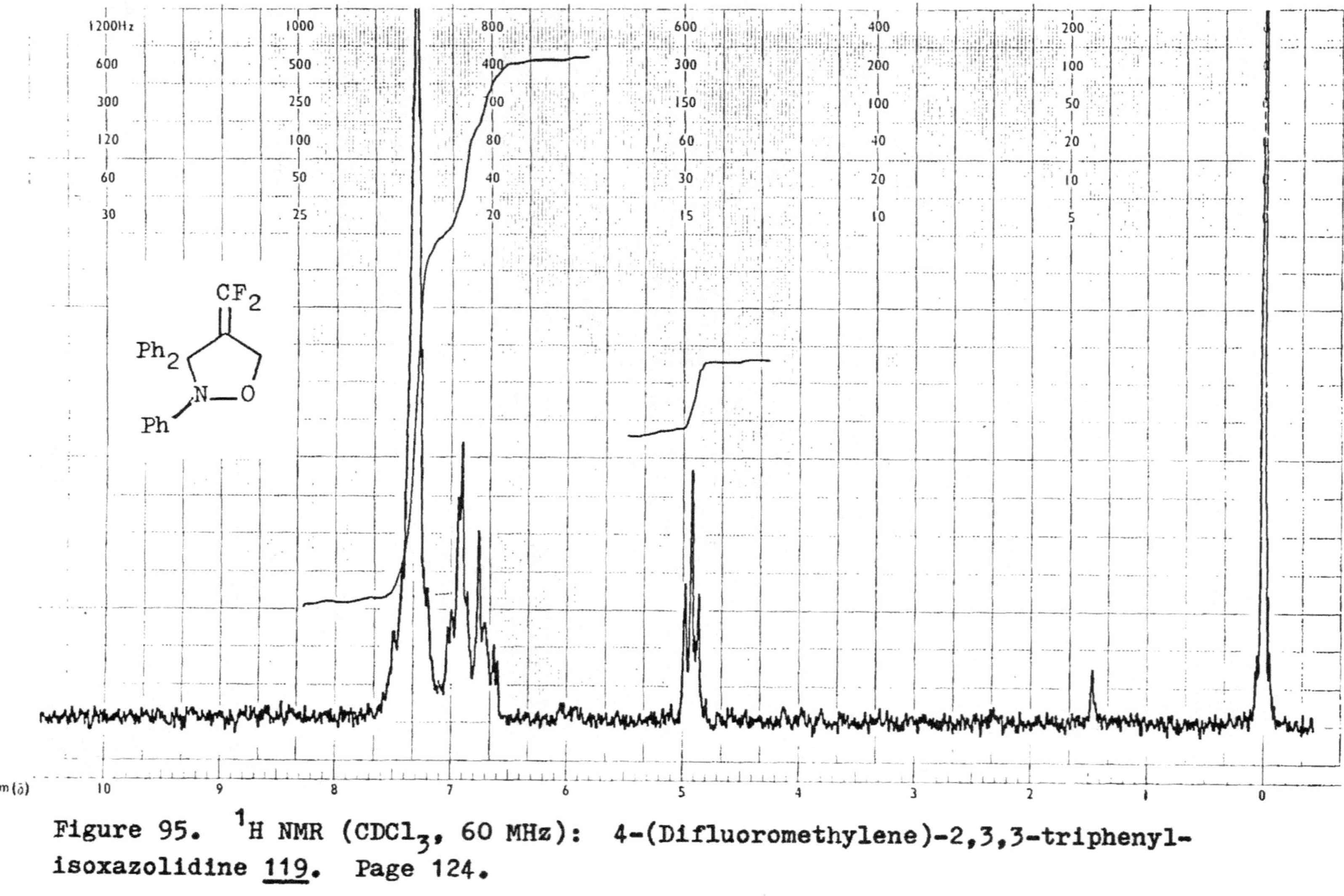

Figure 95. ^{1}H NMR ($CDCl_3$, 60 MHz): 4-(Difluoromethylene)-2,3,3-triphenylisoxazolidine <u>119</u>. Page 124.

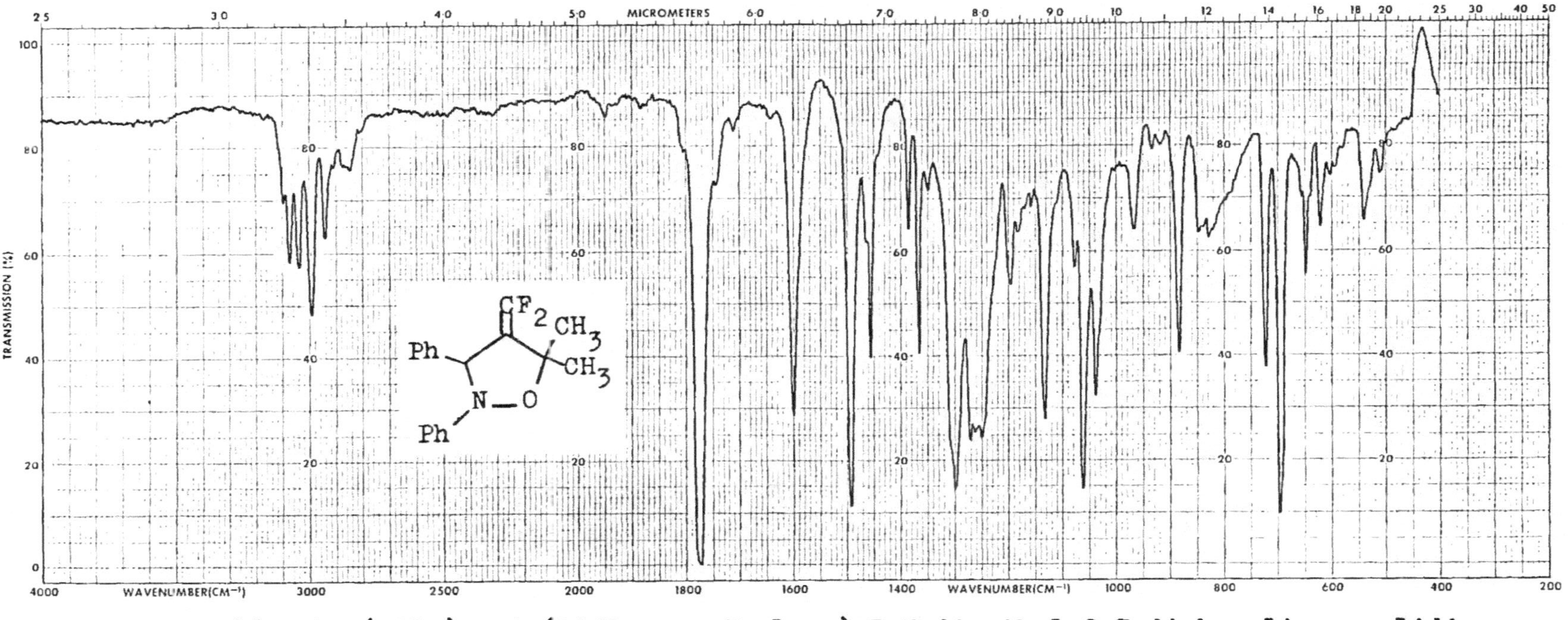

Figure 96. IR (CCl_4): 4-(Difluoromethylene)-5,5-dimethyl-2,3-diphenylisoxazolidine 120. Page 124.

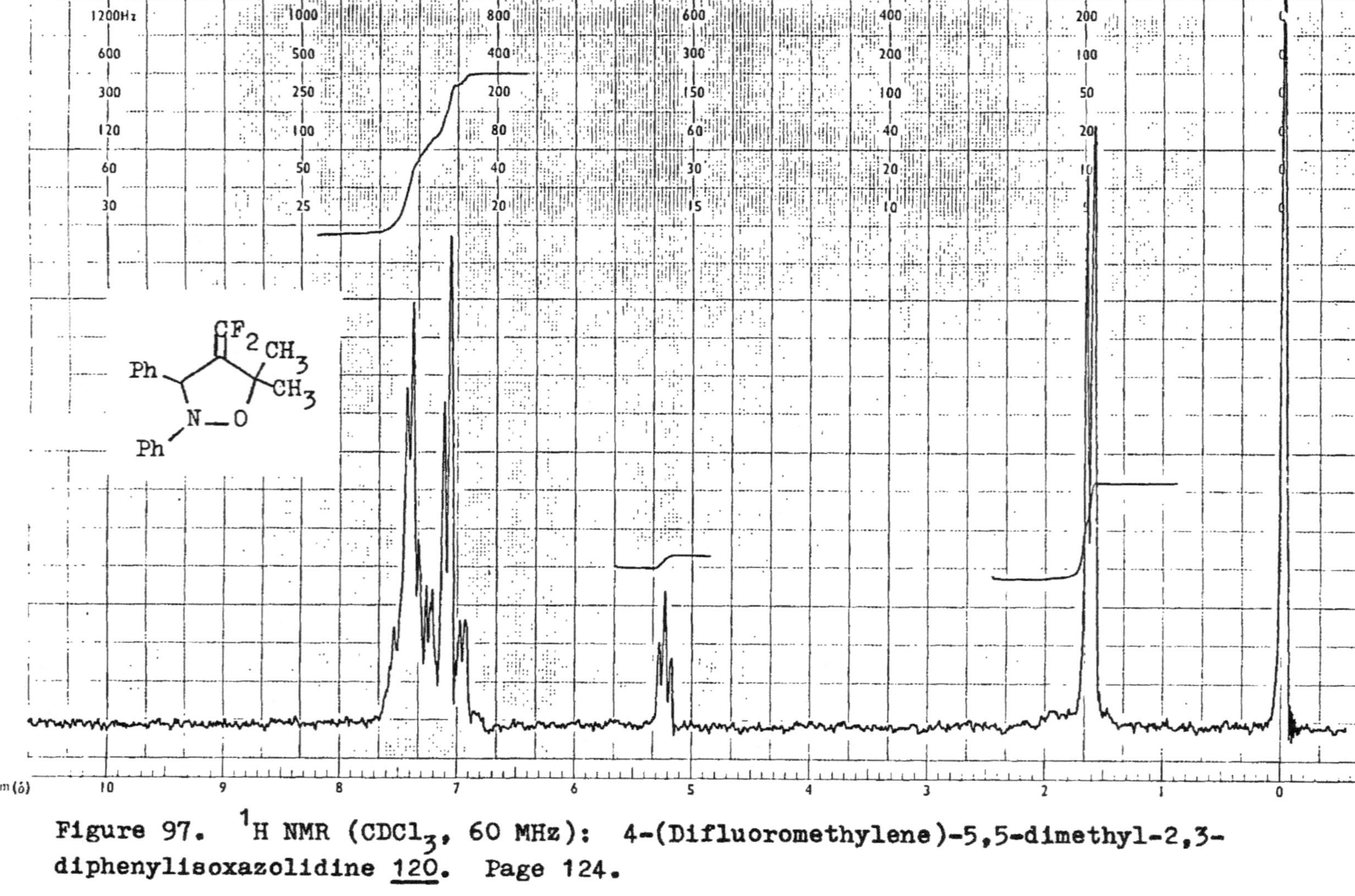

Figure 97. ^{1}H NMR ($CDCl_3$, 60 MHz): 4-(Difluoromethylene)-5,5-dimethyl-2,3-diphenylisoxazolidine <u>120</u>. Page 124.

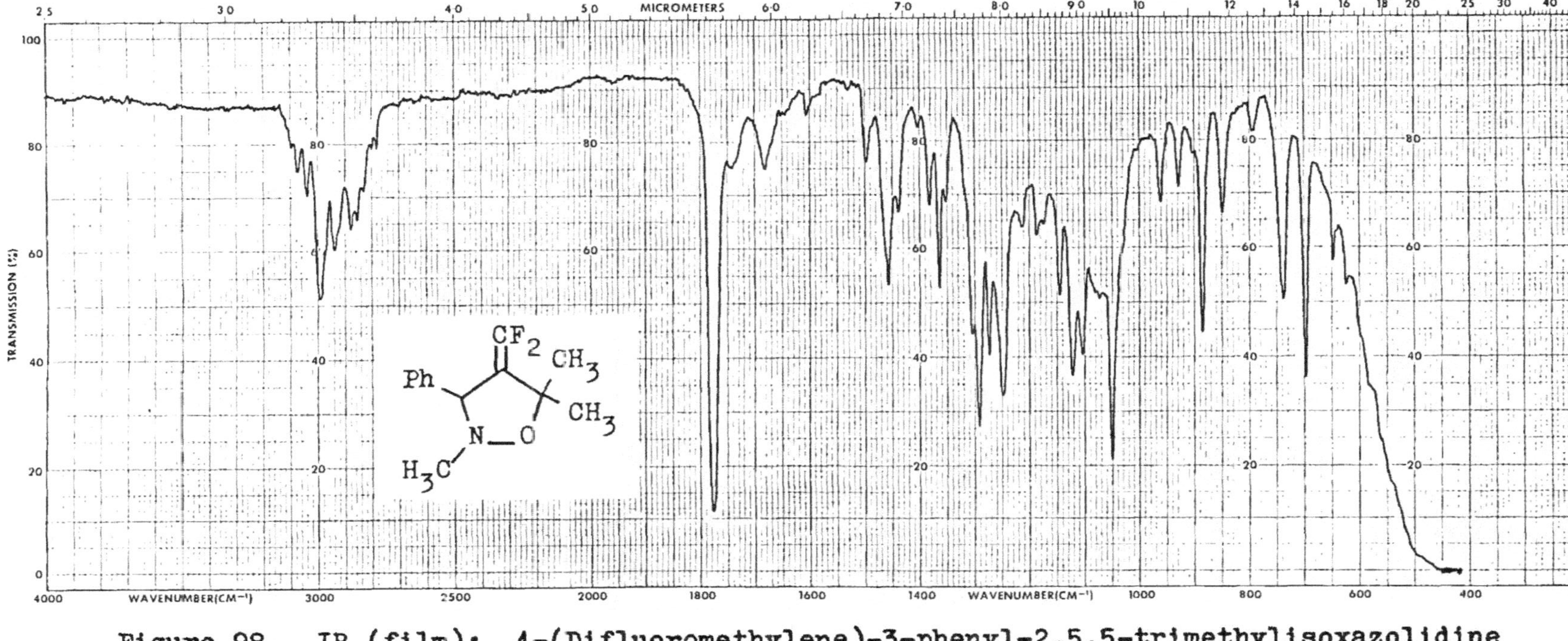

Figure 98. IR (film): 4-(Difluoromethylene)-3-phenyl-2,5,5-trimethylisoxazolidine <u>121</u>. Page 125.

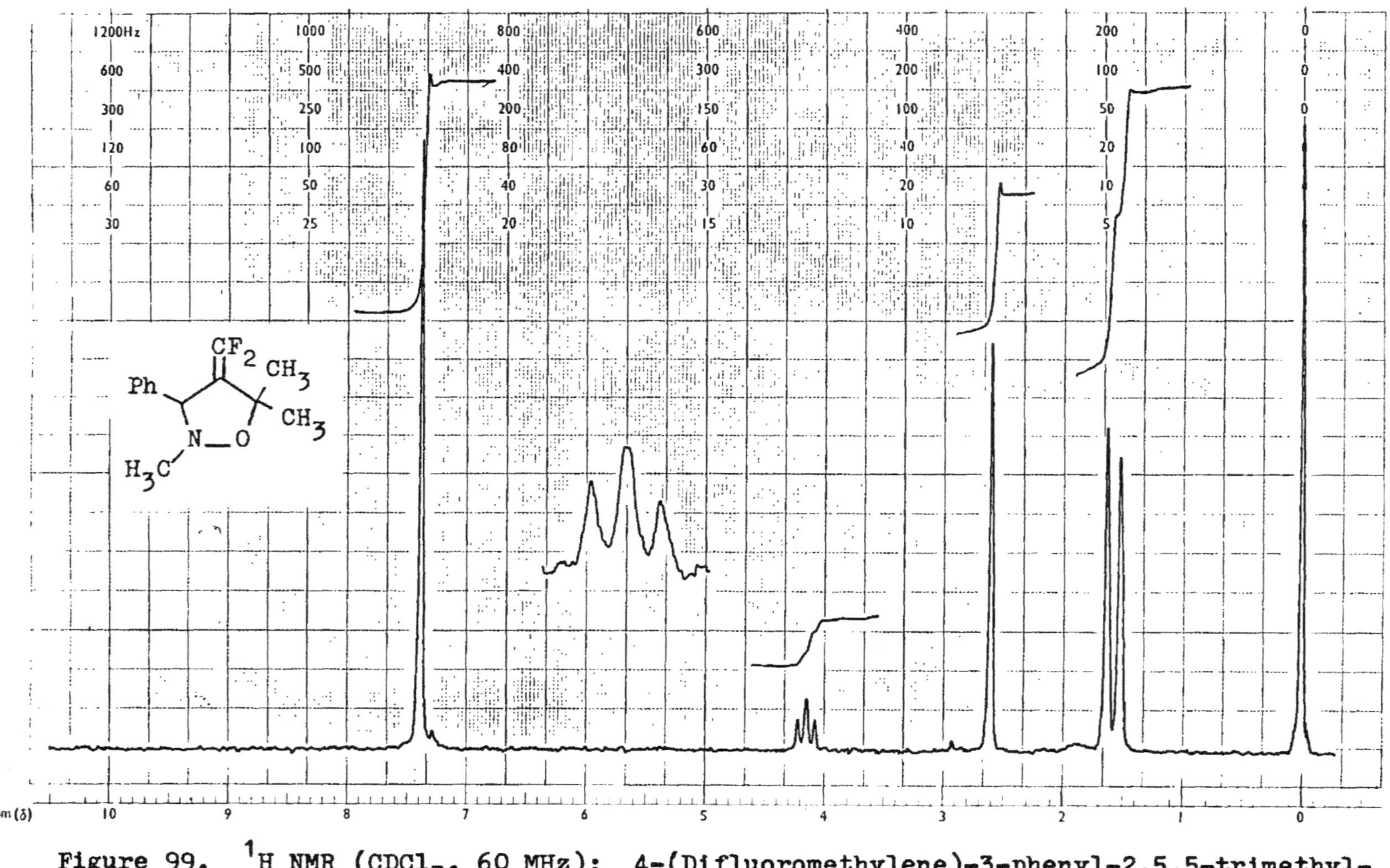

Figure 99. ^{1}H NMR ($CDCl_3$, 60 MHz): 4-(Difluoromethylene)-3-phenyl-2,5,5-trimethylisoxazolidine 121. Page 125.

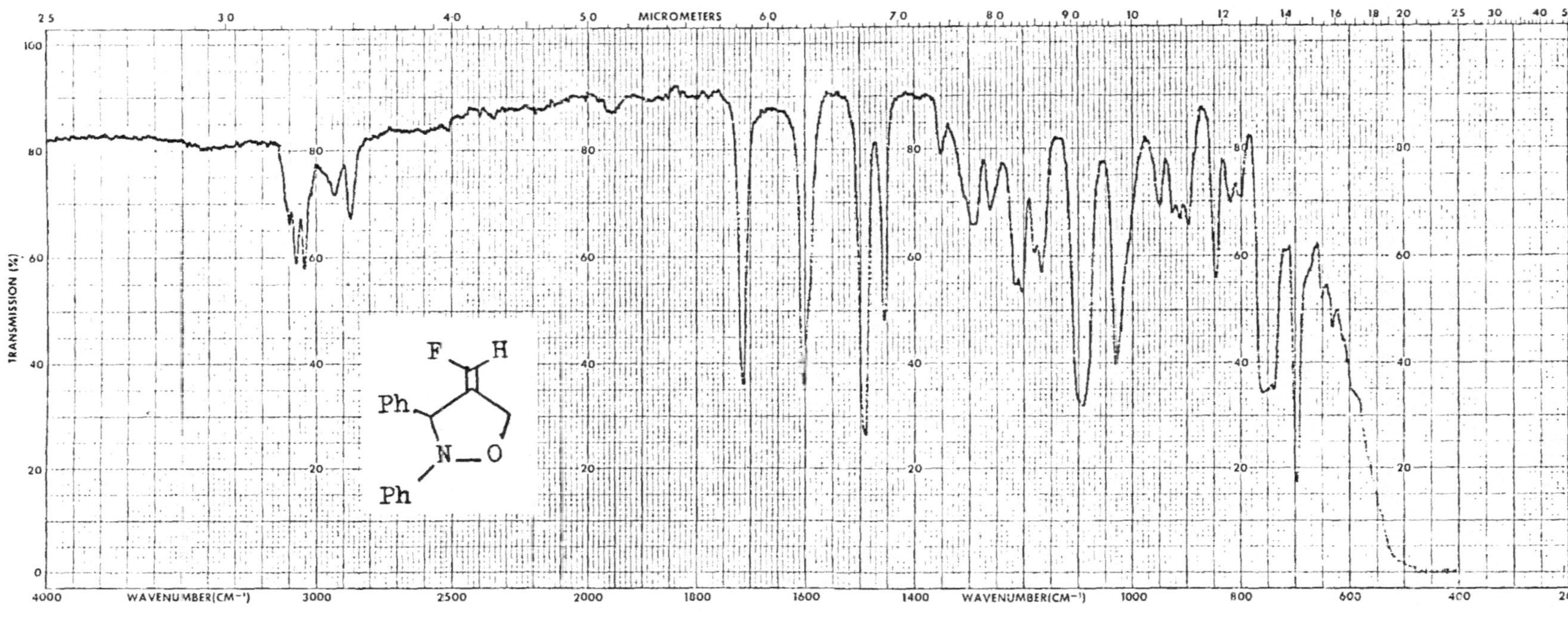

Figure 100. IR (film): (E)-2,3-Diphenyl-4-(fluoromethylene)isoxazolidine <u>122</u>. Page 126.

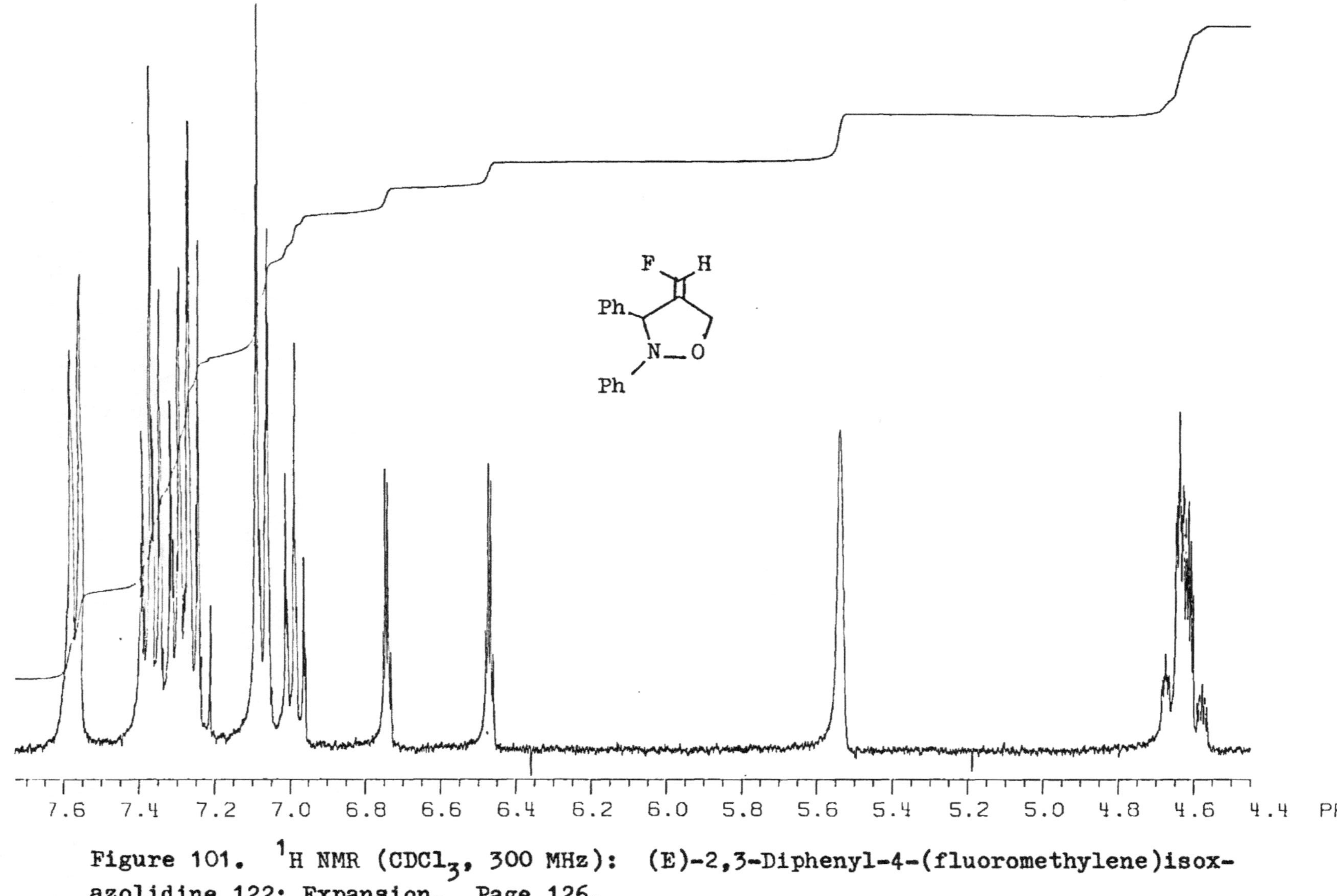

Figure 101. ^{1}H NMR ($CDCl_3$, 300 MHz): (E)-2,3-Diphenyl-4-(fluoromethylene)isoxazolidine <u>122</u>; Expansion. Page 126.

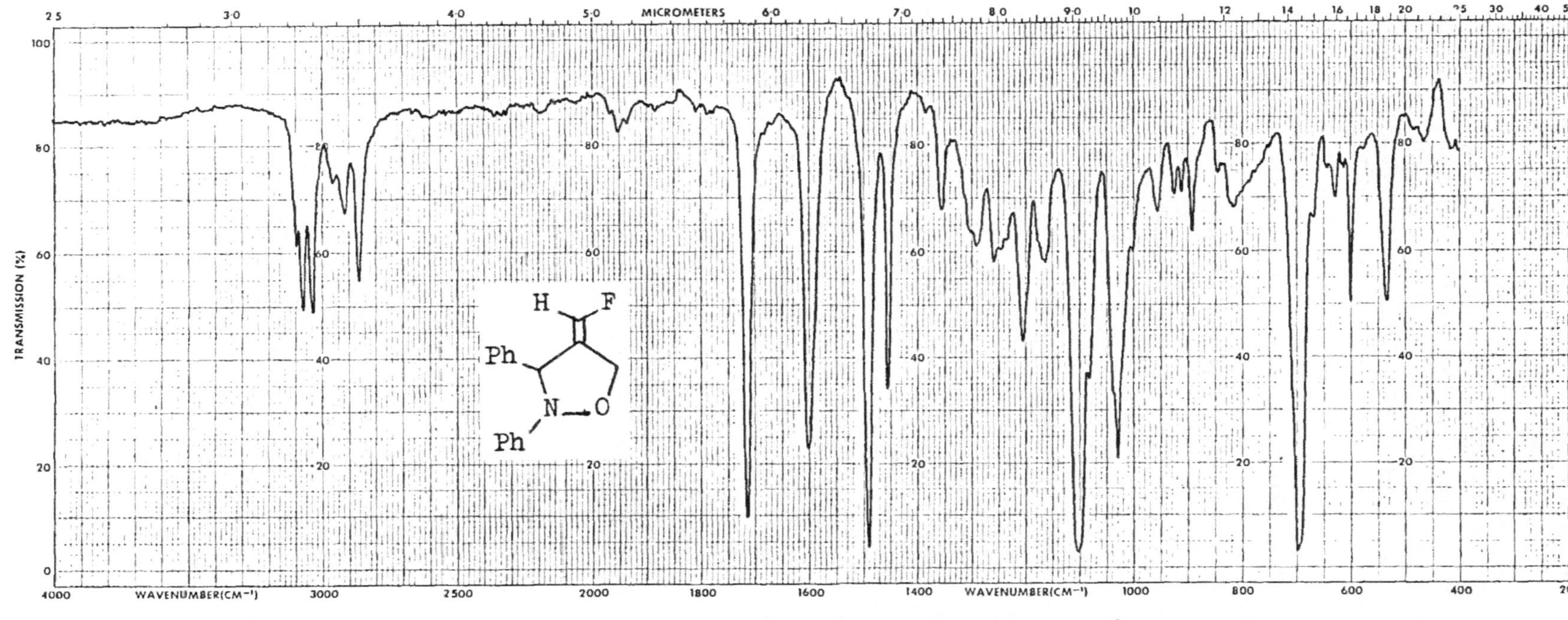

Figure 102. IR (CCl_4): (Z)-2,3-Diphenyl-4-(fluoromethylene)isoxazolidine <u>123</u>. Page 126.

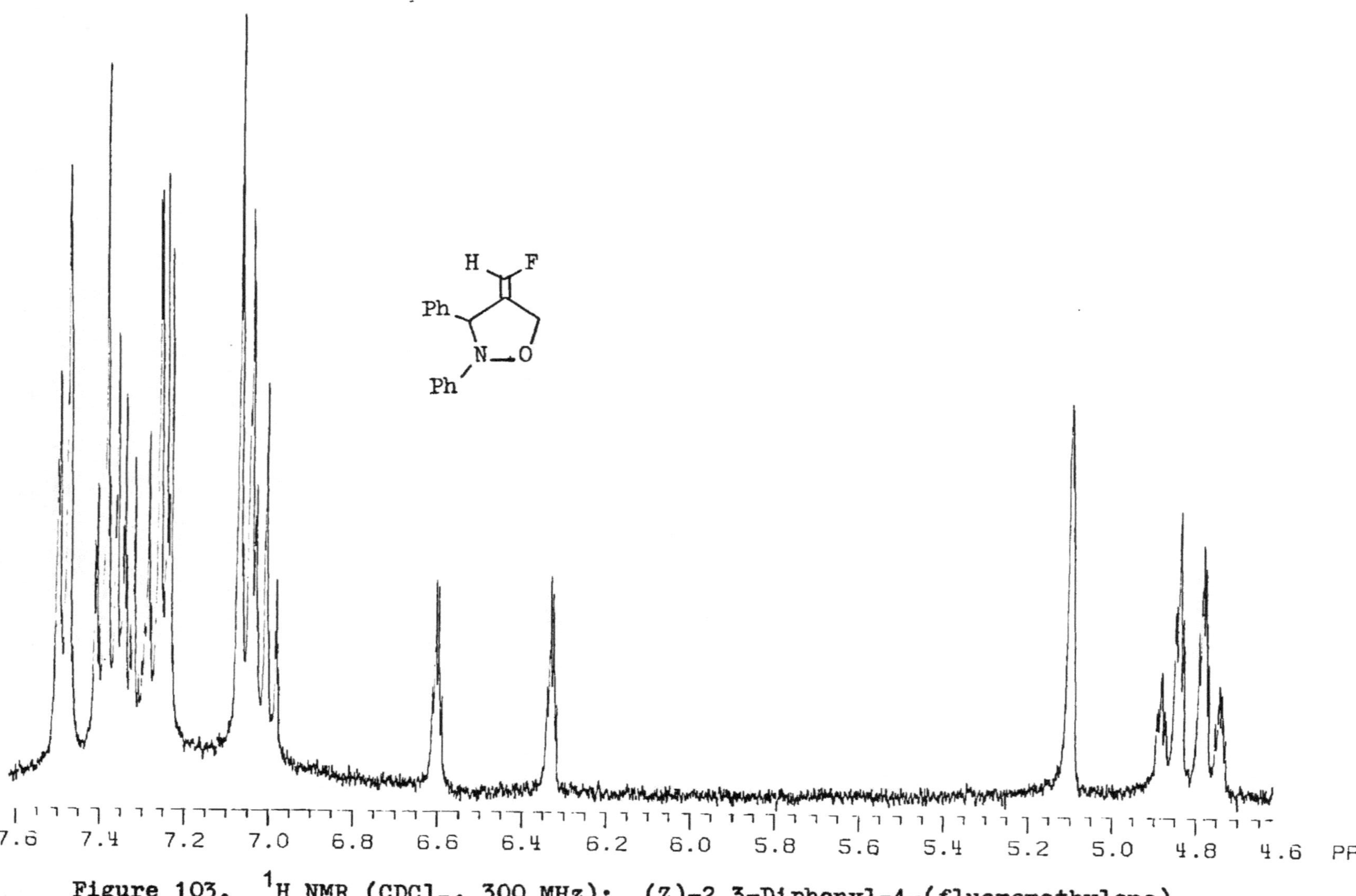

Figure 103. ^{1}H NMR ($CDCl_3$, 300 MHz): (Z)-2,3-Diphenyl-4-(fluoromethylene)-isoxazolidine <u>123</u>; Expansion. Page 126.

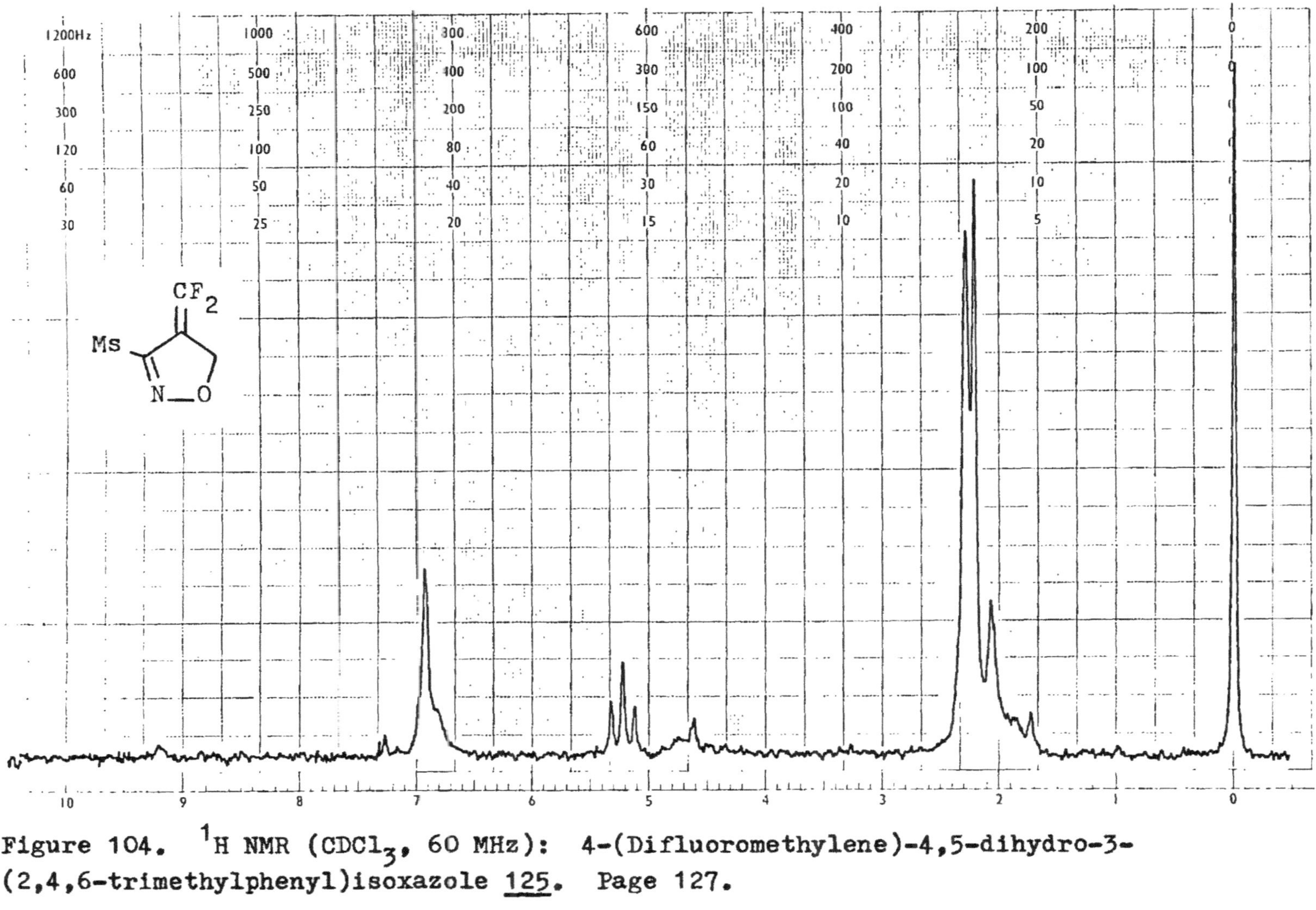

Figure 104. ^{1}H NMR ($CDCl_3$, 60 MHz): 4-(Difluoromethylene)-4,5-dihydro-3-(2,4,6-trimethylphenyl)isoxazole <u>125</u>. Page 127.

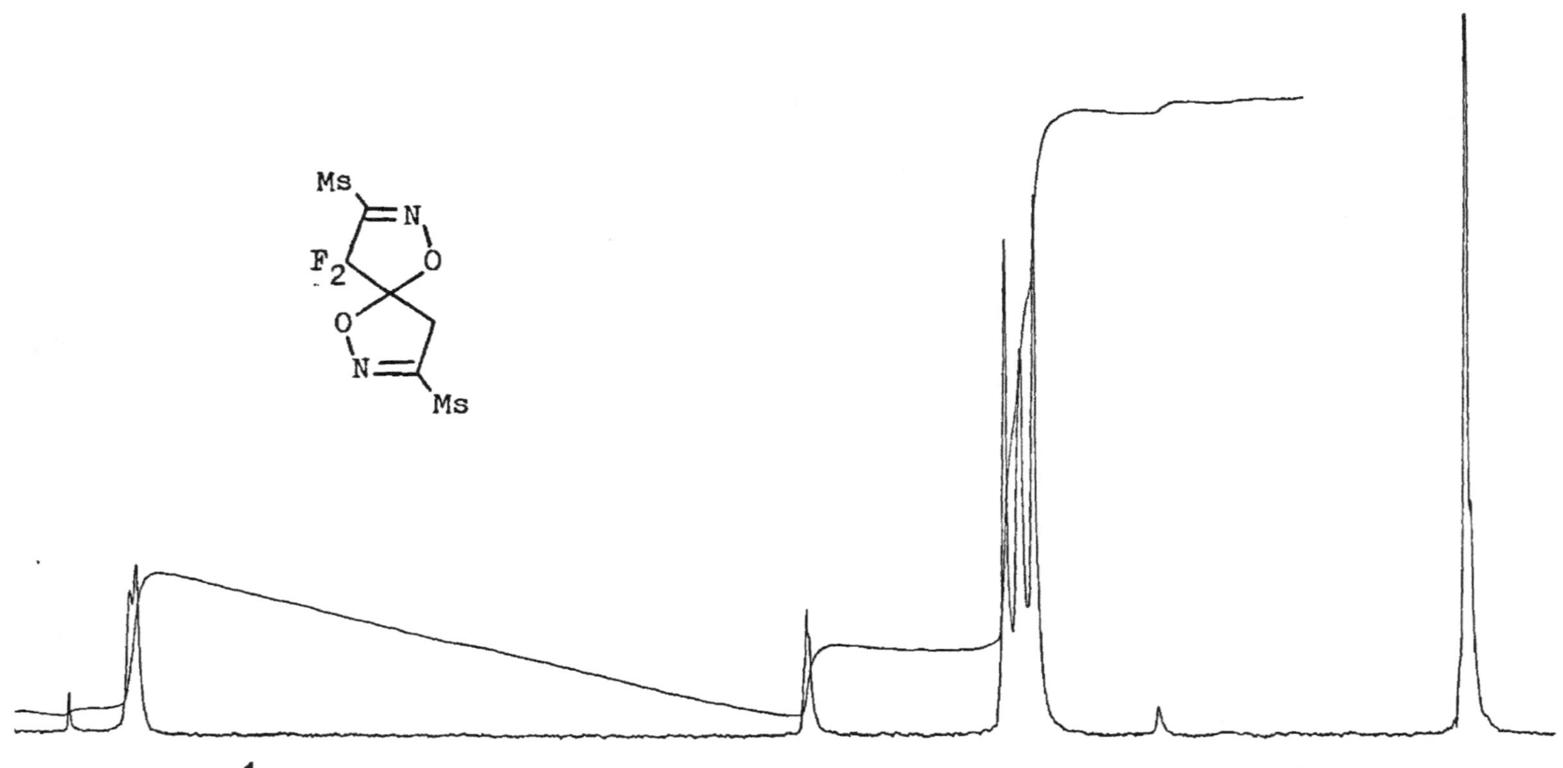

Figure 105. ^{1}H NMR ($CDCl_3$, 100 MHz): 3,8-Bis-(2,4,6-trimethylphenyl)-4,4-difluoro-1,6-dioxa-2,7-diazaspiro[4.4]nona-2,7-diene <u>126</u>. Page 127.

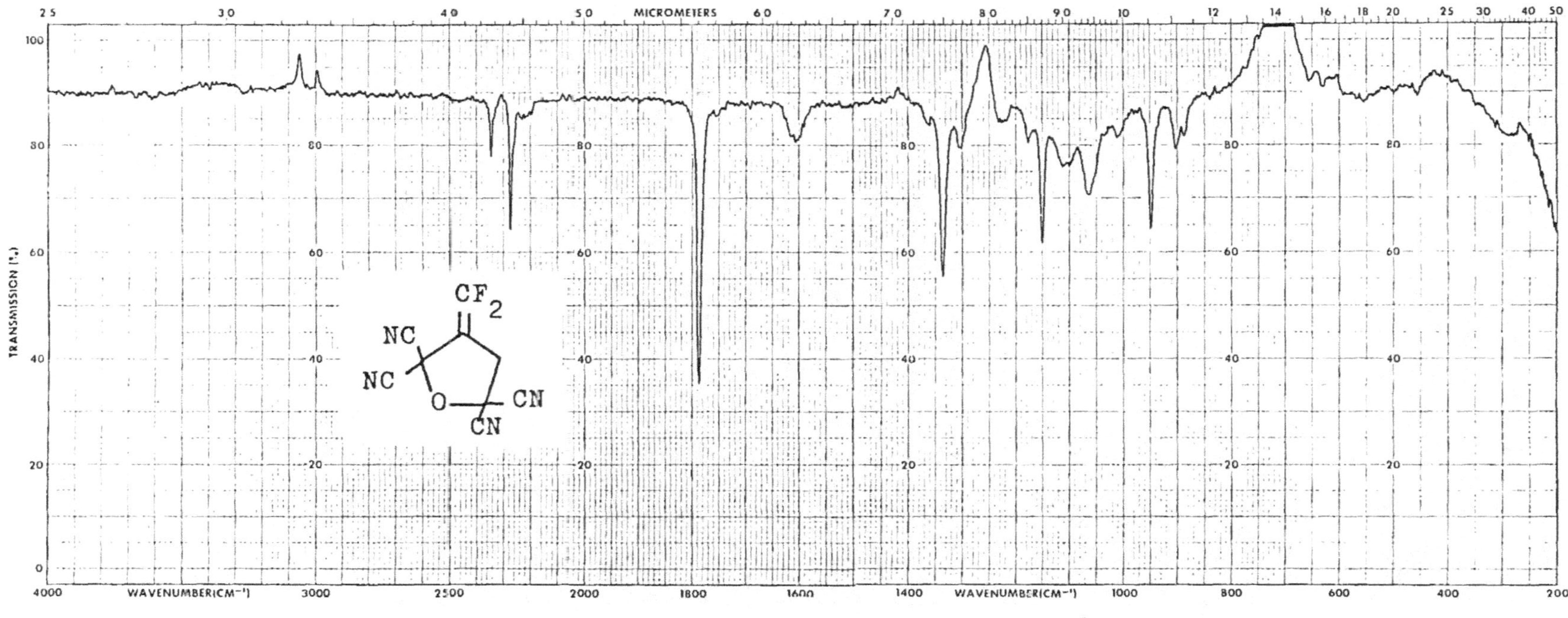

Figure 106. IR (CH_2Cl_2): Tetrahydro-3-(difluoromethylene)-2,2,5,5-furantetracarbonitrile 130. Page 128.

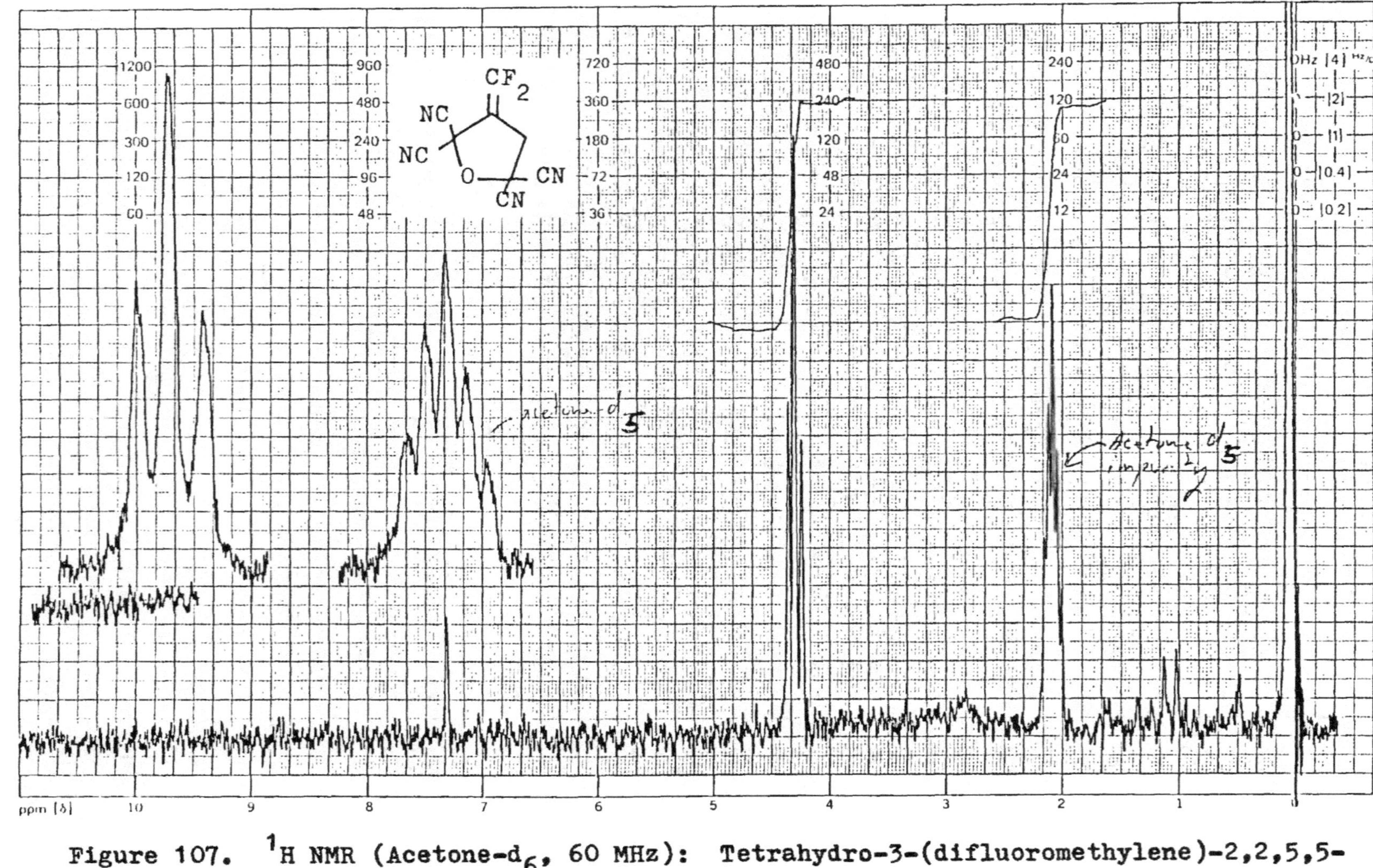

Figure 107. ^{1}H NMR (Acetone-d_6, 60 MHz): Tetrahydro-3-(difluoromethylene)-2,2,5,5-furantetracarbonitrile 130. Page 128.

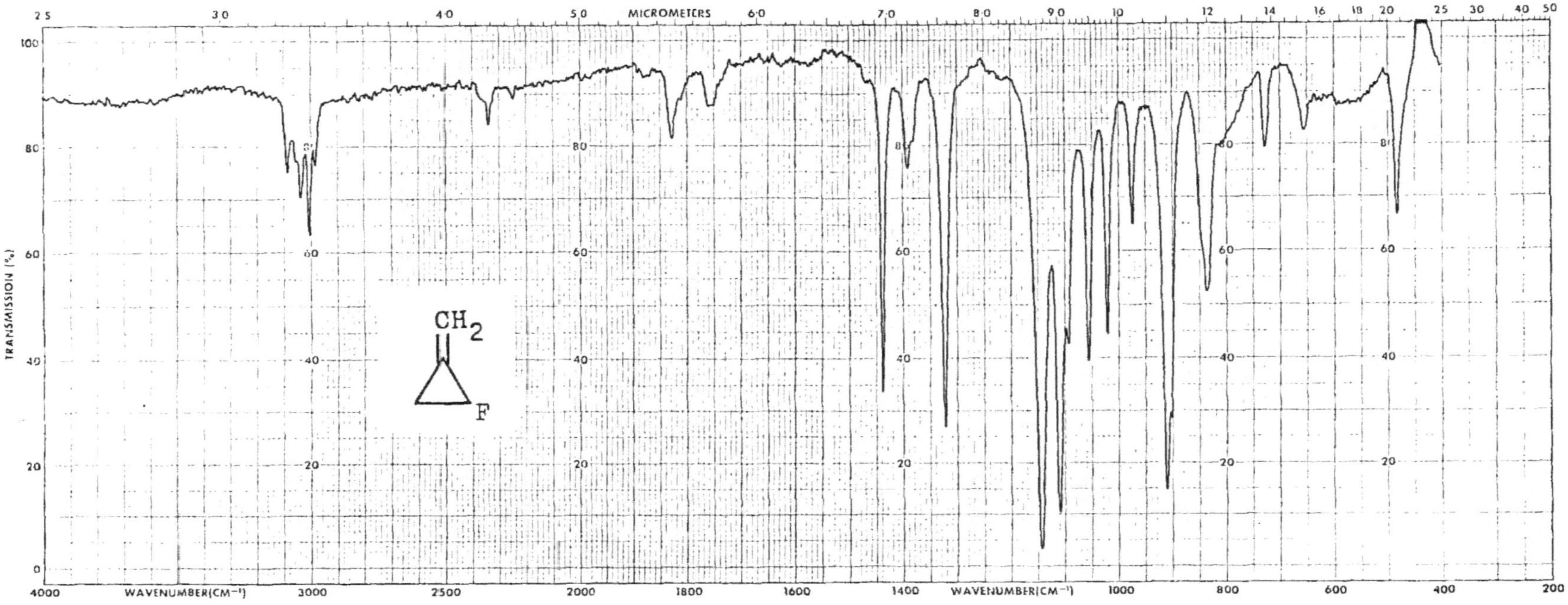

Figure 108. IR (CCl_4): 1-Fluoro-2-methylenecyclopropane <u>136</u>. Page 129.

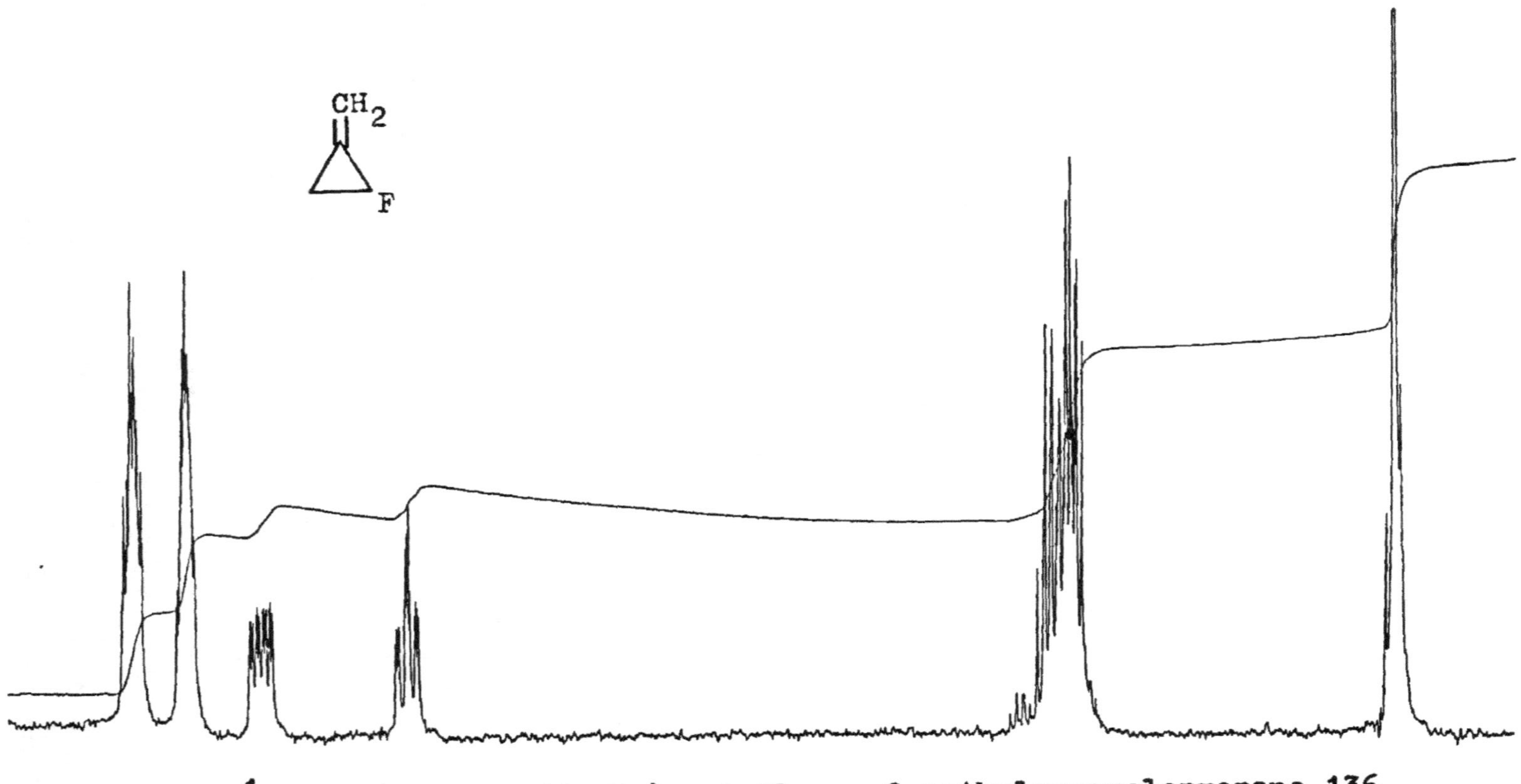

Figure 109. ^{1}H NMR ($CDCl_3$, 100 MHz): 1-Fluoro-2-methylenecyclopropane <u>136</u>. Page 129.

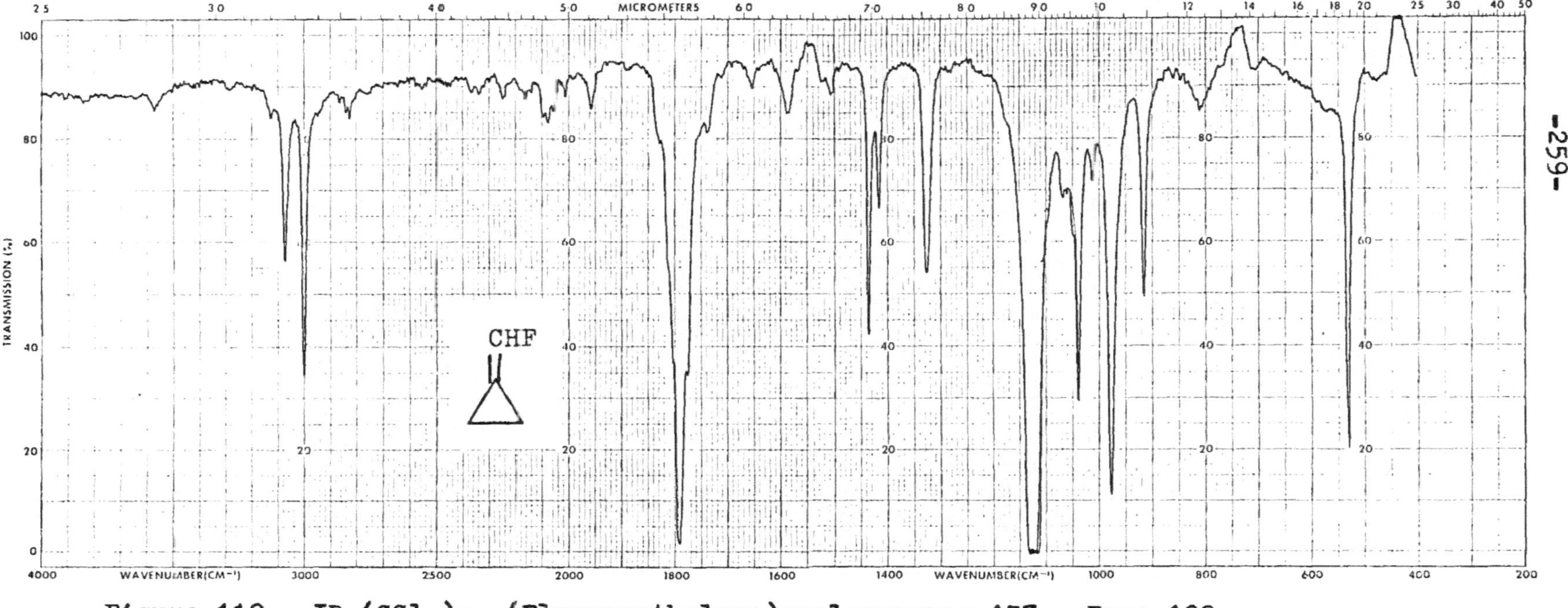

Figure 110. IR (CCl_4): (Fluoromethylene)cyclopropane **137**. Page 129.

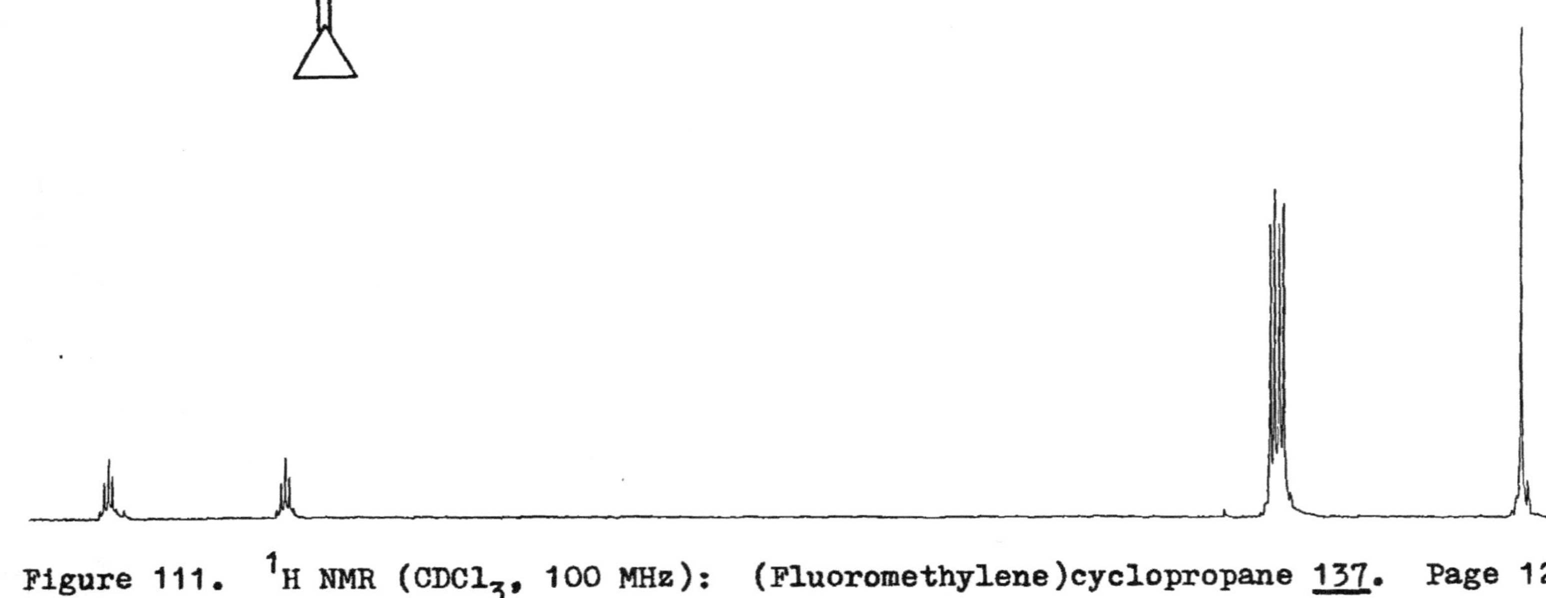

Figure 111. ^{1}H NMR ($CDCl_3$, 100 MHz): (Fluoromethylene)cyclopropane <u>137</u>. Page 129.

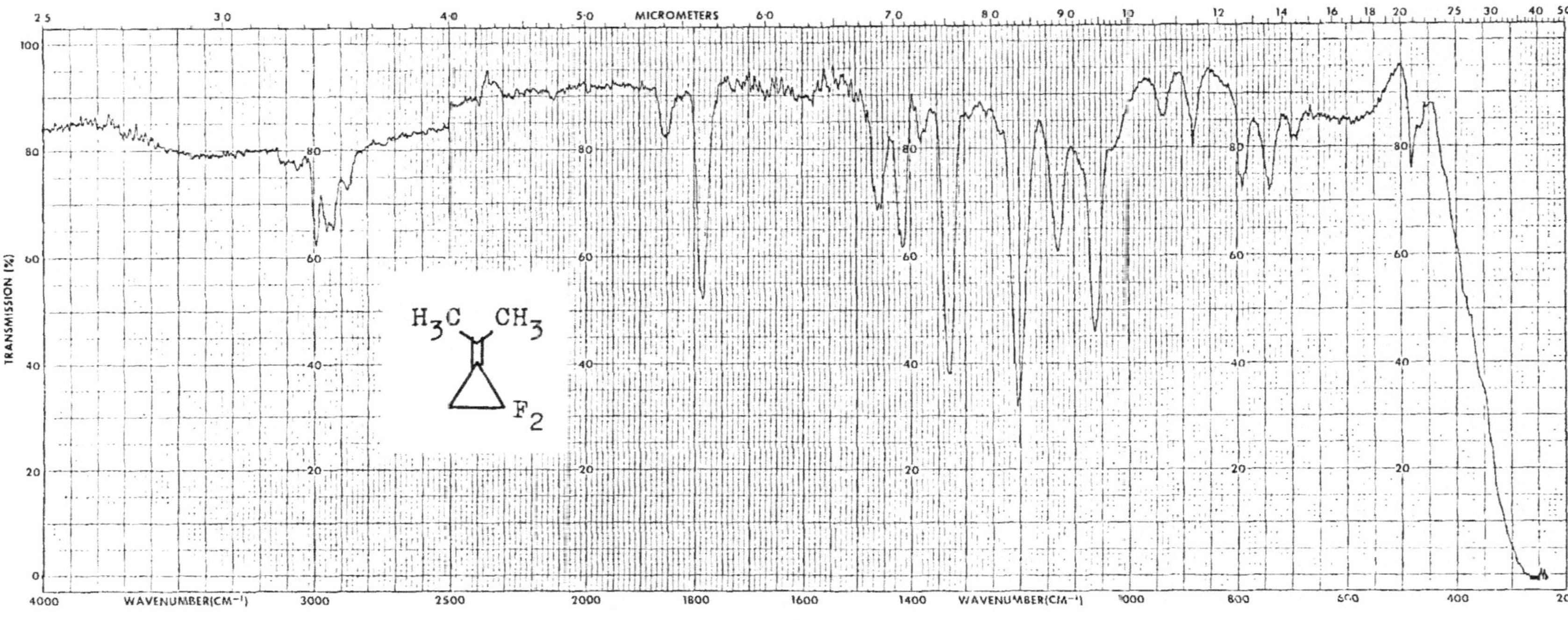

Figure 112. IR (gas): 1,1-Difluoro-2-(1-methylethylidene)cyclopropane <u>138</u>. Page 130.

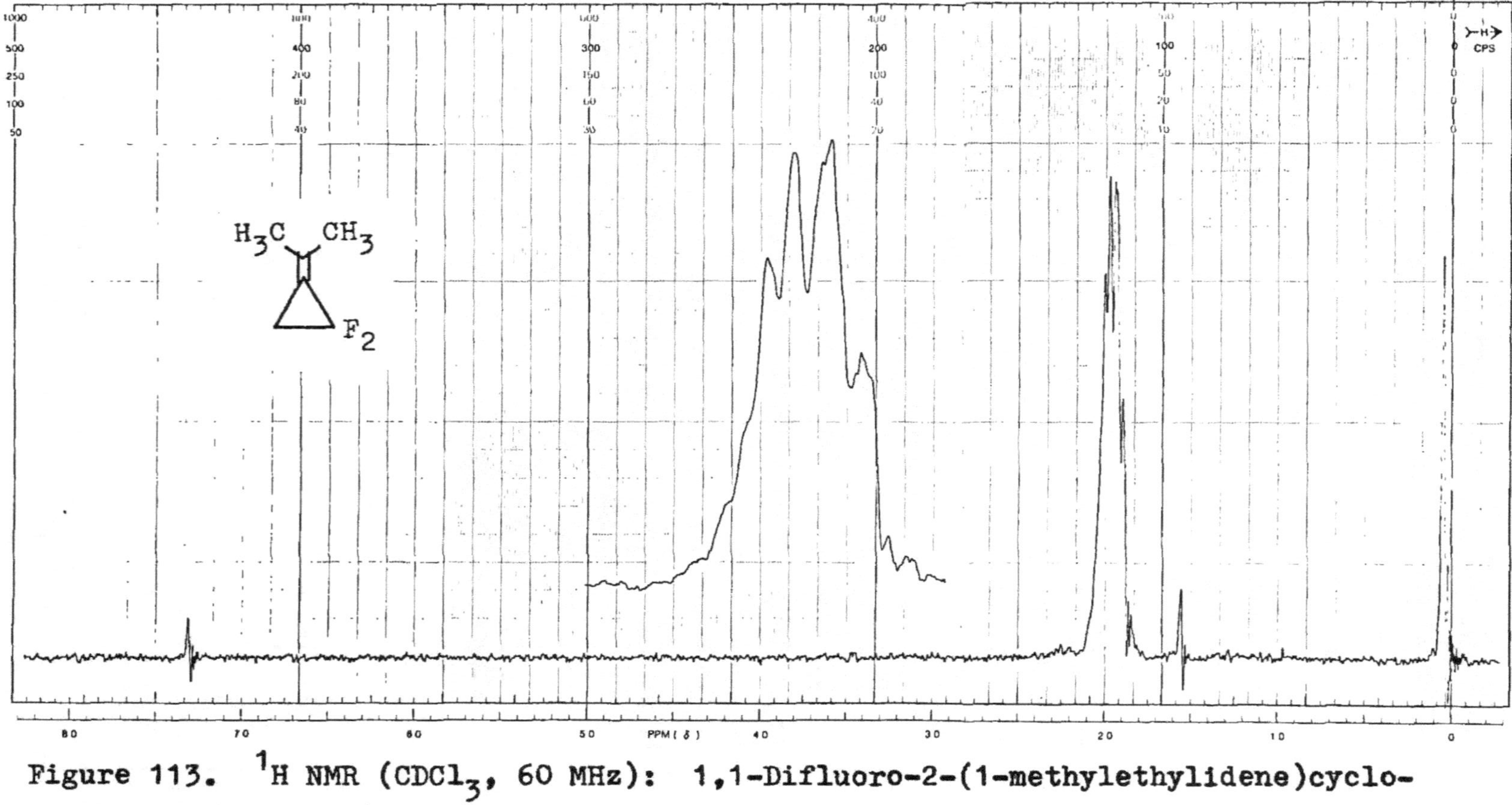

Figure 113. ^{1}H NMR ($CDCl_3$, 60 MHz): 1,1-Difluoro-2-(1-methylethylidene)cyclopropane <u>138</u>. Page 130.

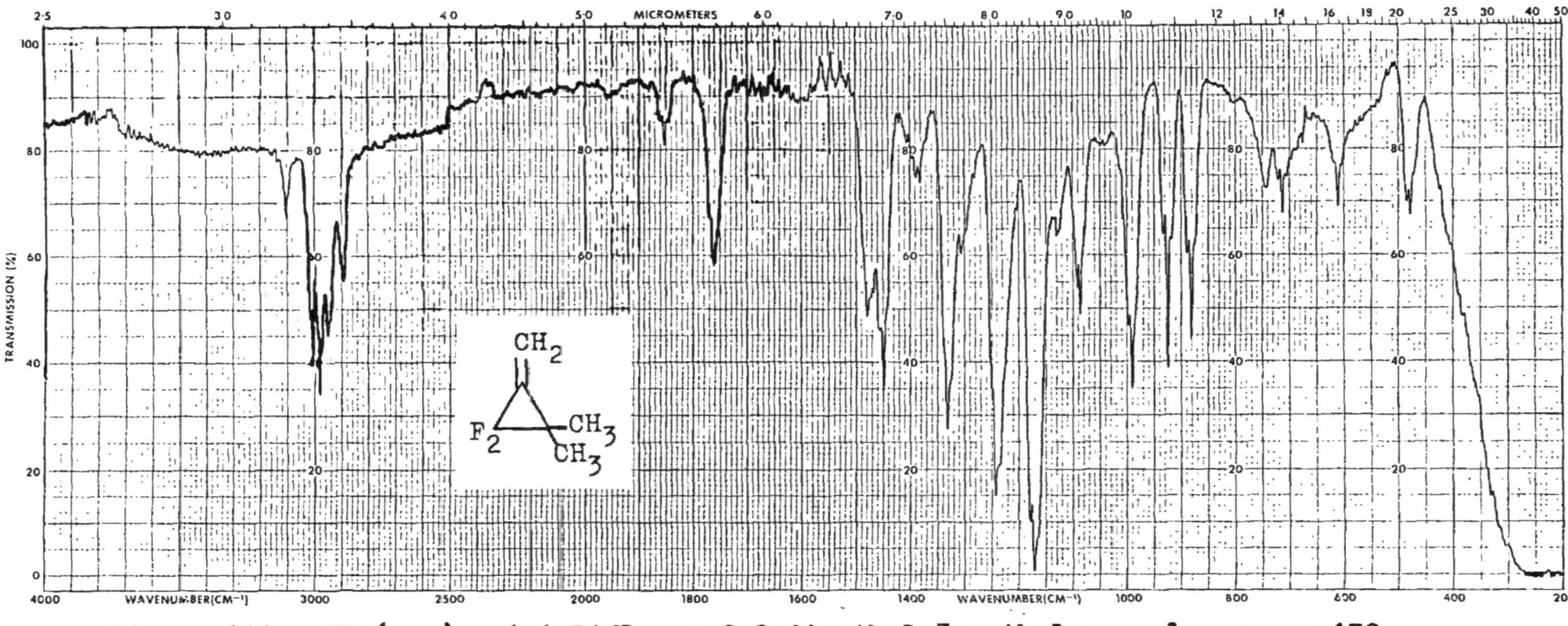

Figure 114. IR (gas): 1,1-Difluoro-2,2-dimethyl-3-methylenecyclopropane <u>139</u>. Page 131.

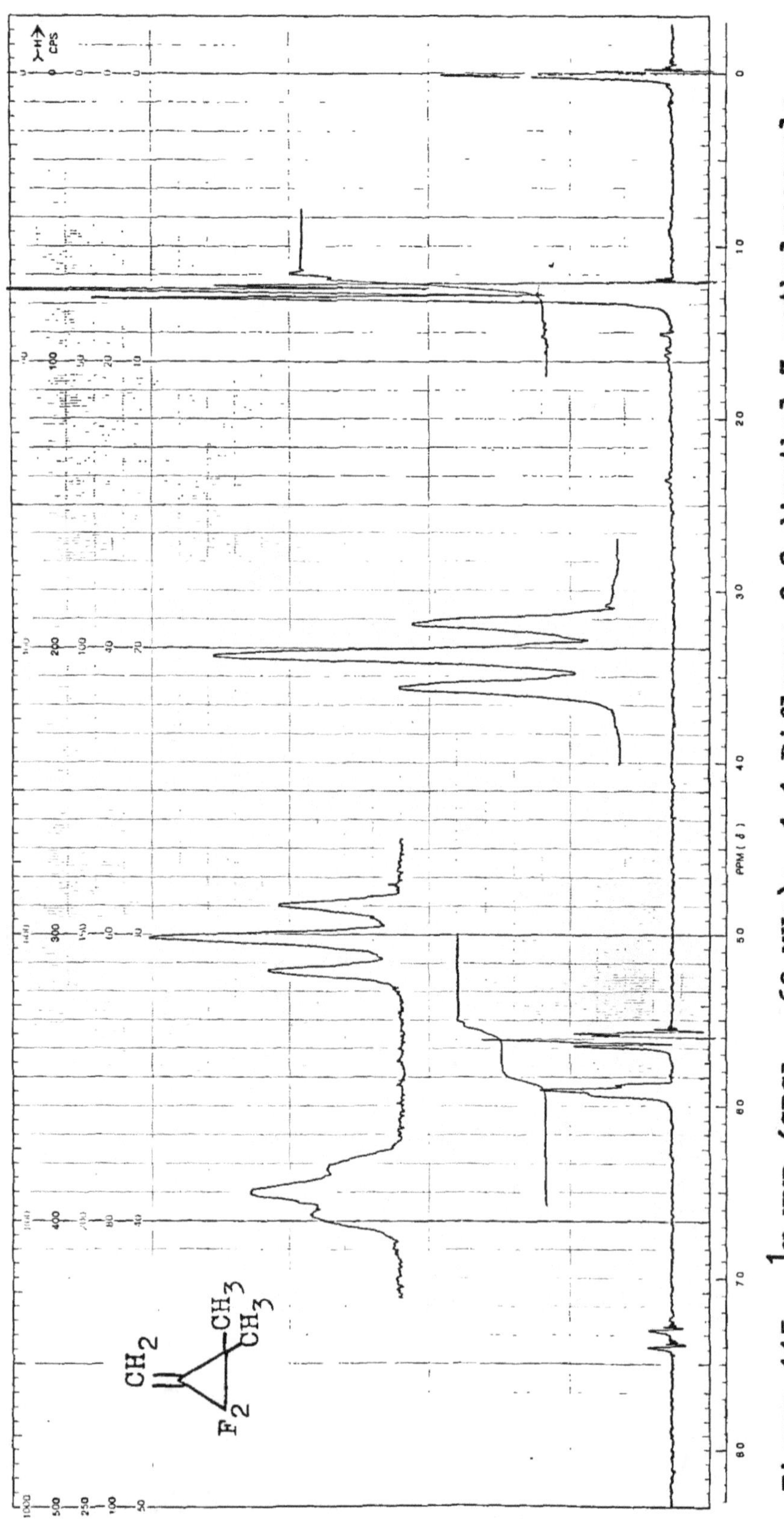

Figure 115. ^{1}H NMR ($CDCl_3$, 60 MHz): 1,1-Difluoro-2,2-dimethyl-3-methylenecyclopropane 139. Page 131.

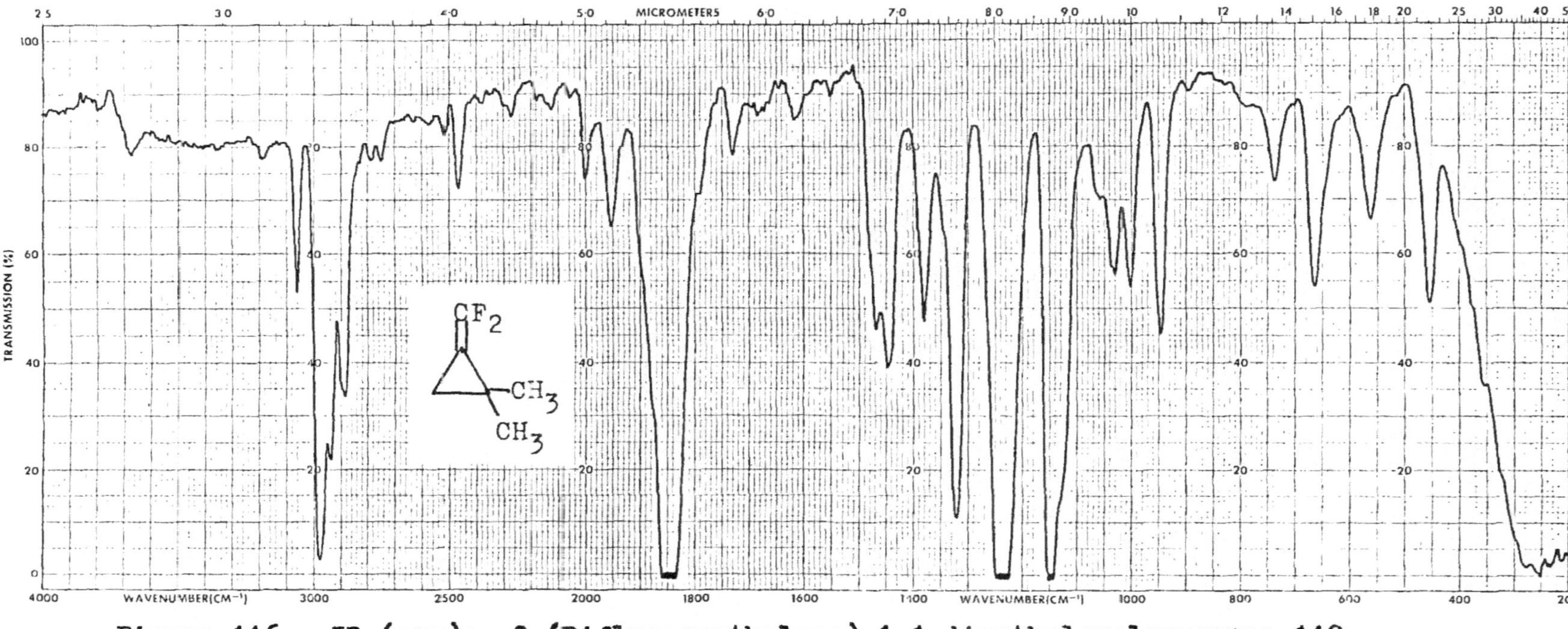

Figure 116. IR (gas): 2-(Difluoromethylene)-1,1-dimethylcyclopropane 140.
Page 130.

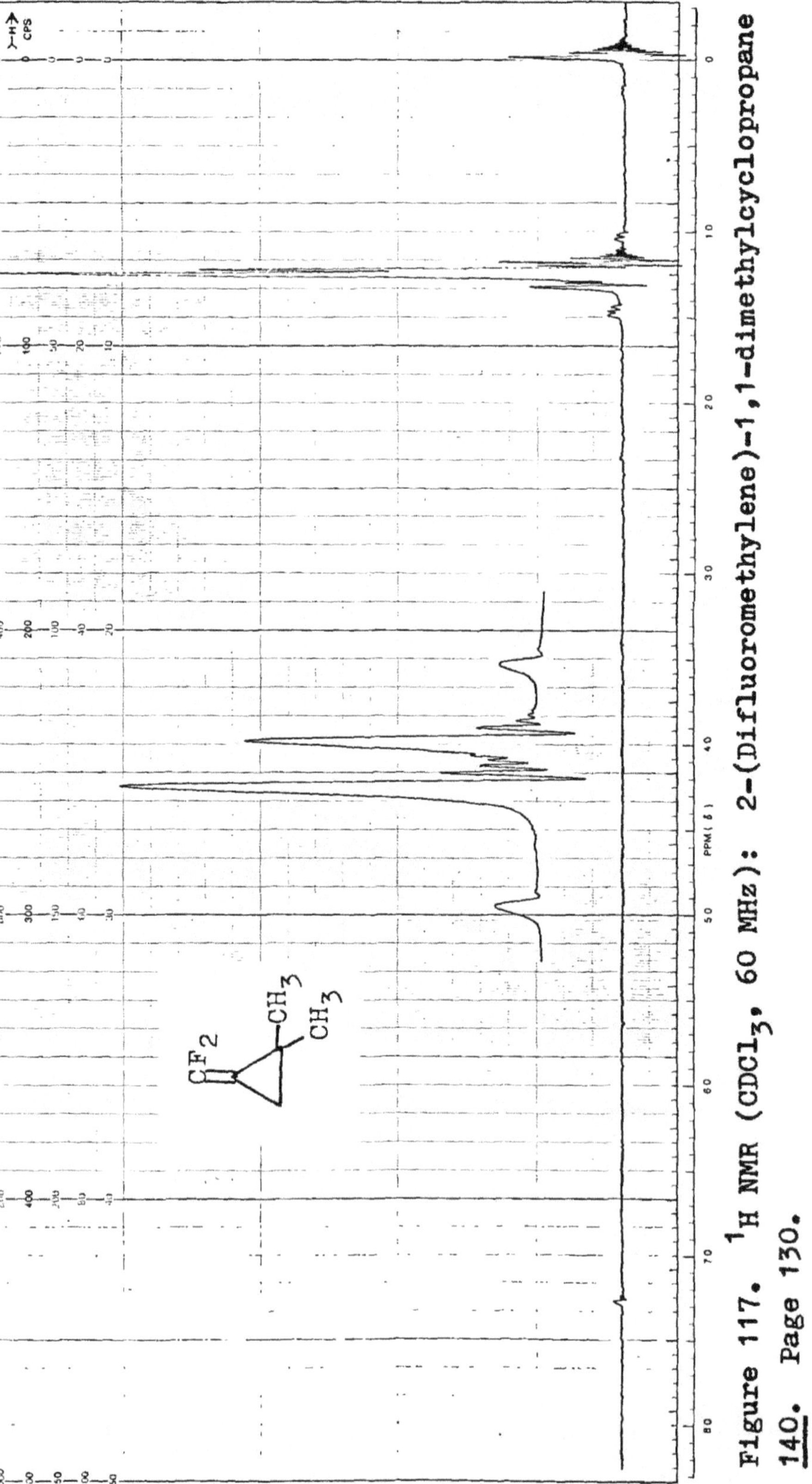

Figure 117. ^{1}H NMR ($CDCl_3$, 60 MHz): 2-(Difluoromethylene)-1,1-dimethylcyclopropane 140. Page 130.

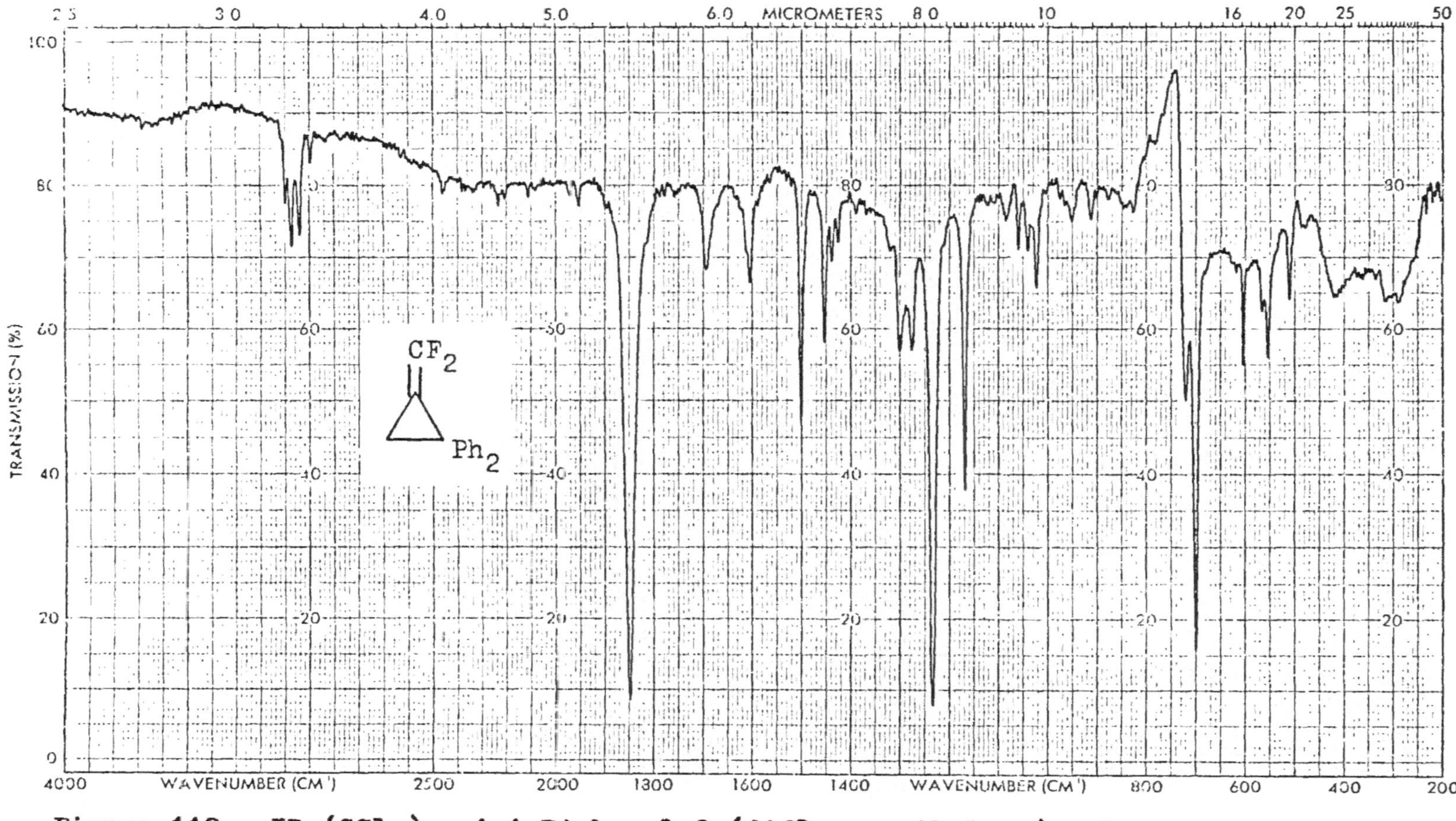

Figure 118. IR (CCl_4): 1,1-Diphenyl-2-(difluoromethylene)cyclopropane <u>142</u>. Page 132.

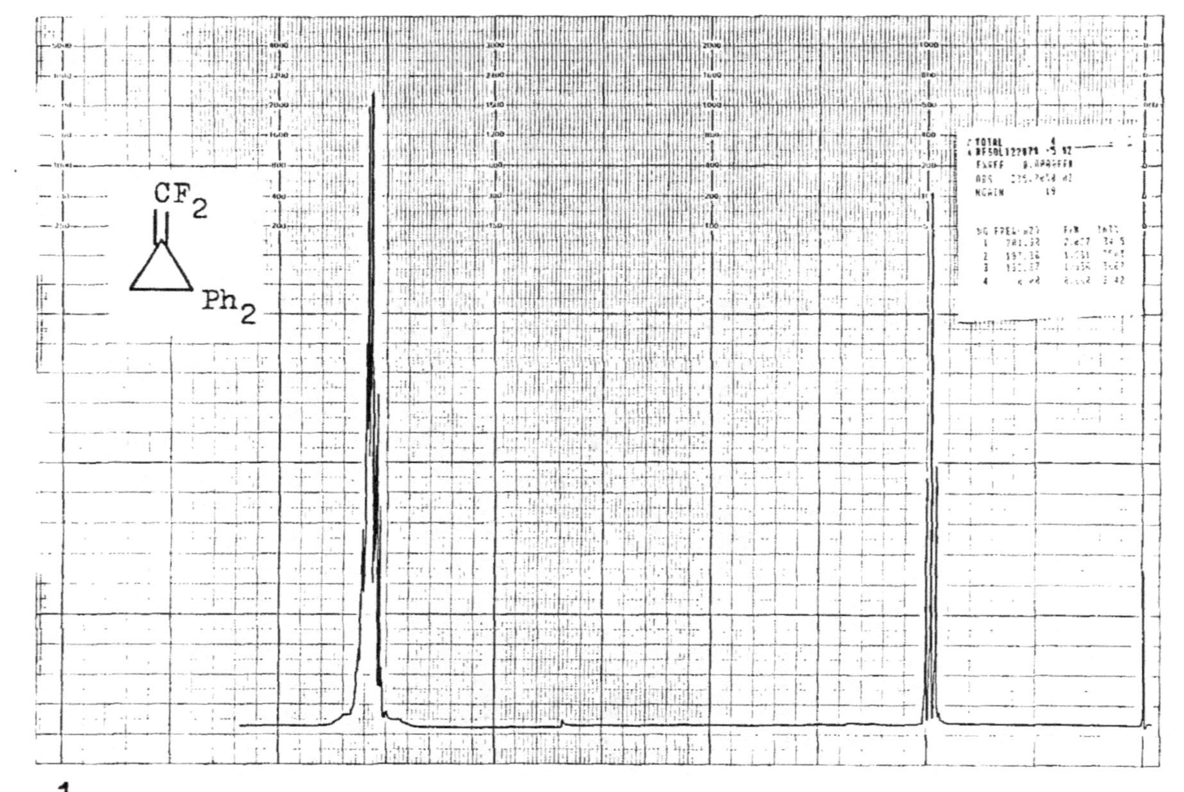

Figure 119. ^{1}H NMR ($CDCl_3$, 100 MHz): 1,1-Diphenyl-2-(difluoromethylene)cyclopropane <u>142</u>. Page 132.

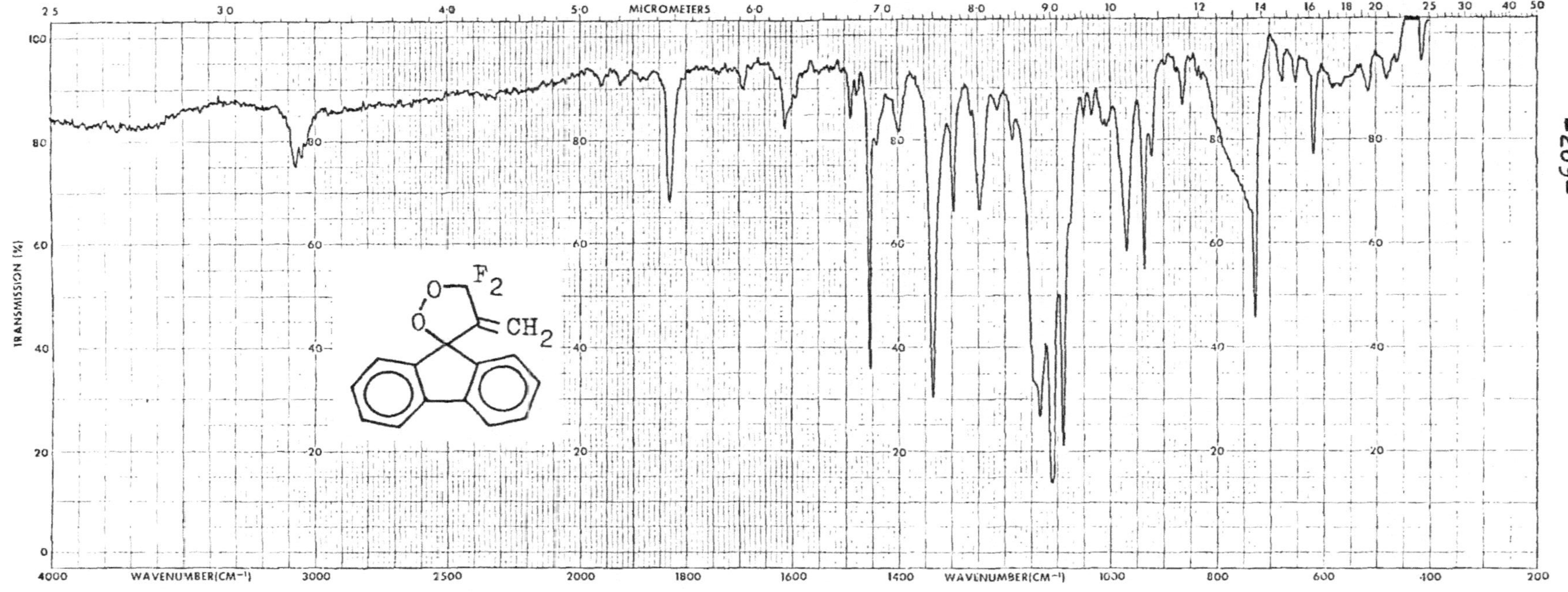

Figure 120. IR (CCl_4): 5,5-Difluoro-4-methylenespiro[1,2-dioxolane-3,9'-[9H]-fluorene] <u>145</u>. Page 134.

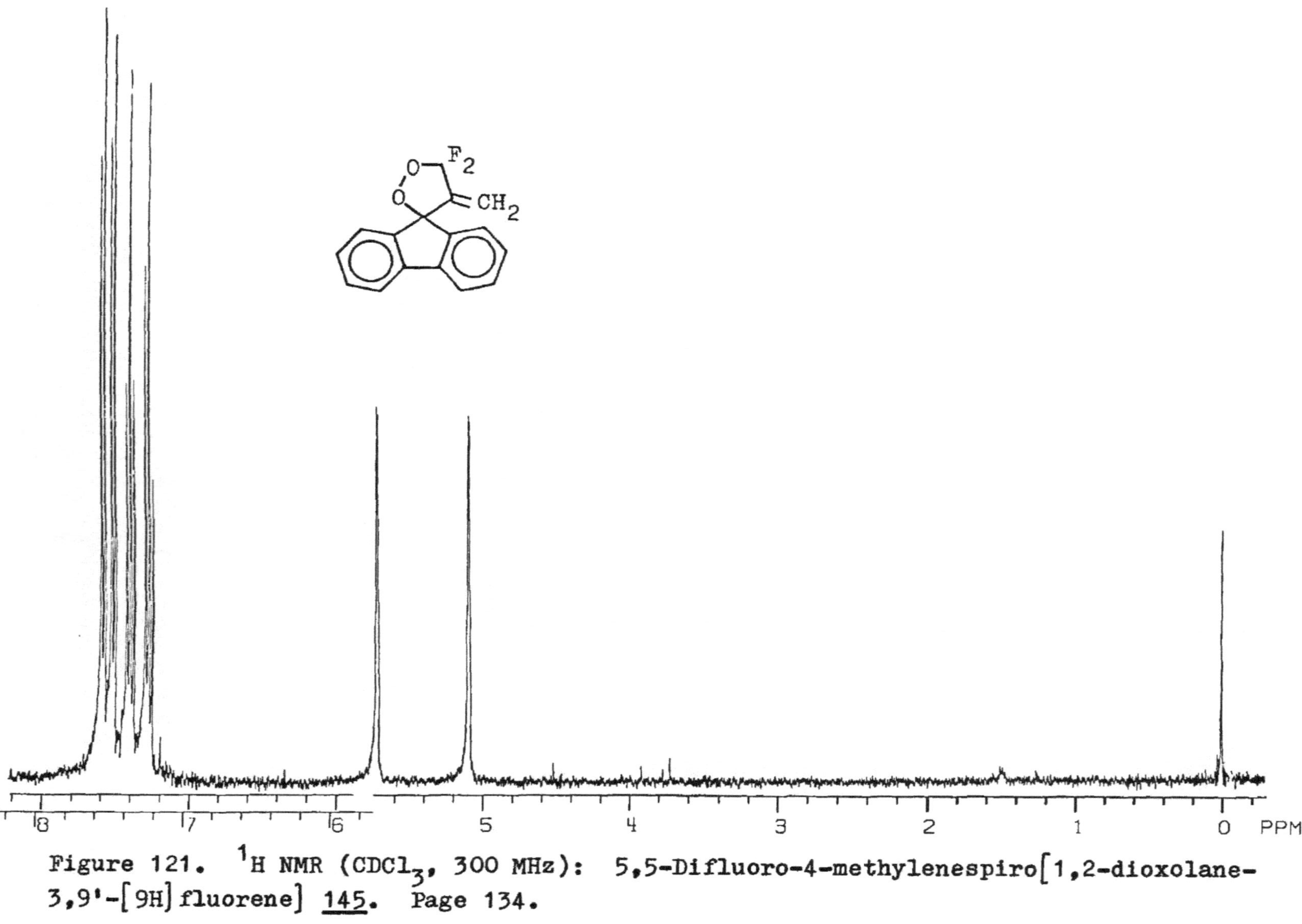

Figure 121. ^{1}H NMR ($CDCl_3$, 300 MHz): 5,5-Difluoro-4-methylenespiro[1,2-dioxolane-3,9'-[9H]fluorene] <u>145</u>. Page 134.

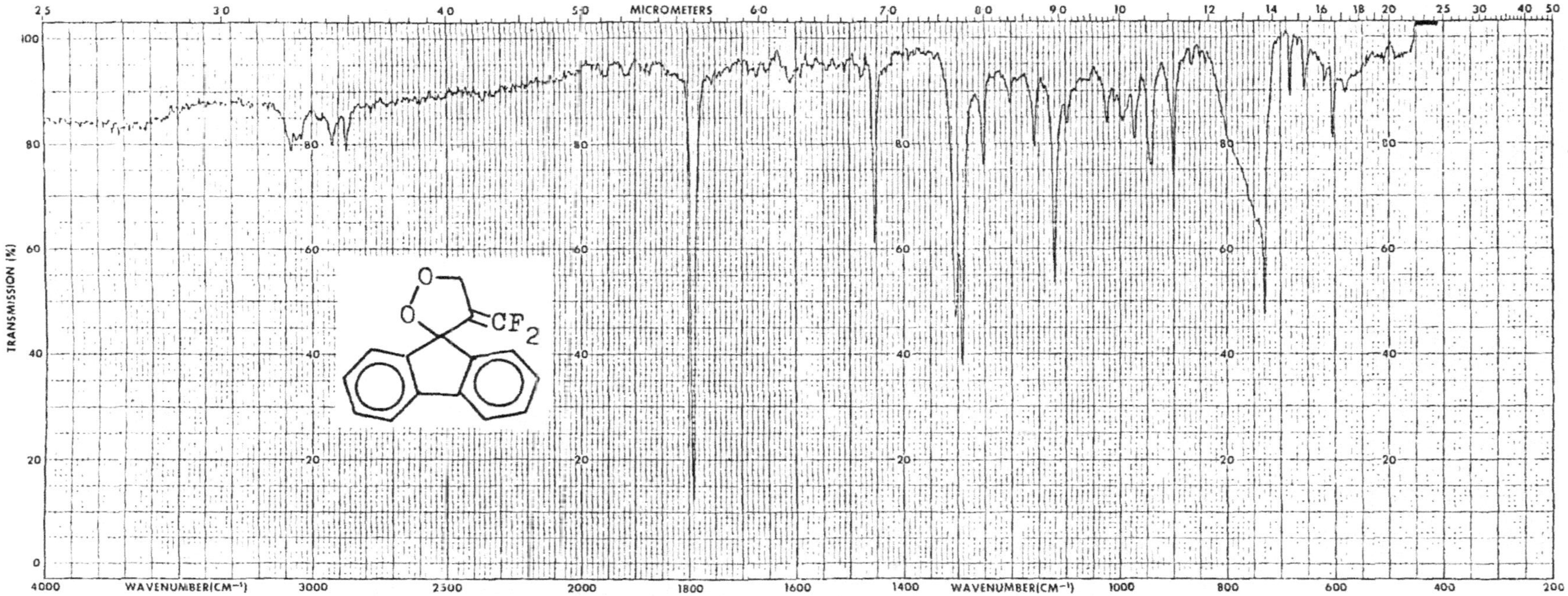

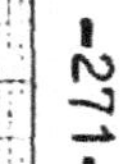

Figure 122. IR (CCl_4): 4-(Difluoromethylene)spiro[1,2-dioxolane-3,9'-[9H]fluorene] 146. Page 134.

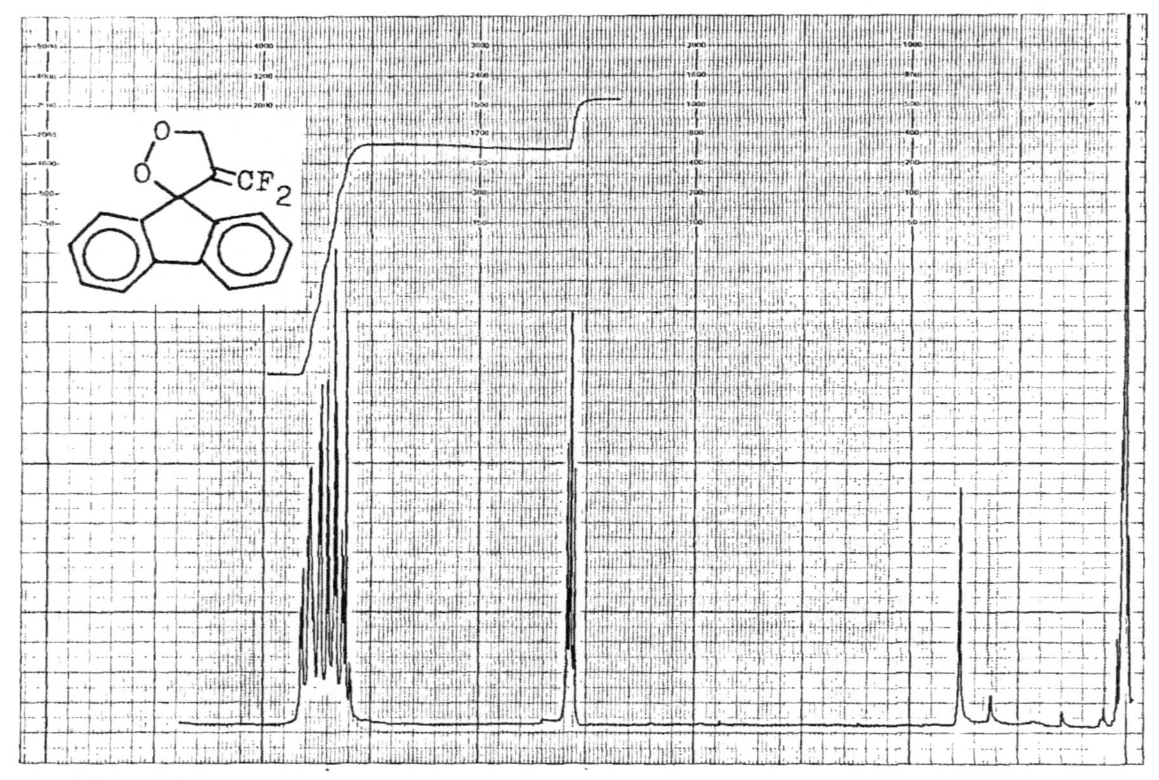

Figure 123. ^{1}H NMR ($CDCl_3$, 100 MHz): 4-(Difluoromethylene)-spiro[1,2-dioxolane-3,9'-[9H]fluorene] 146. Page 134.

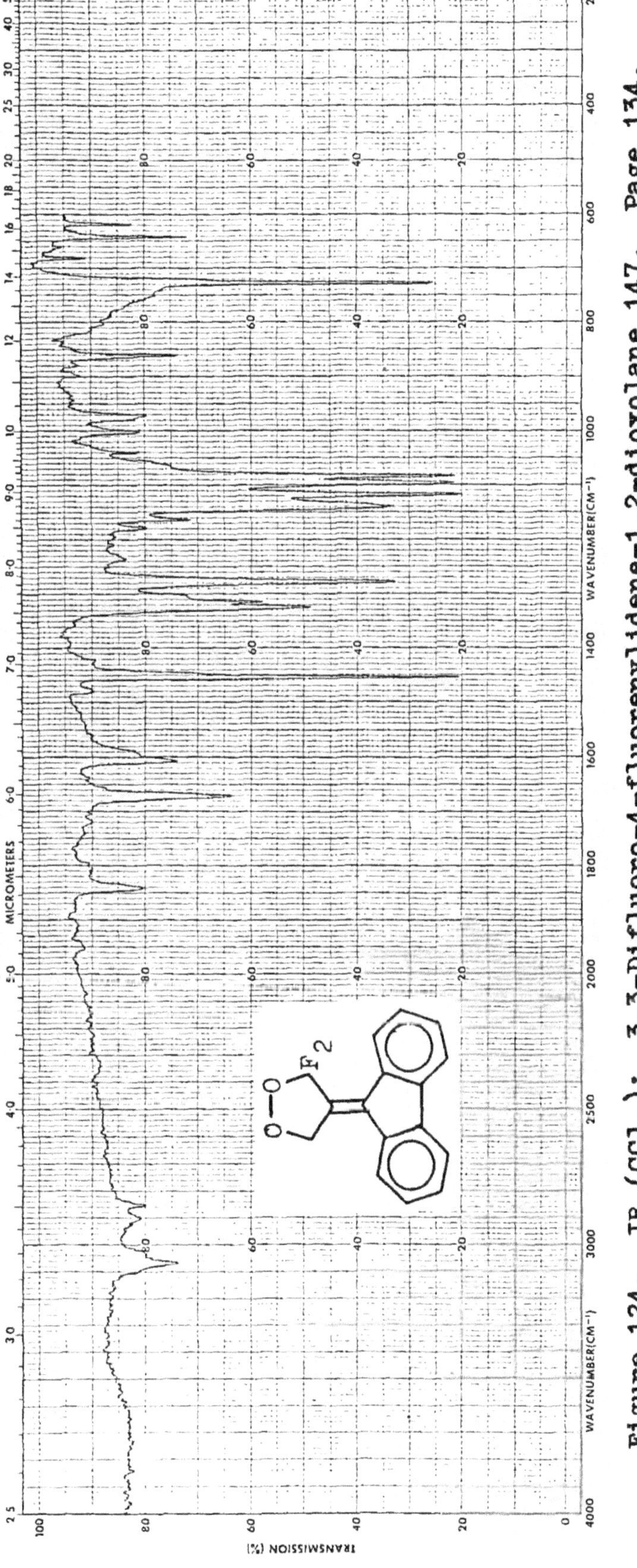

Figure 124. IR (CCl_4): 3,3-Difluoro-4-fluorenylidene-1,2-dioxolane <u>147</u>. Page 134.

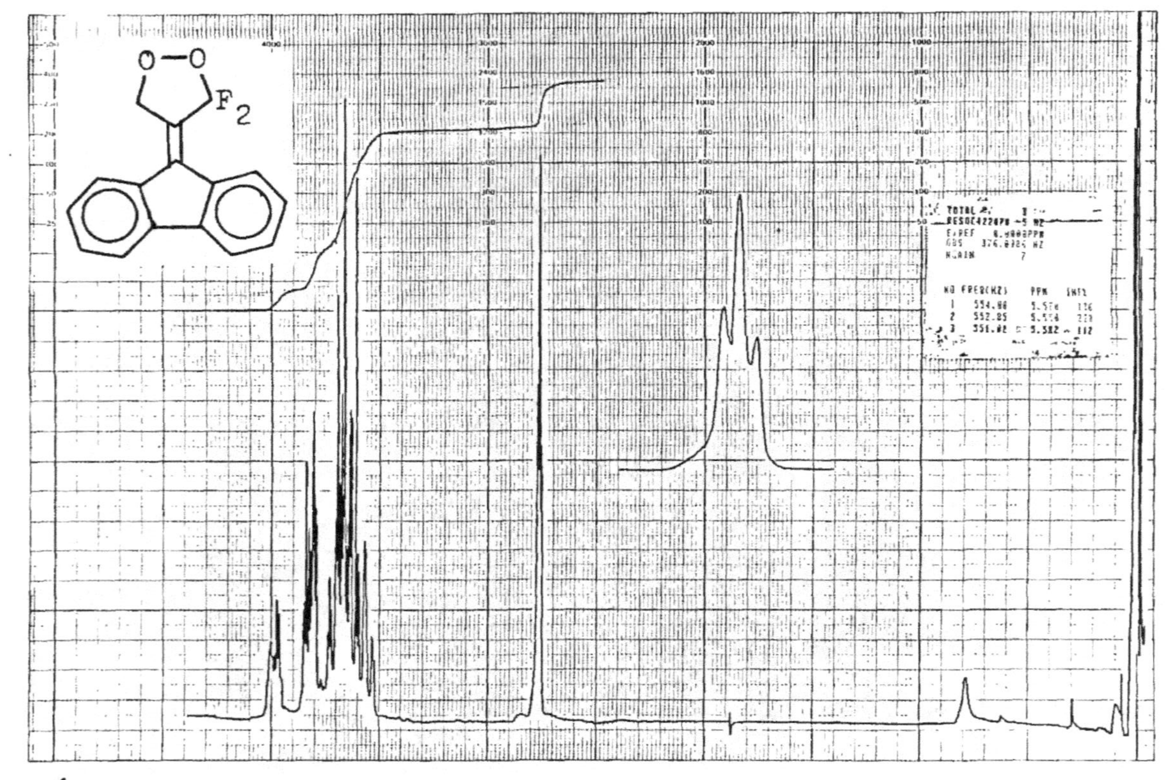

Figure 125. ^{1}H NMR ($CDCl_3$, 100 MHz): 3,3-Difluoro-4-fluorenylidene-1,2-dioxolane <u>147</u>. Page 134.

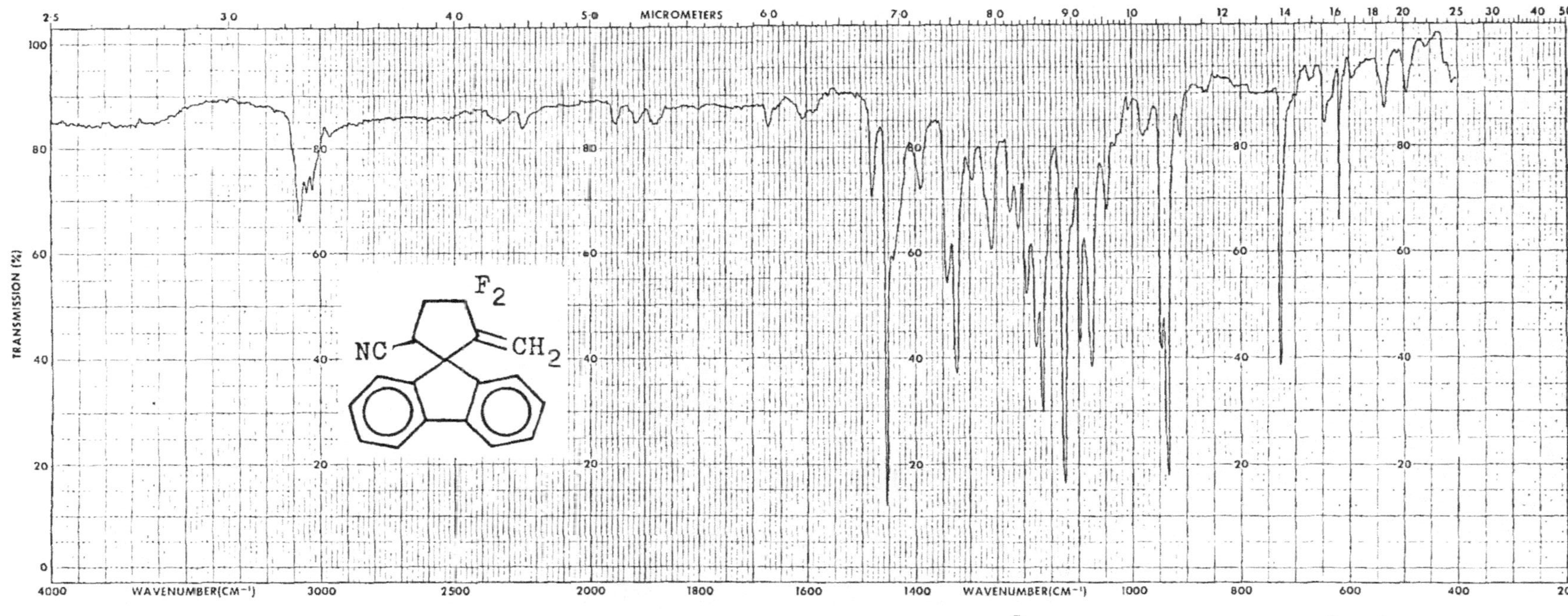

Figure 126. IR (CCl_4): 4,4-Difluoro-5-methylenespiro[cyclopentane-1,9'-[9H]-fluorene]-2-carbonitrile <u>152</u>. Page 136.

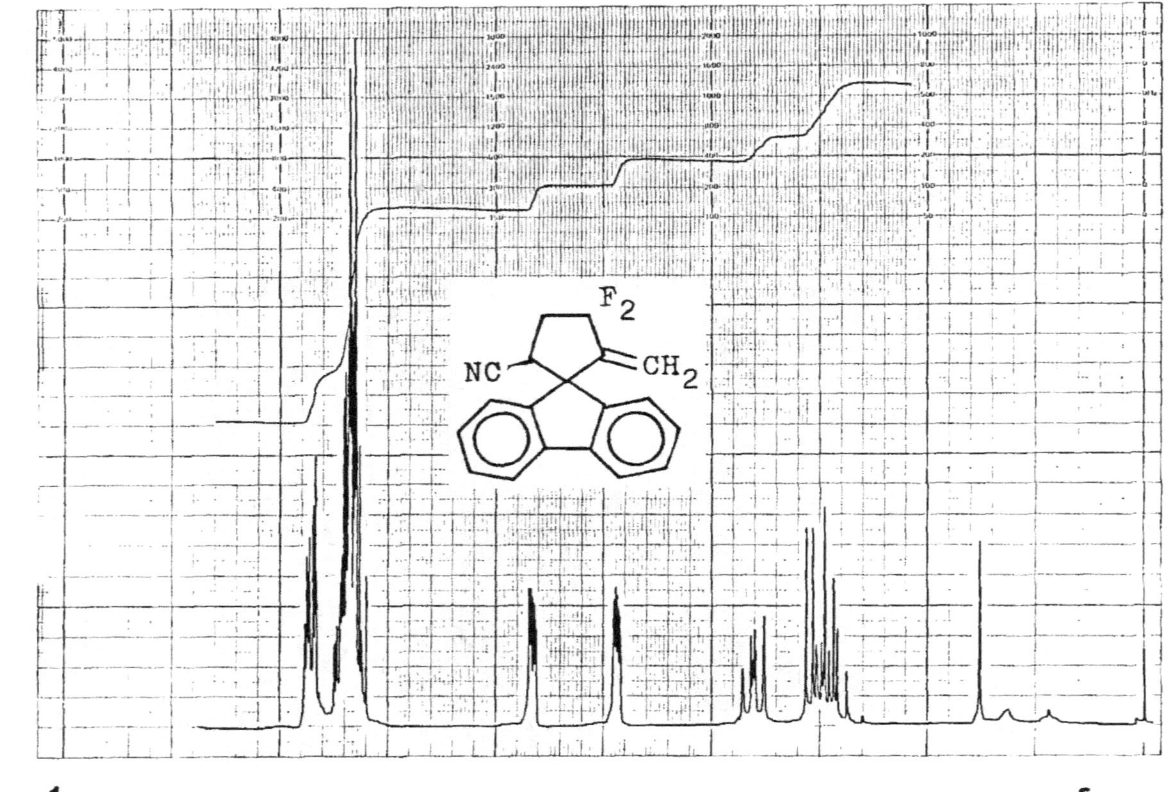

Figure 127. ^1H NMR ($CDCl_3$, 100 MHz): 4,4-Difluoro-5-methylenespiro[cyclopentane-1,9'-[9H]fluorene]-2-carbonitrile <u>152</u>. Page 136.

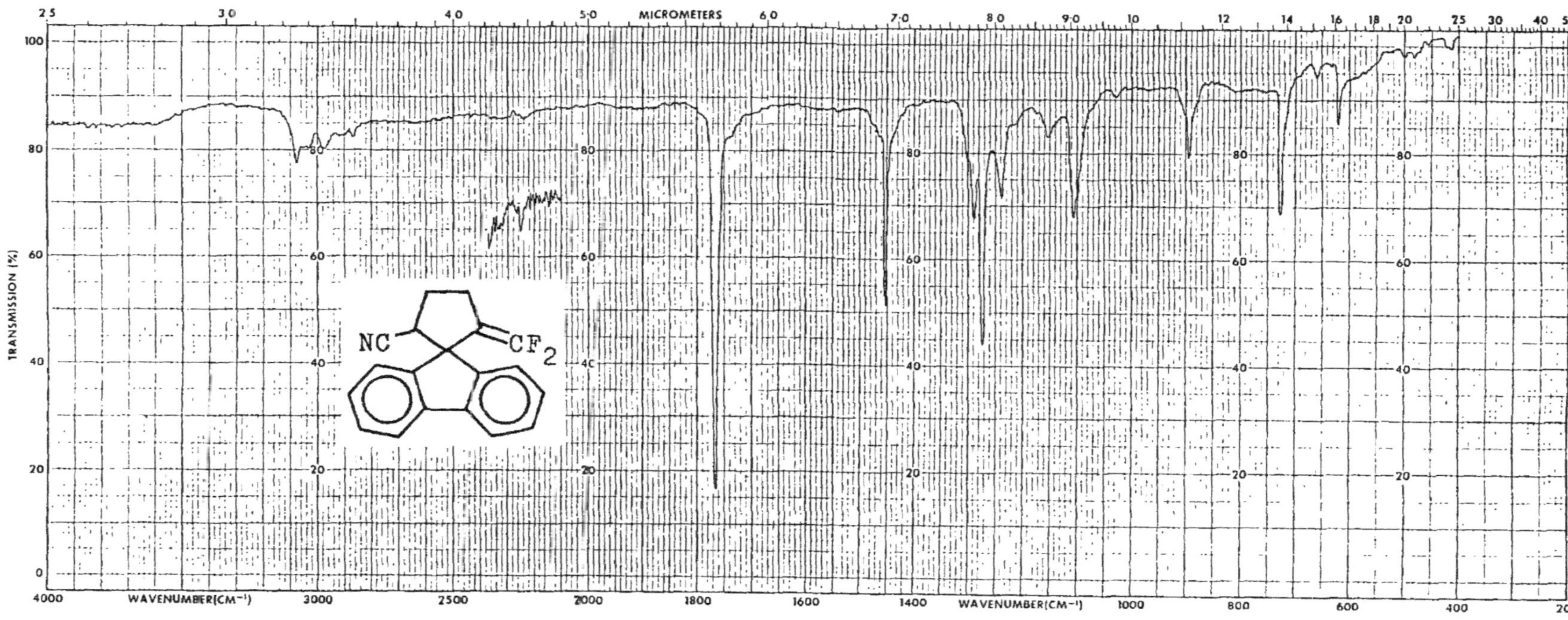

Figure 128. IR (CCl_4): 5-(Difluoromethylene)spiro[cyclopentane-1,9'-[9H]fluorene]-2-carbonitrile <u>153</u>. Page 136.

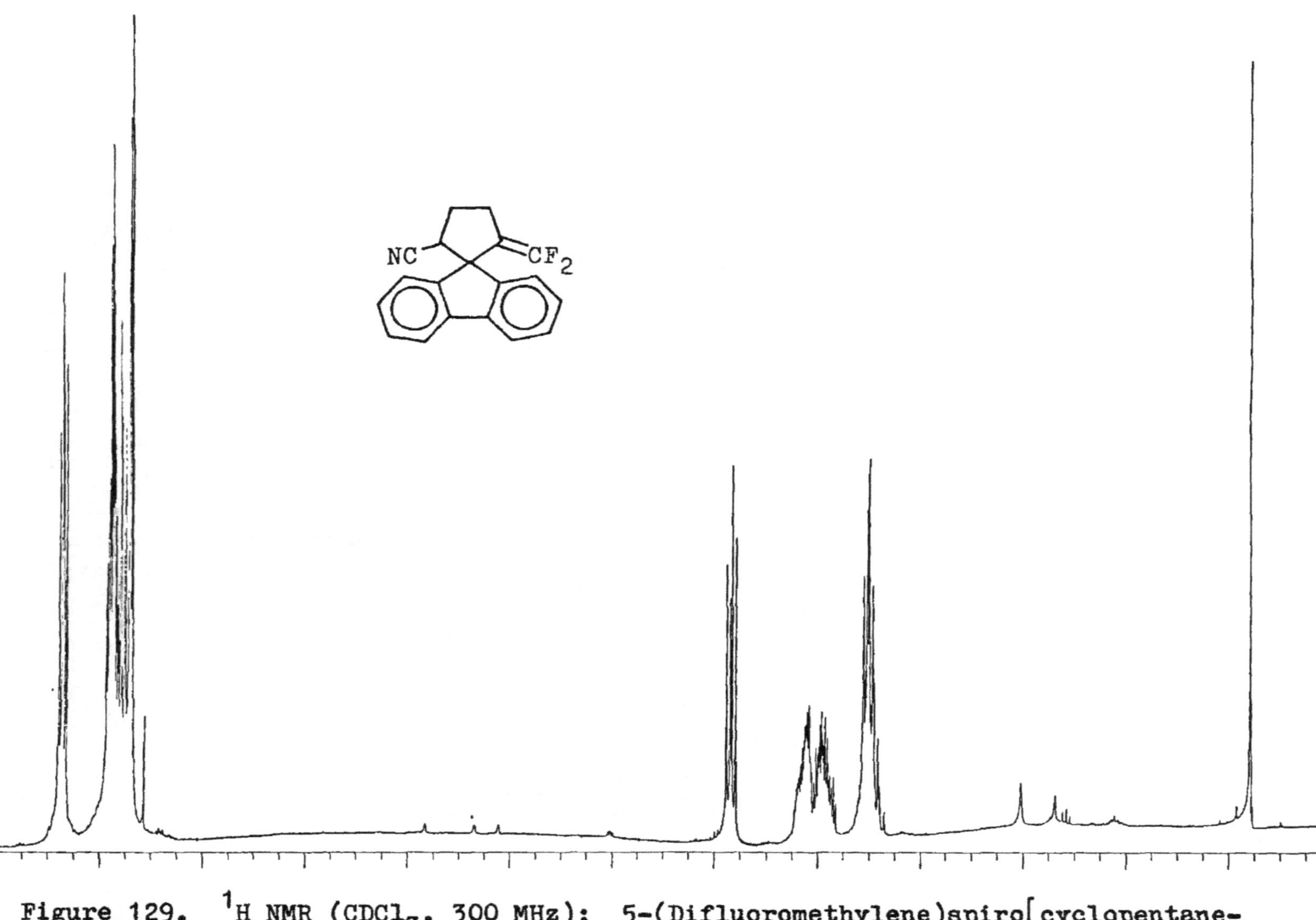

Figure 129. ^{1}H NMR ($CDCl_3$, 300 MHz): 5-(Difluoromethylene)spiro[cyclopentane-1,9'-[9H]fluorene]-2-carbonitrile <u>153</u>. Page 136.

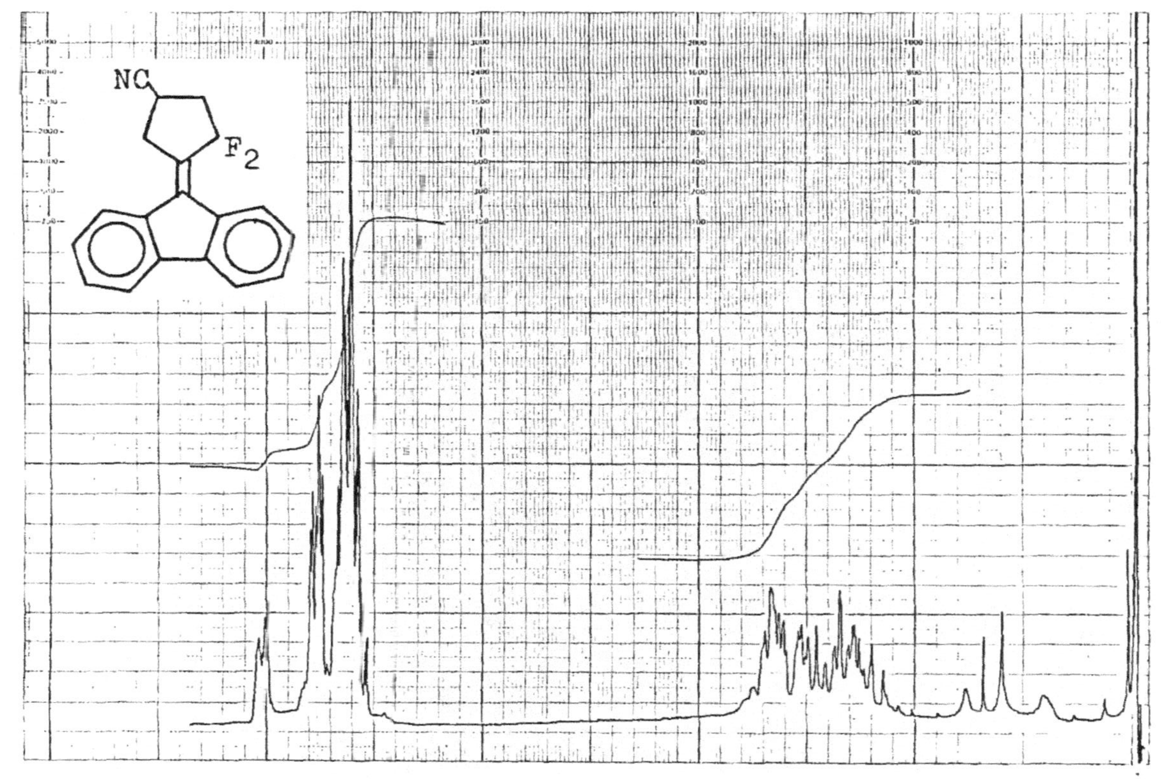

Figure 130. ^{1}H NMR ($CDCl_3$, 100 MHz): 3,3-Difluoro-4-fluorenylidenecyclopentanecarbonitrile <u>154</u>. Page 136.

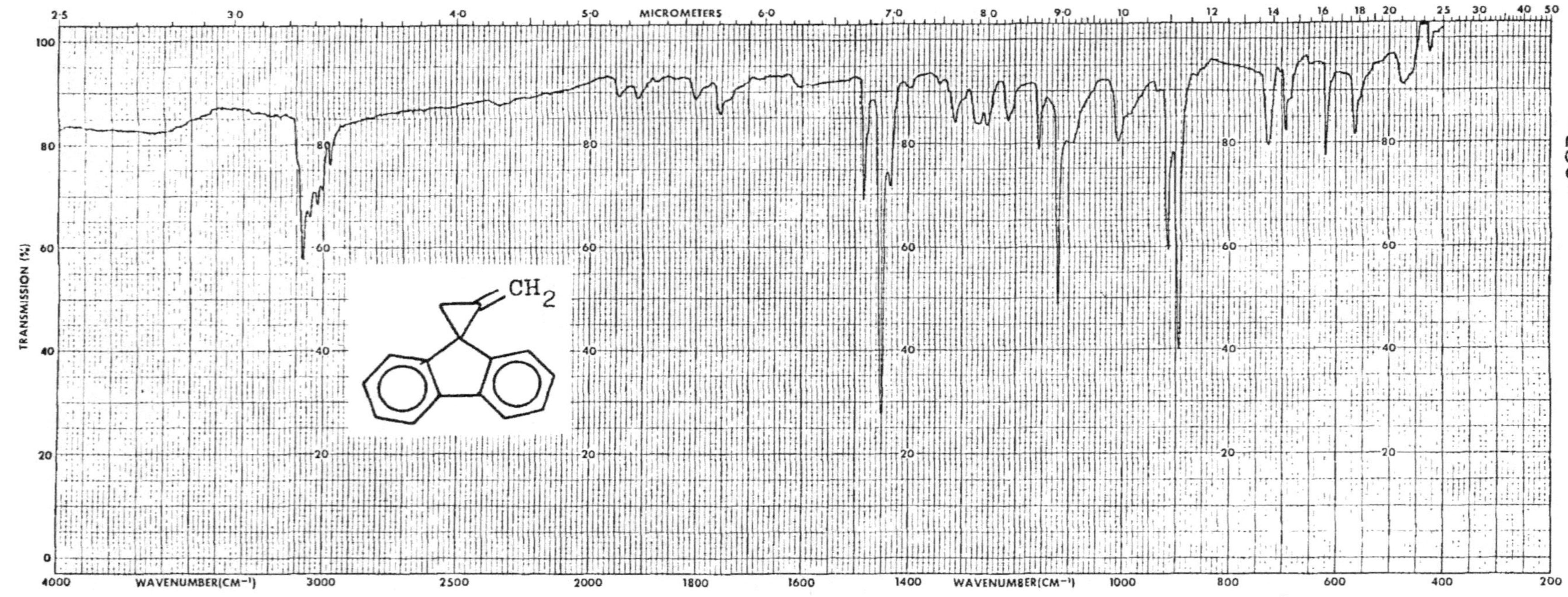

Figure 131. IR (CCl_4): 2-Methylenespiro[cyclopropane-1,9'-[9H]fluorene] <u>157</u>.
Page 138.

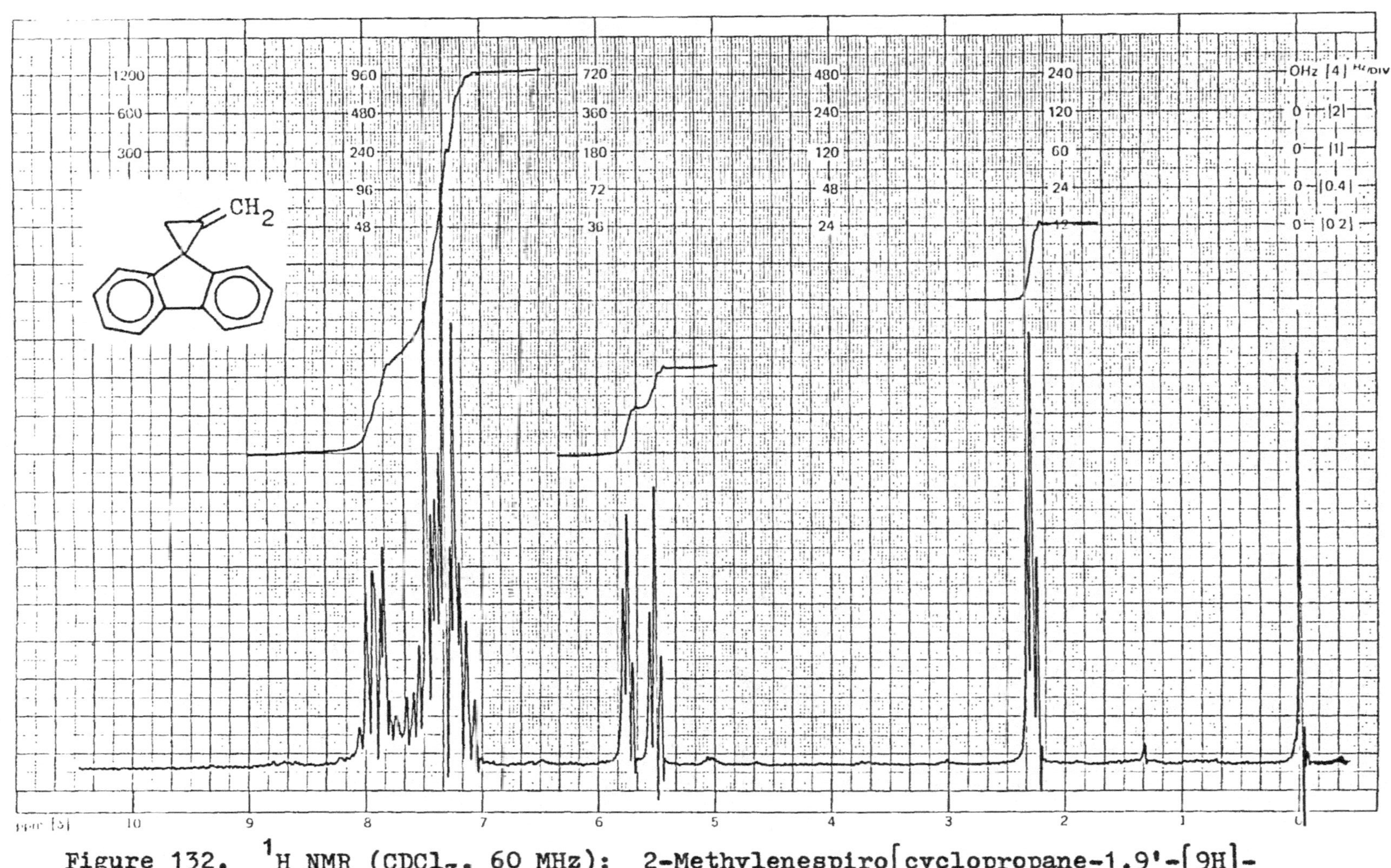

Figure 132. ^{1}H NMR ($CDCl_3$, 60 MHz): 2-Methylenespiro[cyclopropane-1,9'-[9H]-fluorene] 157. Page 138.

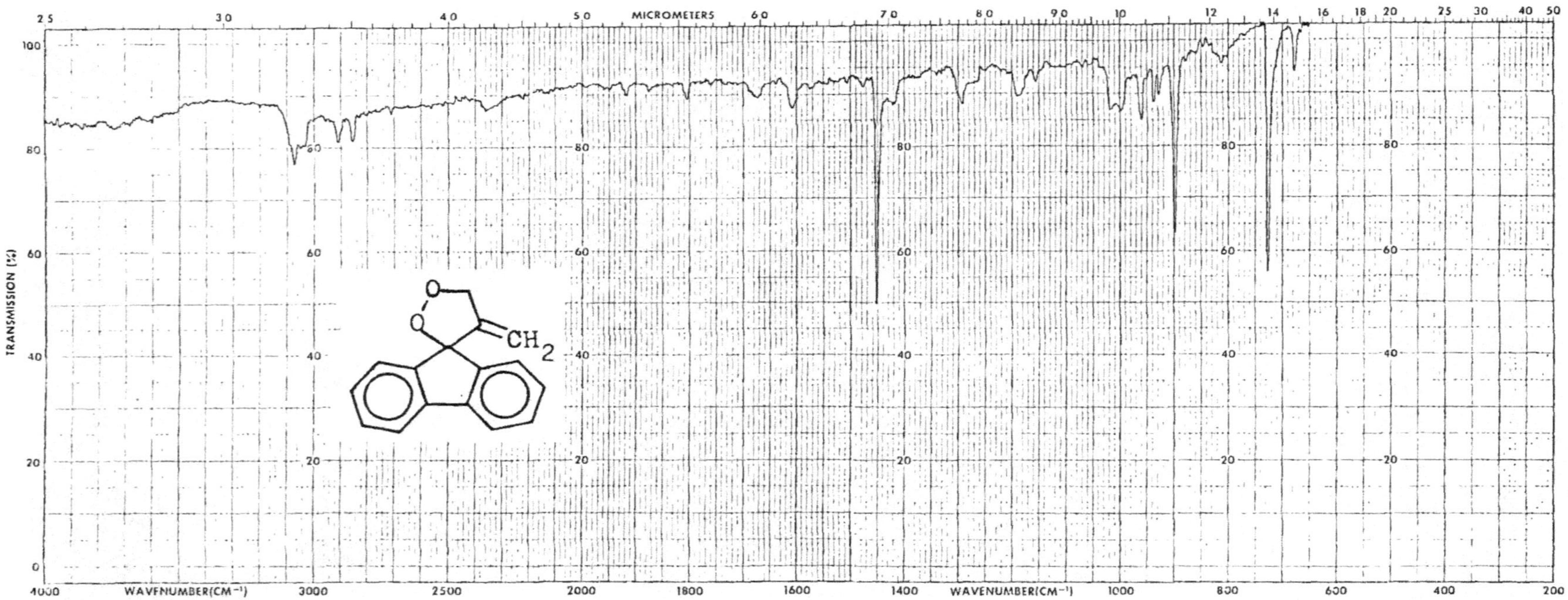

Figure 133. IR (CCl_4): 4-Methylenespiro[1,2-dioxolane-3,9'-[9H]fluorene] 158.
Page 139.

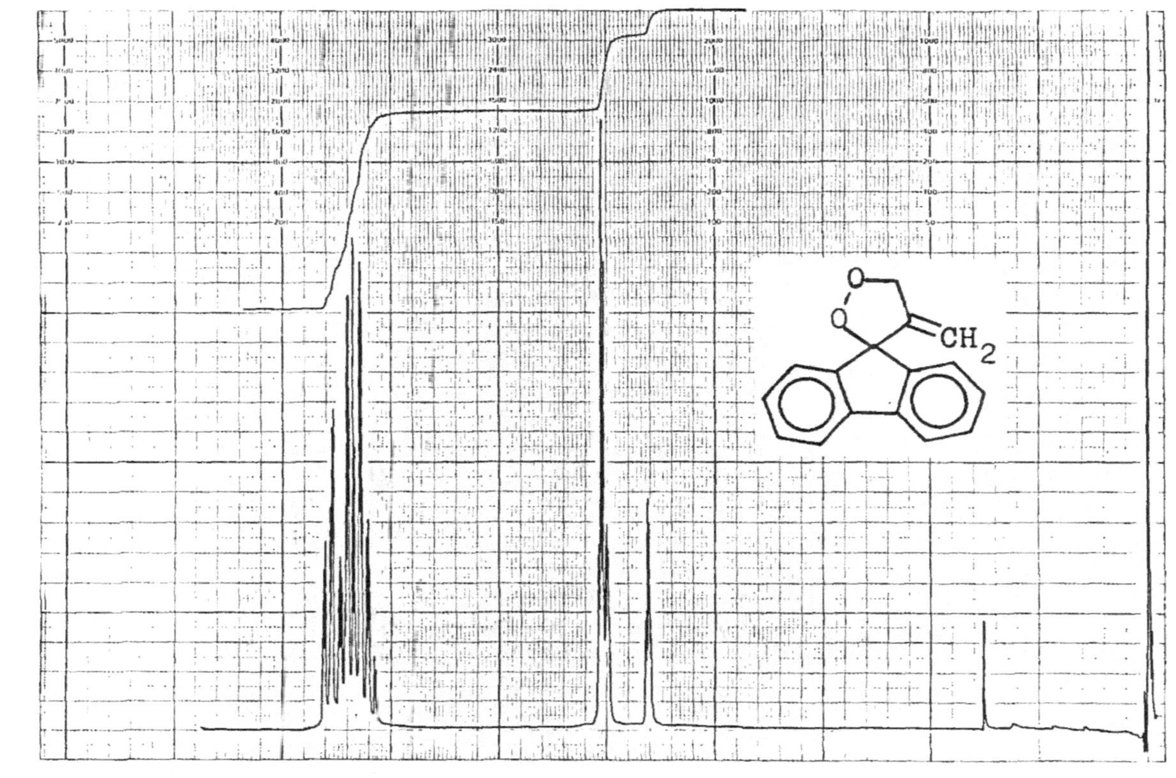

Figure 134. ^{1}H NMR ($CDCl_3$, 100 MHz): 4-Methylenespiro[1,2-dioxolane-3,9'-[9H]-fluorene] 158. Page 139.

REFERENCES

1. J.K. Williams and W.H. Sharkey, J. Am. Chem. Soc., 81, 4296 (1959).

2. W.R. Dolbier and S. Dai, J. Am. Chem. Soc., 92, 1774 (1970).

3. W.T. Borded, L. Sharpe, and I.L. Reich, J. Chem. Soc., Chem. Commun., 461 (1970).

4. S.R. Byrn, E. Maverick, J. Muscio, K.N. Trueblood, and T.L. Jacobs, J. Am. Chem. Soc., 93, 6680 (1971).

5. T.L. Jacobs and O.J. Muscio, Tetrahedron Lett., 4829 (1970).

6. S. Dai and W.R. Dolbier, J. Org. Chem., 37, 950 (1972).

7. H.N. Cripps, J.K. Williams, and W.H. Sharkey, J. Am. Chem. Soc., 81, 2723 (1959).

8. S. Dai and W.R. Dolbier, J. Am. Chem. Soc., 94, 3946 (1972).

9. D.R. Taylor, M.R. Warburton, and D.B. Wright, J. Chem. Soc., Perkin Trans. 1, 1365 (1972).

10. D.R. Taylor, M.R. Warburton, and D.B. Wright, J. Chem. Soc. C, 385 (1971).

11. H. Pledger, J. Org. Chem., 25, 278 (1960).

12. W.R. Dolbier and S. Dai, J. Am. Chem. Soc., 90, 5028 (1968).

13. S. Kunichika, T. Okamoto, and K. Yoshikawa, Bull. Inst. Chem. Res., Kyoto Univ., 49, 109 (1971).

14. W.T. Miller, U.S. 2,668,182, Feb. 2, 1954; Chem. Abstr., 49, P2479a.

15. T.L. Jacobs and R.S. Bauer, J. Am. Chem. Soc., 81, 606 (1959).

16. J.D. McCullough, R.S. Bauer, and T.L. Jacobs, Chem. Ind. (London), 706 (1957).

17. R.E. Banks, R.N. Haszeldine, and D.R. Taylor, J. Chem. Soc., 978 (1965).

18. R.E. Banks, W.R. Deem, R.N. Haszeldine, and D.R. Taylor, J. Chem. Soc. C, 2051 (1966).

19. G.B. Blackwell, R.N. Haszeldine, and D.R. Taylor, J. Chem. Soc., Perkin Trans. 1, 1 (1983).

20. A.P. Zens, P.D. Ellis, and R. Ditchfield, J. Am. Chem. Soc., 96, 1309 (1974).

21. J.R. Durig, Y.S. Li, J.D. Witt, A.P. Zens, and P.D. Ellis, Spectrochim. Acta, Part A, 33, 529 (1977).

22. W.H. Knoth and D.D. Coffman, J. Am. Chem. Soc., 82, 3873 (1960).

23. C. Piedrahita (Univ. of Florida Dissertation), Diss. Abstr. Int. B, 39, 3337 (1979).

24. W.R. Dolbier and C. Piedrahita, Tetrahedron Lett., 2231 (1978).

25. P.J. Gorton and R. Walsh, J. Chem. Soc., Chem. Commun., 783 (1972).

26. H.E. O'Neal and S.W. Benson, J. Phys. Chem., 72, 1866 (1968).

27. P.D. Bartlett, G.E. Wallbillich, A.S. Wingrove, J.S. Swenton, L.K. Montgomery, and B.D. Kramer, J. Am. Chem. Soc., 90, 2049 (1968).

28. J.G. Aston, G. Szasz, H.W. Woolley, and F.G. Brickwedde, J. Chem. Phys., 14, 67 (1946).

29. J.S. Swenton and P.D. Bartlett, J. Am. Chem. Soc., 90, 2056 (1968).

30. L.N. Domelsmith, K.N. Houk, C. Piedrahita, and W.R. Dolbier, J. Am. Chem. Soc., 100, 6908 (1978).

31. W.R. Dolbier, K.S. Medinger, A. Greenberg, and J.F. Liebman, Tetrahedron, 38, 2415 (1982).

32. W.R. Dolbier, C.R. Burkholder, and C.A. Piedrahita, J. Fluorine Chem., 20, 637 (1982).

33. W.R. Dolbier and C.R. Burkholder, Tetrahedron Lett., 21, 785 (1980).

34. P.D. Bartlett, A.S. Wingrove, and R. Owyang, J. Am. Chem. Soc., 90, 6067 (1968).

35. N.J. Turro and P.D. Bartlett, J. Org. Chem., 30, 1849 (1965).

36. P. Hemmersbach, M. Klessinger, and P. Bruckmann, J. Am. Chem. Soc., 100, 6344 (1978).

37. T. Hayashi and T. Nakajima, Bull. Chem. Soc. Jap., 48, 980 (1975).

38. A.T. Blomquist and J.A. Verdol, J. Am. Chem. Soc., 77, 1806 (1955).

39. J.W. van Straten, J.J. van Norden, T. van Schaik, G. Franke, W. de Wolf, and F. Bickelhaupt, Rec. Trav. Chim. Pays-Bas, 97, 105 (1978).

40. J.J. Gajewski and C.N. Shih, J. Am. Chem. Soc., 94, 1675 (1972).

41. Y. Slobodin and A.P. Khitrov, Zh. Obshch. Khim., 33, 2819 (1963).

42. W.T. Borden, L Reich, L. Sharpe, R. Weinberg, and H. Reich, J. Org. Chem., 40, 2438 (1975).

43. W. von E. Doering and W.R. Dolbier, J. Am. Chem. Soc., 89, 4534 (1967).

44. D. Bellus, K. von Bredow, H. Sauter, and C.D. Weis, Helv. Chim. Acta, 56, 3004 (1973).

45. D. Bellus and C.D. Weis, Tetrahedron Lett., 999 (1973).

46. P.C. Hiberty and G. Ohanessian, J. Am. Chem. Soc., 104, 66 (1982).

47. R. Huisgen, Angew. Chem., Int. Ed. Engl., 2, 633 (1963).

48. R. Huisgen, J. Org. Chem., 33, 2291 (1968).

49. P. Battioni, L. VoQuang and Y. VoQuang, Bull. Soc. Chim. Fr., (Part II), 401 (1978).

50. P. Battioni, L. VoQuang, and Y. VoQuang, Bull. Soc. Chim. Fr., (Part II), 415 (1978).

51. R.A. Firestone, J. Org. Chem., 33, 2285 (1968).

52. R.A. Firestone, Tetrahedron, 33, 3009 (1977).

53. K.N. Houk, J. Sims, R.E. Duke, R.W. Strozier, and J.K George, J. Am. Chem. Soc., 95, 7287 (1973).

54. K.N. Houk, J. Sims, C.R. Watts, and L.J. Luskus, J. Am. Chem. Soc., 95, 7301 (1973).

55. P. Caramella, R.W. Gandour, J.A. Hall, C.G. Deville, and K.N. Houk, J. Am. Chem. Soc., 99, 385 (1977).

56. P.G. DeBenedetti, C. DeMicheli, R. Gandolfi, P. Gariboldi, and A. Rastelli, J. Org. Chem., 45, 3646 (1980).

57. R.J. Crawford, D.M. Cameron, and H. Tokunaga, Can. J. Chem., 52, 4025 (1974).

58. M. Schneider, O. Schuster, and H. Rau, Chem. Ber., 110, 2180 (1977).

59. J.R. Durig, Y.S. Li, C.C. Tong, A.P. Zens, and P.D. Ellis, J. Am. Chem. Soc., 96, 3805 (1974).

60. W.C. Herndon, Chem. Rev., 72, 157 (1972).

61. L. Salem, J. Am. Chem. Soc., 90, 543, 553 (1968).

62. J. Geittner, R. Huisgen, and R. Sustmann, Tetrahedron Lett., 881 (1977).

63. R. Huisgen, H. Seidl, and I. Bruning, Chem. Ber., 102, 1102 (1969).

64. M.D. Gordon, P.V. Alston, and A.R. Rossi, J. Am. Chem. Soc., 100, 5701 (1978).

65. G.P. Newsoroff and S. Sternhell, Aust. J. Chem., 25, 1669 (1972).

66. J.W. Emsley, L. Phillips, and V. Wray, "Fluorine Coupling Constants," Pergamon Press, Oxford, New York, Toronto, Sydney, Paris, and Frankfurt, 196-207, 421-446 (1977).

67. C. Grundmann and J.M. Dean, J. Org. Chem., 30, 2809 (1965).

68. K. Kurita, N. Hirakawa, T. Dobashi, and Y. Iwakura, J. Polym. Sci., Polym. Chem. Ed., 17, 2567 (1979).

69. M. Christl and R. Huisgen, Chem. Ber., 106, 3345 (1973).

70. W.J. Linn and R.E. Benson, J. Am. Chem. Soc., 87, 3657 (1965).

71. J.A. Berson, Acc. Chem. Res., 11, 446 (1978).

72. P. Dowd and M. Chow, J. Am. Chem. Soc., 99, 6438 (1977).

73. W.R. Dolbier and T.H. Fielder, J. Am. Chem. Soc., 100, 5577 (1978).

74. P. Engel, Chem. Rev., 80, 99 (1980).

75. R.J. Crawford, C.M. Cameron, and H. Tokunaga, Can. J. Chem., 52, 4025 (1974).

76. R.J. Crawford and H. Tokunaga, Can. J. Chem., 52, 4033 (1974).

77. R.J. Crawford, H. Tokunaga, L. Schrijver, and J.C. Godard, Can. J. Chem., 56, 998 (1978).

78. M.H. Chang and R.J. Crawford, Can. J. Chem., 59, 2556 (1981).

79. R.J. Crawford and M.H. Chang, Tetrahedron, 38, 837 (1982).

80. G.D. Andrews and J.E. Baldwin, J. Am. Chem. Soc., 99, 4853 (1977).

81. W.R. Dolbier and C.R. Burkholder, Tetrahedron Lett., 24, 1217 (1983).

82. J.C. Gilbert and J.R. Butler, J. Am. Chem. Soc., 92, 2168 (1970).

83. G.C. Stevens, R.M. Clarke, and E.J. Hart, J. Phys. Chem., 76, 3863 (1972).

84. B. Hickel, J. Phys. Chem., 79, 1054 (1975).

85. J. Rabani, M. Pick, and M. Simic, J. Phys. Chem., 78, 1049 (1974).

86. A.L. Aleksandrov, E.M. Pliss, and V.F. Shuralov, Izv. Akad. Nauk. SSSR, Ser. Khim., 11, 2446 (1979).

87. R.M. Wilson and F. Geiser, J. Am. Chem. Soc., 100, 2225 (1978).

88. L.R. Corwin, D.M. McDaniel, R.J. Bushby, and J.A. Berson, J. Am. Chem. Soc., 102, 276 (1980).

89. J.A. Mondo and J.A. Berson, J. Am. Chem. Soc., 105, 3340 (1983).

90. K.C. Stueben, J. Polymer Sci., Part A-1, 4, 829 (1966).

91. M.S. Newman, J. Org. Chem., 26, 2630 (1961).

92. O.R. Pierce, J.C. Siegle, and E.T. McBee, J. Am. Chem. Soc., 75, 6324 (1953).

93. S.D. Andrews, A.C. Day, P. Raymond, and M.C. Whiting, "Organic Syntheses," R. Breslow, ed., John Wiley and Sons, Inc., New York, 50, 27 (1970).

94. J.B. Miller, J. Org. Chem., 24, 560 (1959).

95. H. Staudinger and O. Kupfer, Chem. Ber., 44, 2197 (1911).

96. I. Bruning, R. Grashey, H. Hauck, R. Huisgen, and H. Seidl, "Organic Syntheses," E.J. Corey, ed., John Wiley and Sons, Inc., New York, 46, 127 (1966).

97. O.L. Brady, J. Chem. Soc., 2390 (1926).

98. H. Staudinger and K. Miescher, Helv. Chim. Acta, 2, 554 (1919).

99. W.J. Linn, "Organic Syntheses," K.B. Wiberg, ed., John Wiley and Sons, Inc., New York, 49, 103 (1969).

100. D. Seyferth and P. Hopper, J. Org. Chem., 37, 4070 (1972).

BIOGRAPHICAL SKETCH

Conrad Burkholder was born in Washington, D.C., on November 11, 1954. In June of 1972, he graduated from Wellesley High School in Wellesley, Massachusetts. He received the Bachelor of Science degree in chemistry from the University of Massachusetts at Amherst in June of 1976. He entered graduate school in the fall of 1976 at the University of Florida in Gainesville. He received the Master of Science degree in chemistry from the University of Florida in June of 1980. He received the Doctor of Philosophy degree in chemistry from the University of Florida in April of 1984.

I certify that I have read this study and that in my opinion it conforms to acceptable standards of scholarly presentation and is fully adequate, in scope and quality, as a dissertation for the degree of Doctor of Philosophy.

William R. Dolbier, Jr.
Chairman
Professor of Chemistry

I certify that I have read this study and that in my opinion it conforms to acceptable standards of scholarly presentation and is fully adequate, in scope and quality, as a dissertation for the degree of Doctor of Philosophy.

William M. Jones
Professor of Chemistry

I certify that I have read this study and that in my opinion it conforms to acceptable standards of scholarly presentation and is fully adequate, in scope and quality, as a dissertation for the degree of Doctor of Philosophy.

Martin Vala
Professor of Chemistry

I certify that I have read this study and that in my opinion it conforms to acceptable standards of scholarly presentation and is fully adequate, in scope and quality, as a dissertation for the degree of Doctor of Philosophy.

James A. Deyrup
Professor of Chemistry

I certify that I have read this study and that in my opinion it conforms to acceptable standards of scholarly presentation and is fully adequate, in scope and quality, as a dissertation for the degree of Doctor of Philosophy.

E.P. Goldberg
Professor of Materials Science
and Engineering

This dissertation was submitted to the Graduate Faculty of the Department of Chemistry in the College of Liberal Arts and Sciences and to the Graduate School, and was accepted as partial fulfillment of the requirements for the degree of Doctor of Philosophy.

April 1984

Dean for Graduate Studies
and Research

CPSIA information can be obtained
at www.ICGtesting.com
Printed in the USA
BVHW052106090619
550552BV00002B/233/P